The Architecture and Biology of Soils

Life in Inner Space

FSC
www.fsc.org
MIX
Paper from
responsible sources
FSC® C013604

This book is dedicated to the memory of
Bev Kay, scientist and gentleman

The Architecture and Biology of Soils

Life in Inner Space

Edited by

Karl Ritz

Cranfield University, UK

and

Iain Young

*School of Environmental & Rural Science,
Armidale, Australia*

www.cabi.org

CABI is a trading name of CAB International

CABI
Nosworthy Way
Wallingford
Oxfordshire OX10 8DE
UK

CABI
875 Massachusetts Avenue
7th Floor
Cambridge, MA 02139
USA

Tel: +44 (0)1491 832111
Fax: +44 (0)1491 833508
E-mail: cabi@cabi.org
Website: www.cabi.org

Tel: +1 617 395 4056
Fax: +1 617 354 6875
E-mail: cabi-nao@cabi.org

A catalogue record for this book is available from the British Library, London, UK.

Library of Congress Cataloging-in-Publication Data

The Architecture and biology of soils: life in inner space / edited by Karl Ritz, Iain Young.
        p. cm.
  Includes bibliographical references and index.
  ISBN 978-1-84593-532-0 (alk. paper)
1. Soil structure.   2. Soil biology.   3. Soil ecology.   I. Ritz, K. (Karl)
II. Young, Iain, 1962-   III. Title.

  S593.2.A73 2011
  631.4--dc23

                                                                    2011021526

ISBN-13: 978 1 84593 532 0

Commissioning editor: Sarah Hulbert
Production editor: Simon Hill

Typeset by AMA DataSet, Preston, UK.
Printed and bound by CPI Group (UK) Ltd, Croydon, CR0 4YY.

Contents

Contributors

Mark Bartlett, National Soil Resources Institute, Department of Environmental Science and Technology, School of Applied Sciences, Cranfield University, Cranfield MK43 0AL, UK. *Current address*: The Scotts Miracle-Gro Company, Levington Research Station, Bridge Road, Levington, Ipswich IP10 0NE, UK.

Richard J.F. Bewley, URS Corporation Ltd, Brunel House, 54 Princess Street, Manchester M1 6HS, UK.

Claire Chenu, AgroParisTech, UMR Bioemco, Bâtiment EGER, 78850 Thiverval Grignon, France.

Diego Cosentino, AgroParisTech, UMR Bioemco, Bâtiment EGER, 78850 Thiverval Grignon, France. *Current address*: Depto. de Recursos Naturales y Ambiente, Facultad de Agronomía, Universidad de Buenos Aires, C1417DSE BS. As. Argentina.

Michael S. Fitzsimons, Biosciences Division, Argonne National Laboratory, Argonne, IL 60439, USA.

Dmitry Grinev, School of Engineering Sciences, University of Southampton, University Road, Southampton SO17 1BJ, UK.

Christian Hartmann, Institut de Recherche pour le Développement (IRD), International Water Management Institute (IWMI), National Agronomy and Forestry Research Institute – UMR 211 BIOEMCO, BP 06, Vientiane, Lao PDR. *Current address*: Institut de Recherche pour le Développement (IRD)-UMR 211 BIOEMCO, 32, av. H. Varagnat, 93143 Bondy cedex, France.

Simon Hockin, Environmental Reclamation Services, Ltd, Westerhill Road, Bishopbriggs, Glasgow G64 2QH, UK.

Patricia A. Holden, Bren School of Environmental Science & Management, 2400 Bren Hall, University of California Santa Barbara, Santa Barbara, CA 93106-5131, USA.

Iain James, National Soil Resources Institute, Department of Environmental Science and Technology, School of Applied Sciences, Cranfield University, Cranfield MK43 0AL, UK.

Bev. D. Kay (deceased), University of Guelph, School of Environmental Sciences, Guelph, ON, N1G 2W1 Canada.

Jean-Luc Maeght, Institut de Recherche pour le Développement (IRD), International Water Management Institute (IWMI), National Agronomy and Forestry Research Institute – UMR 211 BIOEMCO, BP 06, Vientiane, Lao PDR. *Current address*: de Recherche pour le

Développement (IRD), International Water Management Institute (IWMI), National Agronomy and Forestry Research Institute – UMR 211 BIOEMCO, P.O. Box 4199, Vientiane, Lao PDR.

David A.C. Manning, School of Civil Engineering and Geosciences, Newcastle University, Newcastle upon Tyne NE1 7RU, UK.

R. Michael Miller, Biosciences Division, Argonne National Laboratory, Argonne, IL 60439, USA.

Lars J. Munkholm, Århus University, Department of Agroecology, Blichers Allé 20, PO Box 50, DK-8830, Tjele, Denmark.

Loïc Pagès, Institut National de la Recherche Agronomique (INRA), Centre d'Avignon UR 1115 PSH, Site Agroparc, 84914 Avignon cedex 9, France.

Alain Pierret, Institut de Recherche pour le Développement (IRD), International Water Management Institute (IWMI), National Agronomy and Forestry Research Institute – UMR 211 BIOEMCO, BP 06, Vientiane, Lao PDR. *Current address*: Institut de Recherche pour le Développement (IRD)-UMR 211 BIOEMCO, 32, av. H. Varagnat, 93143 Bondy cedex, France.

Karl Ritz, National Soil Resources Institute, Department of Environmental Science and Technology, School of Applied Sciences, Cranfield University, Cranfield MK43 0AL, UK.

Kejian Wu, Institute of Petroleum Engineering, Heriot-Watt University, Edinburgh EH14 4AS, UK.

Iain M. Young, The School of Environmental and Rural Sciences, University of New England, NSW 2351, Australia.

Xiaoxiang Zhang, SIMBIOS Centre, the University of Abertay Dundee, Dundee DD1 1HK, UK. *Current address*: School of Engineering, University of Liverpool, Liverpool L69 3BX, UK.

Preface

Soils are a fundamental feature of our contemporary planet, and form the basis of all past and present civilizations. The same word is used in many languages to name both planet Earth, and the earth that supports the majority of terrestrial habitats.

Soils are highly complex materials, constituted of an extremely wide variety of mineral and organic components. These constituents are arranged in space such that soils are essentially porous materials, but with remarkable properties. Soil pore networks are structurally diverse across a huge range of size scales, imparting particular and important properties in terms of how gases, water, solutes, colloids and organisms are held within the matrix and move through it. There is an intriguing paradox in that arguably the most important feature of soils is where the solid material is literally absent, i.e. the labyrinthine voids of the pore networks. Soil structure regulates the dynamic processes underpinning soil function, providing the physical framework in and through which all soil processes occur.

Whilst the soil biomass comprises a tiny proportion of the total soil mass, it plays a dominant role in delivering the wide range of ecosystem services provided by soil systems. The soil biota create physical structure within their habitat, affect the resultant dynamics of many soil processes and are themselves affected by such structure. This interplay between the spatial constructs of the soil matrix and the life within it is appositely conceptualized in terms of an *architecture of the soil*, and the pore networks therein as a form of planetary *inner space*. That then is the theme of this book, which aims to explore these concepts from a variety of perspectives and reveal how, ultimately, the functioning of the terrestrial components of the Earth system fundamentally depends upon the quite remarkable spatial organization of the soil.

Karl Ritz
Iain Young

1 Views of the Underworld: *in situ* Visualization of Soil Biota

Karl Ritz[*]

1 Introduction

The unaided human eye and visual cortex are restricted to being able to detect only a relatively narrow waveband of light, and have a lower spatial resolution of the order of one millimetre. Beyond the absolute surface, soils are opaque to such a waveband and hence humans cannot see into, or through, the soil matrix. Furthermore, whilst the fresh biomass in soils typically exceeds many tonnes per hectare, the greater majority of such life is microscopic and hence invisible to the unaided human eye, even when the soil fabric is disintegrated. David Coleman captured this notion appositely in the title of his seminal review 'Through a ped darkly' (Coleman, 1985), which captures the essence of the challenge in visualizing soil organisms whilst in their natural habitat. Visualizing the spatial organization of soil systems, and the life therein, requires adoption of particular techniques and technologies in order to see through a ped *rather less darkly*, and provide forms of looking glass to reveal the inhabited architecture of the underground.

In this chapter, such methods are briefly reviewed, with consideration confined to the observation of the organisms themselves, since other chapters in this volume deal with the biotic interactions with physical structure of the soil *per se*.

2 Tools for the Job

Given that the majority of soil biota are microscopic, magnification systems need to be adopted in order to visualize them. Microscopy systems afford a wide range of spatial resolution, essentially contingent on whether photons or electrons are used to illuminate the specimens. Within visible spectra, basic magnifying lenses, typically up to ×20, can sometimes reveal a surprising range of soil biota, notably insect and arachnid fauna, and some fungal mycelia, but in general more sophisticated lens configurations are needed to provide adequate magnification. Stereo (dissecting) microscopes offer a range of ×1–100 magnification, with the advantage that three-dimensional (3D) visualization is instantaneous to the viewer and can, in principle at least, be captured photographically (published stereographs of soils visualized in this manner are rarely encountered). Compound microscopes can provide a resolution down to

*k.ritz@cranfield.co.uk

300 μm, and be used in both transmitted-light modes (suited to thin sections) or via surface-illuminated epifluorescence systems. Confocal laser scanning systems can be effective with the use of appropriate stains, but to date have rarely been used for *in situ* soil visualization, being largely confined to either rhizospheres in non-natural circumstances (e.g. Assmus *et al.*, 1997) or roots extracted from soil-grown circumstances (e.g. Mogge *et al.*, 2000; Buddrus-Schiemann *et al.*, 2010; Hassen and Labuschagne, 2010). Scanning electron microscopes (SEM) operate on the basis of reflected electrons and offer a wide spatial range of high-resolution imaging (down to c. 20 nm). Transmission electron microscopes (TEM) provide still greater resolution, down to the sub-nm scale. In general, increased spatial resolution is offset by the maximum physical size of the sample that can be accommodated by the instrument, as is the instantaneous field-of-view during imaging. Digital-based imaging systems are starting to remove the constraints of the latter, in that contiguous fields-of-view can be combined to single multi-scale images with increasing precision and capacity (e.g. Nunan *et al.*, 2001; Herrmann *et al.*, 2007).

The fundamental requirement of any *in situ* observation of soil biota is that the fabric of the soil is left as undisturbed as possible whilst realizing the opportunity to visualize any organisms. This is an important principle, since it is understood that soils function by virtue of being spatially connected and integrated systems, and that any disintegration of them will affect their functioning and, in experimental or observational terms, their apparent properties (Young and Ritz, 2005). For visualization, disturbance that results in a physical disruption of the soil matrix is therefore particularly to be avoided.

3 Field-based Observations

In the field, intact soil systems can be observed by excavation of trenches or pits into the soil and creating 'profile walls' that are moderately undisturbed and can be observed directly. Such exposed faces can be covered by glass or plastic to prevent further disruption, and allow observations over extended periods. This concept has been applied particularly in the context of observation of plant roots, hence the construct is often termed 'rhizotron' (e.g. Huck and Taylor, 1982; Vandegeijn *et al.*, 1994; Hendrick and Pregitzer, 1996), but studies focusing on soil biota have also adopted such systems (e.g. Ermakov and Stepanova, 1992; Lussenhop and Fogel, 1993; Wilson *et al.*, 1995; McCormack *et al.*, 2010). In some remarkable pioneering work, Kubiena (1938) scrutinized such exposed soil profiles using a compound microscope system configured for such application and was able directly to visualize a wide variety of soil life, including protozoa swimming in water films within pore networks (Fig. 1.1). His observations – and arguably his prose – may be unsurpassed to this day:

> The best conditions for their [protozoa] development occur when spaces of higher order rich in plant or animal residues are entirely filled with water. By evaporation the water content is decreased, and the soil solution retires to the side cracks, small worm holes and empty plant root channels. These water residues become very crowded by the extremely active population of protozoa remaining. With great rapidity and surety they move along all bends and corners of the channels until they are stopped at the ends at which their paths are closed by the surface meniscus of the water. Their behaviour at the menisci is particularly interesting. Two or three times they would approach the barrier in rapid succession and then quickly turn around and pursue their way in an opposite direction ...
> (Kubiena, 1938, pp. 233–234).

Endoscopes, originally developed for medical application, can also be inserted into the soil, via transparent observation tubes pre-inserted into the fabric (Gijsman *et al.*, 1991; Lopez *et al.*, 1996; Kautz and Kopke, 2010). These have the advantage that pits need not necessarily be dug, they can be adapted at a variety of sizes and are amenable to replication. Disadvantages include the fact that the insertion process can be disruptive, and the resultant interface

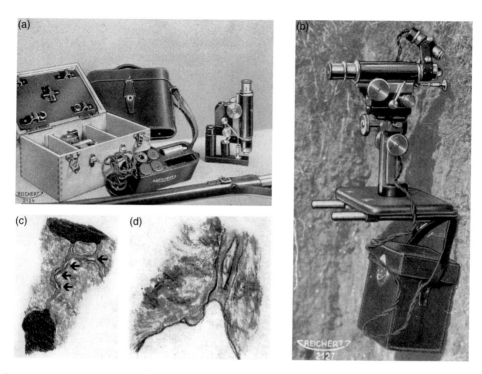

Fig. 1.1. Early application of soil micropedology. (a) Microscope system produced by Reichert Company specifically for soil micropedological application; (b) microscope system installed on soil profile wall; (c) drawing of soil pore with water films and protozoa (arrowed) therein, retiring to side channels of the pore with decreasing water content of the soil; (d) higher magnification drawing of a pore filled with water containing a protozoan cell confined by meniscus (from Kubiena, 1938).

between the observation tube wall and the soil can result in artefacts.

4 Out of the Field

One of the most common methods adopted to preserve structural integrity when sampling soils is to use coring systems, where circular or rectangular sleeves are inserted into the soil using minimal force and then excised by excavation around their periphery. Material confined within the cores is thus subjected to minimal disturbance. Structural preservation is then ensured by subsequently embedding the excised soil in synthetic resin formulations, which can be perfused through the soil matrix in liquid form and subsequently solidified via a process of polymerization. This is akin to the biological histology approach and is standard prac-

tice in the sub-discipline of soil micromorphology (e.g. Fitzpatrick, 1984; Murphy, 1986; Vepraskas and Wilson, 2008). Such resin-embedded soil blocks can then be cut and polished to create surfaces that may be observed directly, or thin soil sections created by adhering such polished blocks to glass slides, cutting the bulk of the material away with geological saws and grinding and polishing the resultant slices to the desired thickness using lapidary pastes. The resultant thin sections can then be observed using a variety of imaging techniques. Since the resins used in such processes are extremely hydrophobic, it is imperative that the soil being impregnated contains minimal amounts of free water, and hence samples need to be desiccated. Air-drying tends to be disruptive to the physical structure of the soil, but certainly affects biological material to a considerable extent. Desiccation based upon solvent exchange is therefore adopted, typically by perfusing the soil with a graded series of acetone:water mixtures of decreasing water concentration. However, even desiccation by such procedures can still disrupt biological tissues (Tippkötter *et al.*, 1986), and in the early application of soil micromorphology only biological material relatively resistant to desiccation, such as woody tissues, some fungal hyphae and thick-walled spores, was observed in an apparently intact state (see, for example, images in Fitzpatrick, 1993). However, if soils are pretreated with biological fixatives such as glutaraldehyde or paraformaldehyde, i.e. chemicals that bind with biomolecules and render them structurally resistant, then the majority of the soil biota are apparently preserved in an intact state following micromorphological preparations. Examples of the preparation of such 'biological thin sections' at the core scale are described by Tippkötter *et al.* (1986), Postma and Altemüller (1990), Tippkötter and Ritz (1996) and Nunan *et al.* (2001).

As well as taking intact cores from soils in the field situation, structurally intact systems can be constructed experimentally for the express purpose of observing soil biota. Such micro- or meso-cosms can include observation windows or walls (Fig. 1.2), observation tubes or be of a volume conducive to direct resin impregnation and micromorphological preparation.

Cryopreservation techniques, where samples are frozen in liquid nitrogen, can also be used effectively to preserve soil samples and their associated biology, but the volumes of soil that can be handled are relatively small, and such approaches are, in practice, best suited to SEM (Campbell and Porter, 1982; Chenu *et al.*, 2001; McCully *et al.*, 2009; Chapter 3).

Visualizing soil biota at the highest spatial resolution by transmission electron microscopy requires concomitantly small sample volumes, and is technically very challenging (e.g. Kilbertus and Proth, 1979; Foster, 1994), particularly since ultra-thin sections are required (sub-μm), which can only be attained by use of a microtome (i.e. a

(a)

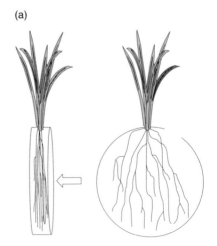

(b)

Fig. 1.2. Microcosm-scale observation chamber suitable for direct observation of soil biota in undisturbed state. (a) Configuration of chamber, fabricated from polystyrene Petri dish (diameter 12 cm) filled with soil; two holes are cut in the top of the dish – one to allow plants to emerge and the adjacent one for watering; arrow shows line of sight for (b) image of roots of *Plantago lanceolata* and associated fungal hyphae growing in observation chamber, photograph taken using dissecting microscope. Scale bar = 1 mm (from Ritz, 1984).

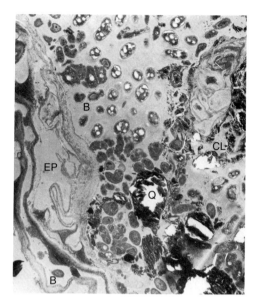

Fig. 1.3. Transmission electron micrograph (× 10,000) of mature rhizosphere of clover. EP, epidermal cells; B, bacteria; CL, clay; Q, quartz grain (from Foster *et al.*, 1983).

cutting/slicing system using a sharp edge); it is not possible to prepare sufficiently thin sections by lapidary. Mineral constituents in the soil are not conducive to being cut as such, and hence, even with the use of diamond microtome knives, it is very difficult to obtain intact sections, particularly in soils containing quartz material. None the less, a number of pioneering studies, notably those by Ralph Foster *et al.*, have revealed the remarkable extent of spatial structuring in the rhizosphere at such small scales (e.g. Foster, 1978, 1998; Foster *et al.*, 1983; Fig. 1.3), as well as the morphological diversity of bacterial cells occurring in the natural soil environment (Fig. 1.4), which becomes apparent only through the use of such approaches.

5 Contrasting Matters

Objects are discernible only if there is some form of contrast between them and the background in which they occur, whatever wavelength of radiation is used to image the system. Much of the soil microflora at least (and hence the majority of the soil biomass) is non-pigmented, which challenges any ability to discern it using visible light. Given some contrast, some organisms – or components thereof – can be identified on the basis of their morphology, but this problem is exacerbated by the structurally highly heterogeneous background typically occurring in most soils, arising as a result of the complex mixture of organic and mineral constituents of soil manifest across size scales over several orders of magnitude. Thus staining techniques are needed in order to render most soil biological material discernible. Applications of staining in soil systems can be challenging, due to the often high potential for non-specific binding of such compounds to soil constituents other than the target material. Such stains can be coloured (chromophoric), most suitable for transmission light microscopy, based on fluorescent compounds (Postma and Altemüller, 1990; Tippkötter, 1990; Harris *et al.*, 2002; Li *et al.*, 2004). The latter are likely to be more effective since the stained material tends to show a high degree of contrast to the background, as fluorescence emanates from the materials themselves (Figs 1.5 and 1.6).

Early application of such fluorochromes focused on compounds that were relatively non-specific to soil microbiota, with a range of fluorescent brighteners proving effective. More recently, nucleic acid probes have been successfully applied to specifically visualize eubacterial cells in soil thin sections using the fluorescent *in situ* hybridization (FISH) techniques (Eickhorst and Tippkötter, 2008a, b; Fig. 1.7). This is technically very challenging due to issues of non-specific binding of probes and cells being rendered permeable to the probes, but represents a substantial advance in terms of affording the potential to locate the distribution of specific classes of microbes in the soil fabric. Where more specific probes are used, the relative proportion of bacterial cells that would be marked could be very small and challenging to locate, leading to issues of locating the 'region of interest' at such small spatial scales (Herrmann *et al.*, 2007).

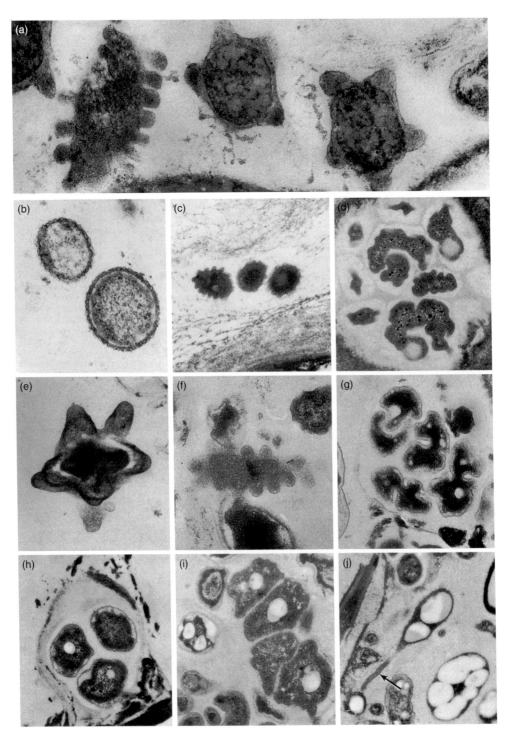

Fig. 1.4. Diverse bacterial morphology apparent in natural rhizosphere soils as visualized by transmission electron microscopy. (a, b) pine; (c, d, e, h, j) wheat; (f, g, i) clover (from Foster *et al.*, 1983).

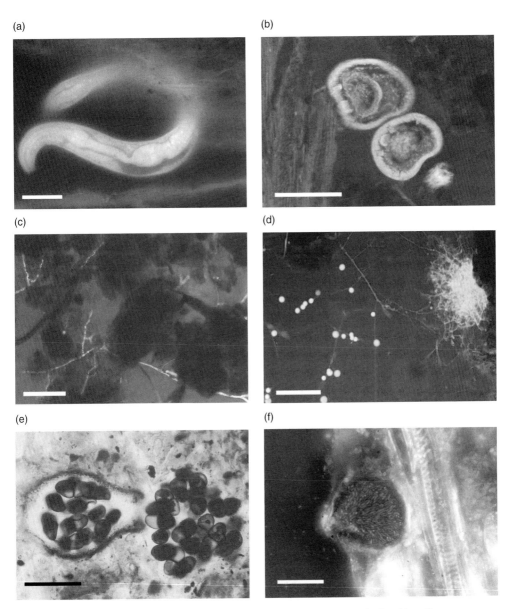

Fig. 1.5. Images of soil biota visualized in biological thin sections of undisturbed arable soils. (a) Bacterial-feeding nematode, constrained by pore dimensions; (b) enchytraeid worm (in cross-section) adjacent to root cortex; (c) hyphal network of mycelium of *Rhizoctonia solani* (sterilized soil); (d) fungal mycelium bridging pore with spore clusters; (e) fungal perithecium in decomposing organic material; (f) fungal perithecium in decomposed cortex of root, with apparently intact stele adjacent. All material stained with fluorescent brighteners and imaged under UV epifluorescence. Scale bars (a, c, d, f) = 20 µm; (b) = 100 µm; (e) = 50 µm (from the author; (c) also used in Harris *et al.*, 2002).

Advances in digital visualization systems are likely to lead to the potential for automatic detection of such signals. For transmission electron microscopy, and associated ultrastructural studies, heavy metal-based compounds suited to visualization via transmission electron microscopy are used (e.g. Robert and Chenu, 1992; Foster, 1994).

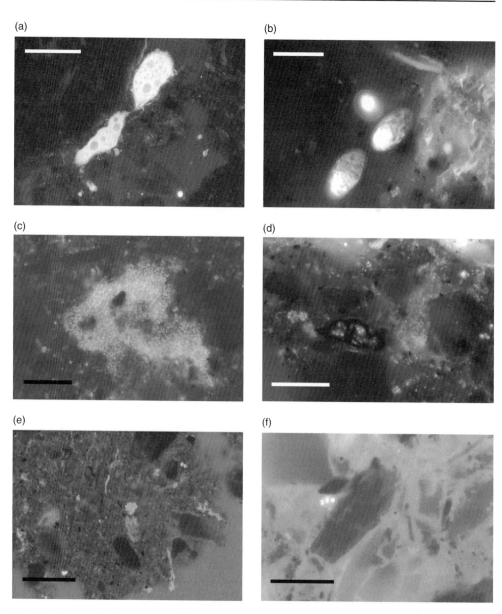

Fig. 1.6. Images of soil protozoa and bacteria visualized in biological thin sections of undisturbed soils. (a) Naked amoeba in arable soil; (b) testate amoebae in arable soil; (c) bacterial colony in arable soil induced by addition of glucose; (d) bacteria in grassland soil; (e) bacterial colony in bulk arable soil; (f) bacterial colony in arable subsoil. All material stained with fluorescent brighteners and imaged under UV epifluorescence. Scale bars all 20 μm (from the author; (c) also used in Nunan *et al.*, 2003).

6 The Next Dimension

Soil constituents may be visualized in 3D non-destructively by X-ray computer-aided tomography (CAT), and across a wide range of spatial resolution, typically down to a few μm (e.g. Aylmore, 1994; Kaestner *et al.*, 2006; Lombi and Susini, 2009; Peth, 2010). Furthermore, the 3D aspect of such visualizations far exceeds that of the stereographic

(a)

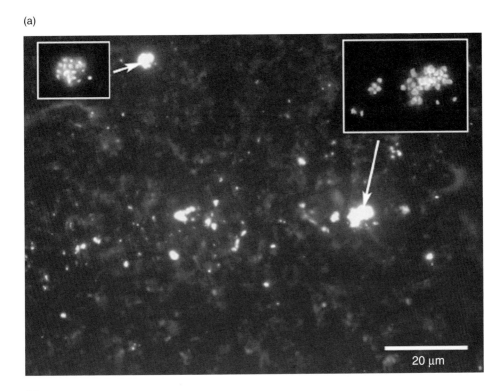

(b)

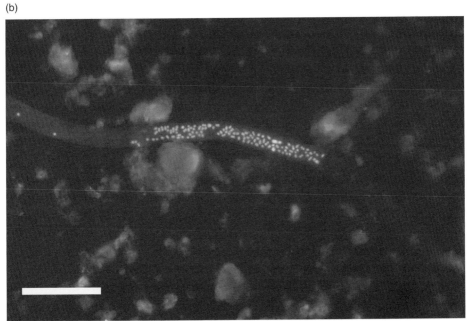

Fig. 1.7.   Fluorescent *in situ* hybridization (FISH) of bacteria in undisturbed soils, using eubacterial-targeted nucleic acid probes. (a) Bacterial colonies visualized in an undisturbed sandy soil; insets show details at reduced exposure times; (b) FISH-stained bacteria colonizing a fungal hypha, imaged using double excitation wavelengths. Scale bars = 20 µm (from Eickhorst and Tippkötter, 2008a (a); Eickhorst and Tippkötter, 2008b (b)).

constructions afforded by stereo light microscopy or SEM, since they are based upon digital reconstructions of tomographic slices and hence can be visualized (or incorporated into modelling frameworks) in any spatial orientation. The base spatial resolution is essentially defined by the overall dimensions of the sample being scanned – the smaller the object, the greater the minimal attainable resolution. Its application to the study of soils is increasing at a great rate as the technology matures, particularly in relation to visualizing pore networks and water flow. Visualizing soil biotic features *per se* using this technique is challenged by the fact that such features tend to attenuate X-rays to the same extent as non-living organic matter, and hence there is often inadequate contrast to enable discrimination of biota from other material. This is true particularly at the microbial scale – bacterial cells are essentially amorphous features compared with

soil organic matter in X-ray scans at the order of 1 µm. Thus, to date, the technique has largely been applied to determine the effects of biota on soil structure rather than visualize the biota *per se*. Some representatives of the larger soil fauna such as earthworms (Joschko *et al.*, 1993; Jegou *et al.*, 2001) and insect larvae (Johnson *et al.*, 2004, 2007) have been successfully imaged, and the non-destructive nature of the technique enables the movement of such animals to be tracked. Considerable progress has been made in relation to visualizing root systems growing in soil (e.g. Gregory *et al.*, 2003; Mooney *et al.*, 2006; Hargreaves *et al.*, 2009), and it is likely that technological and methodological advances will result in other soil biota eventually being visualized by such systems. This will undoubtedly lead to new insights and significant advances in the understanding of the dynamics of soil organisms – to X-rays, the ped is indeed far from dark!

References

Assmus, B., Schloter, M., Kirchhof, G., Hutzler, P. and Hartmann, A. (1997) Improved in situ tracking of rhizosphere bacteria using dual staining with fluorescence-labeled antibodies and rRNA-targeted oligonucleotides. *Microbial Ecology* 33, 32–40.

Aylmore, L.A.G. (1994) Applications of computer assisted tomography to soil–plant-water studies: an overview. In: Anderson, S.H. and Hopmans, J.W. (eds) *Tomography of Soil–Water–Root Processes*. Soil Science Society of America, Madison, WI, pp. 7–15.

Buddrus-Schiemann, K., Schmid, M., Schreiner, K., Welzl, G. and Hartmann, A. (2010) Root colonization by *Pseudomonas* sp DSMZ 13134 and impact on the indigenous rhizosphere bacterial community of barley. *Microbial Ecology* 60, 381–393.

Campbell, R. and Porter, R. (1982) Low-temperature scanning electron-microscopy of microorganisms in soil. *Soil Biology and Biochemistry* 14, 241–245.

Chenu, C., Hassink, J. and Bloem, J. (2001) Short-term changes in the spatial distribution of microorganisms in soil aggregates as affected by glucose addition. *Biology and Fertility of Soils* 34, 349–356.

Coleman, D.C. (1985) Through a ped darkly – an ecological assessment of root soil–microbial–faunal interactions. In: Fitter, A.H., Atkinson, D., Read, D.J. and Usher, M.B. (eds) *Ecological Interactions in Soil*. Blackwell Scientific Publications, Oxford, UK.

Eickhorst, T. and Tippkötter, R. (2008a) Detection of microorganisms in undisturbed soil by combining fluorescence in situ hybridization (FISH) and micropedological methods. *Soil Biology and Biochemistry* 40, 1284–1293.

Eickhorst, T. and Tippkötter, R. (2008b) Improved detection of soil microorganisms using fluorescence in situ hybridization (FISH) and catalyzed reporter deposition (CARD-FISH). *Soil Biology and Biochemistry* 40, 1883–1891.

Ermakov, E.I. and Stepanova, O.A. (1992) Study of microorganisms of the rhizosphere of plants in a rhizotron. *Microbiology* 61, 643–648.

Fitzpatrick, E.A. (1984) *Micromorphology of Soils*. Chapman and Hall, London.

Fitzpatrick, E.A. (1993) *Soil Microscopy and Micromorphology*. John Wiley, Chichester, UK.

Foster, R.C. (1978) Ultramicromorphology of some South Australian soils. In: Emerson, W.W., Bond, R.D. and Dexter, A.R. (eds) *Modification of Soil Structure*. John Wiley, Chichester, UK, pp. 103–109.

Foster, R.C. (1994) The ultramicromorphology of soil biota in situ in natural soils: a review. In: Ringrose-Voase, A.J. and Humphreys, G.S. (eds) *Soil Micromorphology: Studies in Management and Genesis*. Elsevier, Amsterdam, pp. 381–393.

Foster, R.C. (1998) The ultrastructure of the rhizoplane and rhizosphere. *Annual Review of Phytopathology* 24, 211–234.

Foster, R.C., Rovira, A.D. and Cock, T.W. (1983) *Ultrastructure of the Root–Soil Interface*. American Phytopathology Society Press, St Paul, Minnesota.

Gijsman, A.J., Floris, J., Vannoordwijk, M. and Brouwer, G. (1991) An inflatable minirhizotron system for root observations with improved soil tube contact. *Plant and Soil* 134, 261–269.

Gregory, P.J., Hutchison, D.J., Read, D.B., Jenneson, P.M., Gilboy, W.B. and Morton, E.J. (2003) Non-invasive imaging of roots with high resolution X-ray micro-tomography. *Plant and Soil* 255, 351–359.

Hargreaves, C.E., Gregory, P.J. and Bengough, A.G. (2009) Measuring root traits in barley (*Hordeum vulgare* ssp. *vulgare* and ssp. *spontaneum*) seedlings using gel chambers, soil sacs and X-ray microtomography. *Plant and Soil* 316, 285–297.

Harris, K., Crabb, D., Young, I.M., Weaver, H., Gilligan, C.A., Otten, W. *et al.* (2002) *In situ* visualisation of fungi in soil thin sections: problems with crystallisation of the fluorochrome FB 28 (Calcofluor M2R) and improved staining by SCRI Renaissance 2200. *Mycological Research* 106, 293–297.

Hassen, A.I. and Labuschagne, N. (2010) Root colonization and growth enhancement in wheat and tomato by rhizobacteria isolated from the rhizoplane of grasses. *World Journal of Microbiology and Biotechnology* 26, 1837–1846.

Hendrick, R.L. and Pregitzer, K.S. (1996) Applications of minirhizotrons to understand root function in forests and other natural ecosystems. *Plant and Soil* 185, 293–304.

Herrmann, A.M., Ritz, K., Nunan, N., Clode, P.L., Pett-Ridge, J., Kilburn, M.R. *et al.* (2007) Nano-scale secondary ion mass spectrometry – a new analytical tool in biogeochemistry and soil ecology. *Soil Biology and Biochemistry* 39, 1835–1850.

Huck, M.G. and Taylor, H.M. (1982) The rhizotron as a tool for root research. *Advances in Agronomy* 35, 1–35.

Jegou, D., Capowiez, Y. and Cluzeau, D. (2001) Interactions between earthworm species in artificial soil cores assessed through the 3D reconstruction of the burrow systems. *Geoderma* 102, 123–137.

Johnson, S.N., Read, D.B. and Gregory, P.J. (2004) Tracking larval insect movement within soil using high resolution X-ray microtomography. *Ecological Entomology* 29, 117–122.

Johnson, S.N., Crawford, J.W., Gregory, P.J., Grinev, D.V., Mankin, R.W., Masters, G.J. *et al.* (2007) Non-invasive techniques for investigating and modelling root-feeding insects in managed and natural systems. *Agricultural and Forest Entomology* 9, 39–46.

Joschko, M., Muller, P.C., Kotzke, K., Dohring, W. and Larink, O. (1993) Earthworm burrow system – development assessed by means of X-ray computed-tomography. *Geoderma* 56, 209–221.

Kaestner, A., Schneebeli, M. and Graf, F. (2006) Visualizing three-dimensional root networks using computed tomography. *Geoderma* 136, 459–469.

Kautz, T. and Kopke, U. (2010) *In situ* endoscopy: New insights to root growth in biopores. *Plant Biosystems* 144, 440–442.

Kilbertus, G. and Proth, J. (1979) Study of forest soil (Rendzina) using electron-microscope. *Canadian Journal of Microbiology* 25, 943–946.

Kubiena, W.L. (1938) *Micropedology*. Collegiate Press, Ames, Iowa.

Li, Y., Dick, W.A. and Tuovinen, O.H. (2004) Fluorescence microscopy for visualization of soil microorganisms – a review. *Biology and Fertility of Soils* 39, 301–311.

Lombi, E. and Susini, J. (2009) Synchrotron-based techniques for plant and soil science: opportunities, challenges and future perspectives. *Plant and Soil* 320, 1–35.

Lopez, B., Sabate, S. and Gracia, C. (1996) An inflatable minirhizotron system for stony soils. *Plant and Soil* 179, 255–260.

Lussenhop, J. and Fogel, R. (1993) Observing soil biota *in situ*. *Geoderma* 56, 25–36.

McCormack, M.L., Pritchard, S.G., Breland, S., Davis, M.A., Prior, S.A., Runion, G.B. *et al.* (2010) Soil fungi respond more strongly than fine roots to elevated CO_2 in a model regenerating longleaf pine–wiregrass ecosystem. *Ecosystems* 13, 901–916.

McCully, M.E., Canny, M.J. and Huang, C.X. (2009) Cryo-scanning electron microscopy (CSEM) in the advancement of functional plant biology. Morphological and anatomical applications. *Functional Plant Biology* 36, 97–124.

Mogge, B., Loferer, C., Agerer, R., Hutzler, P. and Hartmann, A. (2000) Bacterial community structure and colonization patterns of *Fagus sylvatica* L. ectomycorrhizospheres as determined by fluorescence in situ hybridization and confocal laser scanning microscopy. *Mycorrhiza* 9, 271–278.

Mooney, S.J., Morris, C. and Berry, P.M. (2006) Visualization and quantification of the effects of cereal root lodging on three-dimensional soil macrostructure using X-ray computed tomography. *Soil Science* 171, 706–718.

Murphy, C.P. (1986) *Thin Section Preparation of Soils and Sediments*. AB Academic Publishers, Berkhamsted, UK.

Nunan, N., Ritz, K., Crabb, D., Harris, K., Wu, K., Crawford, J.W. *et al.* (2001) Quantification of the *in situ* distribution of soil bacteria by large-scale imaging of thin-sections of undisturbed soil. *FEMS Microbiology Ecology* 37, 67–77.

Nunan, N., Wu, K., Young, I.M., Crawford, J.W. and Ritz, K. (2003) Spatial distribution of bacterial communities and their relationships with the micro-architecture of soil. *FEMS Microbiology Ecology* 43, 203–215.

Peth, S. (2010) Applications of microtomography in soils and sediments. In: Singh, B. and Gräfe, M. (eds) *Developments in Soil Science*, vol. 34. Elsevier, Amsterdam, pp. 73–101.

Postma, J. and Altemüller, H.-J. (1990) Bacteria in thin soil sections stained with the fluorescence brightener Calcofluor M2R. *Soil Biology and Biochemistry* 22, 89–96.

Ritz, K. (1984) Phosphorus transfer between grassland plants. PhD thesis, University of Bristol, UK.

Robert, M. and Chenu, C. (1992) Interactions between soil minerals and microorganisms. In: Stotzky, G. and Bollag, G. (eds) *Soil Biochemistry*, vol. 7. Marcel Dekker, New York, pp. 307–404.

Tippkötter, R. (1990) Staining of soil microorganisms and related materials with fluorochromes. In: Douglas, L.A. (ed.) *Soil Micromorphology: A Basic and Applied Science*. Elsevier, Amsterdam, pp. 605–611.

Tippkötter, R. and Ritz, K. (1996) Evaluation of polyester, epoxy and acrylic resins for suitability in preparation of soil thin sections for *in situ* biological studies. *Geoderma* 69, 31–57.

Tippkötter, R., Ritz, K. and Darbyshire, J.F. (1986) The preparation of soil thin sections for biological studies. *Journal of Soil Science* 77, 681–690.

Vandegeijn, S.C., Vos, J., Groenwold, J., Goudriaan, J. and Leffelaar, P.A. (1994) The Wageningen Rhizolab – a facility to study soil–root–shoot–atmosphere interactions in crops. 1. Description of main functions. *Plant and Soil* 161, 275–287.

Vepraskas, M.J. and Wilson, M.A. (2008) Soil micromorphology: concepts, techniques and applications. In: Ulery, A.L. and Drees, L.R. (eds) *Methods of Soil Analysis: Mineralogical Methods*, pp. 191–228.

Wilson, K., Gunn, A. and Cherrett, J.M. (1995) The application of a rhizotron to study the subterranean effects of pesticides. *Pedobiologia* 39, 132–143.

Young, I.M. and Ritz, K. (2005) The habitat of soil microbes. In: Bardgett, R.D., Usher, M.B. and Hopkins, D.W. (eds) *Biological Diversity and Function in Soils*. Cambridge University Press, Cambridge, UK, pp. 31–43.

2 Modelling Soil Structure and Processes

Kejian Wu* and Xiaoxiang Zhang

1 Introduction

The spatial arrangement of the solid component in soils is central to soil functions (Young and Crawford, 2004). As visualization in the previous chapter shows, soils are highly complex, characterized by multiple heterogeneity across a wide range of scales. Such heterogeneity, along with the inherent opaqueness of soils, has hampered efforts to quantify soil microstructure despite its fundamental role in all soil processes, such as its conductivity for water flow and chemical transport (Vogel *et al.*, 2005; Zhang and Lv, 2007). These processes combine to impact the spatial distribution of soil organisms and their associated functions, including carbon cycling and uptake of water and nutrients by plant roots (Rappoldt and Crawford, 1999; Crawford *et al.*, 2005). Therefore, the ability to quantify soil microstructure and its responsive changes to the physical and biochemical processes in soil is essential in ensuring the sustainable use of soil resources.

2 Modelling Soil Structure

Traditional methods of quantifying soil structures are mostly indirect and empirical, using a few parameters such as tortuosity and pore size distribution to characterize the capacity of soil to conduct and hold water, chemicals and biological collides (Currie, 1984). Over the past two decades, fractal models have also been found to be an appropriate approach to describe soil pore geometry, as they naturally give a power-law water retention curve, which is consistent with some measurements (Gimenez *et al.*, 1997). Although the fractal model theoretically explains the power-law water retention curve, proof of its ability to describe soil conductivity for water and chemicals is far from satisfactory despite some earlier efforts (Crawford, 1994). One reason is that the fractal model alone cannot describe the connectedness of soil pores over space.

The advent of imaging technology such as X-ray computed tomography (CT) and nuclear magnetic resonance imaging (MRI)

*Corresponding author: kejian.wu@pet.hw.ac.uk

over the past two decades has made direct visualization of opaque materials feasible in three dimensions (3D) (Dunsmoir *et al.*, 1991; Coles *et al.*, 1994; Spanne *et al.*, 1994; Hazlett, 1995; Coker *et al.*, 1996). For example, X-ray CT could provide a measurement of soil structure at resolutions as fine as a few microns, and these techniques have been used in soil science to investigate various processes (Young and Crawford, 2004) although, in spite of their increased use, they have some inherent limitations. For example, it appears that they are unable to separate microbes and fungi from soil minerals and other organic materials, and there is a trade-off between sample size and resolution. On the other hand, the thin-section technique can offer a complement as it is capable of revealing microbes and fungi in soils at the micro scale (Nunan *et al.*, 2003). However, thin-section techniques are limited to two dimensions. Therefore, developing mathematical models to construct 3D images based on 2D measurements can bridge this gap, and this has become an area of interest in both theory and practice over recent years in many fields (Zhang *et al.*, 2005; Wu *et al.*, 2006).

3 Overview of the Quantitative Characterization of Soil Microstructure

As water and chemicals can move only through the void space of soil with water acting as a carrier ensuring dissolved chemicals move in soil, the manner and speed of water moving from one place to another in soil are dictated by soil microstructure. A quantification of soil microstructure and its impact on water flow is hence the basis to understanding other chemical and biological processes. Modelling of soil structure can be dated back to the 1950s, and one of the earlier models is the cut–rejoin capillary concept proposed by Childs and Collis-George (1950). Since then, numerous structure models have been proposed. One shortcoming of these models is their inability to describe heterogeneity, which, in most natural soils, is usually spatially cor-

related (Young and Ritz, 2000; Young *et al.*, 2001). To consider such spatial correlation, Dexter (1976) used a one-dimensional Markov chain to model 2D soil structure. Another model is the fuzzy random concept proposed by Moran and McBratney (1997), in which the pores are treated as a fuzzy system rather than an explicit geometry. Most existing soil structure models are based on a lattice network in which the big pore bodies sit at network nodes with the nodes linked by either cylindrical or irregular throats. Such a network model can be created based on the pore-size distribution estimated from the water-retention curve, and has been applied to investigate the impact of pore geometry on soil transport parameters (Biswal *et al.*, 1999; Yeong and Torquato 1998a, b; Matthews *et al.*, 2006).

Thin sections of soils are easy to obtain and widely available, and therefore how to generate 3D structures from 2D thin sections has become an area of interest since the 1970s. A number of models have been developed based on different theories, ranging from two-point correlation function to Markov process (Joshi, 1974; Quiblier, 1984; Adler *et al.*, 1990, 1992; Hazlett, 1997; Roberts, 1997; Manswart and Hilfer 1998; Yeong and Torquato, 1998a, b; Okabe and Blunt, 2004; Wu *et al.*, 2006). These models normally assume that a soil structure can be fully characterized by a number of parameters and functions, which can be estimated from one or several thin sections. If soil is isotropic, these parameters and functions are also isotropic and independent of directions, and their value can be estimated from a single thin section. When soil is anisotropic, however, these parameters become direction-dependent and their values have to be estimated from several thin sections by cutting the sample from different directions. With these parameters and functions known, the 3D structure can be generated in a stochastic manner in that the generated structure is statistically (rather than geometrically) similar to the original (Dunsmoir *et al.*, 1991; Coles *et al.*, 1994; Hazlett, 1997; Biswal *et al.*, 1999; Manswart *et al.*, 2000; Øren and Bakke, 2003). It is obvious from their assumptions that these models could fail if the structure of a soil sample cannot be statistically characterized by a few parameters and functions. Therefore, instead of assuming statistical similarity, Bakke and Øren (1997) proposed a process-based model by mimicking the sedimentation process of the particles in fluid to generate 3D structures. Such a method might be valid for generating media such as sandstone in the oil industry, but is less applicable for soil as the structures of soils in managed and natural land are subjected to continuous change, and the origins and mechanisms of these changes are too complicated to be simulated in a way like that used in the simulation of sedimentation in water.

Depending on how well a sample can be characterized statistically, the above statistical models could produce a 3D structure differing significantly from its original sample, especially with regard to pore connectivity, and therefore fail as a predictive tool to estimate transport parameters (Hazlett, 1997; Manswart *et al.*, 2000). In this chapter, we present a novel method we have developed over the last decade to characterize and construct 3D structures. The method is still stochastically based, generating a 3D heterogeneous structure using Markov chain Monte Carlo (MCMC) simulation. It consists of two main steps, the first being to calculate the transition probabilities from three 2D thin sections obtained by cutting the sample at three perpendicular directions, and the second step to generate 3D structures based on these estimated transition probabilities. The main difference of the proposed method from existing models is that the MCMC model is able to deal with the substantial levels of anisotropy and heterogeneity existing in most natural soils.

In addition to the MCMC model, we will also present a set of tools to analyse the geometry and topology of soil structure. The numerically generated structure from MCMC can be used as a base to simulate fluid flow in calculating soil transport properties, such as permeability to water flow and dispersion for chemical transport, as well as in water retention to represent the ability of soils to hold water at different suctions.

4 Markov Random Field for Heterogeneous Soils

The MCMC method is developed from image processing analysis where Markov random fields (MRF) are widely used (Geman and Geman, 1984). The core of MRF is to use a small amount of *local* information to predict global features. Each generated image consists of a number of voxels, and the probability of each voxel being in a particular state is determined by the means of a transition matrix (conditional probabilities) and can be determined from an a priori image. For binary soil structure, the possible state of each voxel is either solid or pore, represented by a binary variable with either value equal to one or null. In order to model a heterogeneous soil, the soil porosity and local dependence of the voxels are treated as inputs and their values are estimated from thin sections. To describe spatial heterogeneity effectively, the numbers of the neighbourhood of each voxel should be sufficiently large; for the 3D structure considered here, we use the 15-neighbourhood. Since a high-level neighbourhood is used, a rapid convergence of the chain is vital and a new method is introduced to improve convergence (Wu *et al.*, 2004, 2006).

4.1 3D Markov chain model

The soil structure is assumed to be fully characterized by a probability distribution function (PDF). We represent the structure by an array consisting of n voxels. Let $X = (X_1, X_2 ..., X_n)$ denote the state of the voxels, with $X_i = 0$ representing pore and $X_i = 1$ solid. To reconstruct a 3D structure, we need a full PDF, $p(x)$ (i.e. the dependence of each voxel on all the voxels in the structure). However, this is impracticable, because the 3D sample is too large for a normalizing constant to be calculated directly. Therefore, instead of using the full PDF, we use MRF by assuming that the state of each voxel depends only on the states of its neighbouring voxels. To be more specific, for a particular voxel s, let Λ_{-s} denote the set of all the voxels excluding s. There exists a neighbourhood of s, N_s, in that

$$p(x_s \mid x(\Lambda_{-s})) \approx p(x_s \mid x(N_s)). \quad (2.1)$$

Markov random fields can be viewed as a multi-dimensional version of a Markov chain. In general, the distribution (Eqn 2.1) is difficult to simulate, and existing iteration methods take a long time to converge, thereby limiting its application to 3D. In 2D, a new single-pass simulation has been developed to find the conditional probabilities (Qian and Titterington, 1991a; Wu et al., 2004), and in what follows we extend this to 3D.

The following notations are used for convenience to represent a structure in the 3D Markov chain model. Let $V_{LMN} = \{(l, m, n): 0 < l \le L$ rows, $0 < m \le M$ columns, $0 < n \le N$ layers$\}$ be the structure lattice, (i, j, k) represent the voxel at the intersection at row i, column j and layer k with its associated state being X_{ijk}. V_{ijk} denotes the rectangular parallelopiped array of voxels as shown in Fig. 2.1, with the associated state vector $X(V_{ijk})$. With the single-pass scanning scheme, instead of using the intensive iteration method, we define the 'past' voxel of (a, b, c) as the set $\{(l, m, n): l<a,$ or $m<b$ or $n<c)\}$. The states of the past voxels depend on the direction in which we run the chain. For example, if the chain is run bottom upwards (along the z direction) layer by layer, all voxels in the layers below the current layer k (i.e. $n < k$) are past voxels. Similarly, within the same layer, if the chain runs from inside to outside (along the x direction) and then left to right (along the y direction), the past voxels will be those within $l < i$ and $m < j$. The Markovian assumption for the third-order model is

$$p(x_{ijk} \mid \{x_{lmn} : l < i,\ or\ m < j,$$
$$or\ n < k\}) = p(x_{ijk} \mid x_{i-1,j,k},\ x_{i,j-1,k},\ x_{i,j,k-1}). \quad (2.2)$$

Hence, for $(i, j, k) \in V_{LMN}$, we have the joint probability function as follows:

$$p\left[x(V_{ijk})\right] = \prod_{l=0}^{i} \prod_{m=0}^{j} \prod_{n=0}^{k} \quad (2.3)$$
$$p\left(x_{lmn} \mid x_{l-1,mn}, x_{l,m-1,n}, x_{lm,n-1}\right).$$

From equation (2.1), a general MRF for the lattice would use the conditional probabilities:

$$p\left(x_{ijk} \mid \{x_{lmn} : (l, m, n) \ne (i, j, k)\}\right)$$
$$= p\left(x_{ijk} \mid \{x_{lmn} : (l, m, n) \in N(ijk)\}\right), \quad (2.4)$$

where $N(ijk)$ is one of the neighbourhoods of the voxel at (i, j, k). For the example shown in Fig. 2.1, we set up the MRF and use the neighbourhood that involves only the past voxels. To improve computational efficiency, we use an algorithm that generates two voxels simultaneously: voxel (i, j, k) and voxel $(i, j+1, k)$, similar to that used in the 2D scanning introduced by Wu et al. (2004). Thus, the following 15-neighbourhood is defined in which 13 voxels are past voxels and only two are non-past voxels whose statue needs to be decided, as shown in Fig. 2.2 and

$$N_{15}(ijk; i, j+1, k)$$
$$= \begin{bmatrix} (i-2, j, k-1) & (i-2, j+1, k-1) & (i-1, j, k-1) \\ (i-1, j+1, k-1) & (i, j-1, k-1) & (i, j, k-1) \\ (i, j+1, k-1) & (i-2, j, k) & (i-2, j+1, k) \\ (i-1, j-1, k) & (i-1, j, k) & (i-1, j+1, k) \\ (i, j-1, k) & & \end{bmatrix}$$
$$(2.5)$$

The 15-neighbourhood in 3D is a direct extension of the 6-neighbourhood in 2D (Wu et al., 2004). Details of the scanning using N_{15} as defined in equation (2.5) will be given below in Section 3.3. Therefore, for V_{LMN}, the rows, columns and layers of MRF form a stationary vector Markov chain with dimensions of $(M + 1)(N + 1)$, $(N + 1)(L + 1)$ and $(L + 1)(M + 1)$, respectively. The total number of voxels is $N_v = (M + 1)(N + 1)(L + 1)$. For an h-order model with N_s states and N(h) neighbours, the computational complexity is

$$O\left(N_v \sum_{i=1}^{h} N_s^{N(h)}\right) \quad (2.6)$$

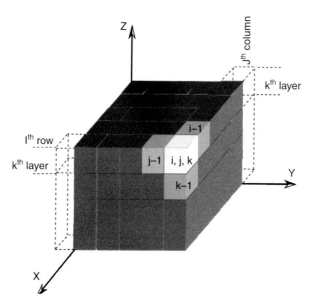

Fig. 2.1.   Definition of V$_{ijK}$ in the 3D Markov chain model.

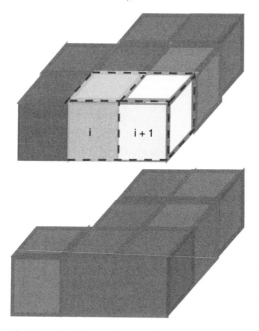

Fig. 2.2.   The 15-neighbourhood system in 3D.

(Qian and Titterington, 1991b). It can be estimated that the computational cost increases exponentially with the number of neighboured voxels. For binary soil structure, the value of N_s is 2.

4.2 3D Markov chain Monte Carlo for soil structure

The core of the above model is to estimate the values of the associated parameters. Theoretically, traditional methods like the maximum likelihood estimation can be used to estimate these parameters, but the computational cost is enormous and beyond current computer power. The MCMC can provide an alternative to improved computational efficiency, and is used here. To accomplish this, we have to construct a 3D Markov chain with state space X, which is straightforward to simulate and has an equilibrium distribution of $p(x)$. The chain is run for a sufficiently long time – i.e. sufficient size in the present discrete spatial context – so that the simulated structure of the chain can reproduce the important statistical features of $p(x)$ of the real material. We constructed a so-called ergodic Markov transition probability matrix; Wu *et al.* (2004) explored this approach in 2D, and here we extend it to 3D.

For a 3D image, the neighbourhood structure (Eqn 2.5) can be directly calculated. What we are interested in here is how to generate the 3D chain based on three 2D thin sections in three perpendicular directions. For each 2D image, a transition matrix can be built using the 6-neighbourhood scheme as described by Wu *et al.* (2004). As in 2D, the 3D 15-neighbourhood shown in Fig. 2.2 is built in two steps. First, an 11-neighbourhood $N_{11}(ijk)$ is constructed to determine the conditional probability of voxel (i, j, k) by combining the 2D 6-neighbourhood $N_{j,6}(ijk)$ and the 5-neighbourhoods $N_{i,5}(ijk)$ and $N_{k,5}(ijk)$, as shown in Fig. 2.3b. To clarify the notations, we use $N_{j,6}(ijk)$, to denote the 2D 6-neighbourhood of (i, j, k) in the plane with constant j and associated conditional probability $p(x_{ijk} | x(N_{j,6}(ijk)))$. Therefore, to determine the probability of voxel $(i, j + 1, k)$, the 6-neighbourhoods $N_{j,6}(i, j + 1, k)$, $N_{j,6}(i, j + 1, k)$ and $N_{k,6}(i, j + 1, k)$ are combined to form the 12-neighbourhood $N_{12}(i, j + 1, k)$, as shown in Fig. 2.3b. Note that the newly determined probability for voxel (i, j, k) is part of $N_{j,6}(i, j + 1, k)$ and $N_{k,6}(i, j + 1, k)$ $N_{11}(ijk)$ and $N_{12}(i, j + 1, k)$ combine to form the 15-neighbourhood $N_{15}(ijk; j + 1, k)$, as shown in Fig. 2.3c.

Hence, this 15-neighbourhood is a direct generalization of the previously developed 2D 6-neighbourhood. Using the above neighbourhood scheme, the Markov chain can be simulated as follows.

At the beginning of the chain, the probability of the first voxel being pore is taken as the value of the porosity. Then, the first row is constructed in a 1D simulation using a 3-neighbourhood, for which the conditional probabilities are derived from the horizontal thin section (constant k). The second row is started with a 4-neighbourhood, also derived from the horizontal section. Then, in one scan, we create the first layer row by row as in the 2D simulation described by Wu *et al.* (2004), using the 6-neighbourhood with probabilities derived from the horizontal thin section.

The first row of the second layer is formed similarly to the second row of the first layer (see step ii above), but using a vertical 4-neighbourhood (with constant i), followed by 6-neighbourhoods in the same plane. The second row of the second layer starts with reduced 7- and 8-neighbourhoods and continues with the reduced 10- and 11-neighbourhoods described above. Then, the inside rows and columns are built using the full 3D 15-neighbourhood described above.

To build the chain, 2- and 3-neighbourhoods in 1D are determined, yielding 2^2 and 2^3 transition possibilities in the binary systems. Similarly, 3- and 4-neighbourhoods in two 2D sections are determined, having the 2×2^3 and 2×2^4 possibilities, with 5- and 6-neighbourhoods in all three 2D sections, adding 3×2^5 and 3×2^6 possibilities. Hence, a total of 348 conditional probabilities need to be determined. The

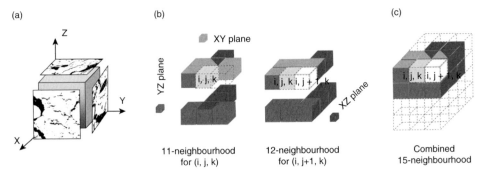

Fig. 2.3. Illustration of combining 2D 5- and 6-neighbourhoods derived from (a) three perpendicular planes in 3D into (b) 11- and 12-neighbourhoods for updating the two voxels (i, j, k) and (i, j + 1, k) simultaneously, resulting in (c), the combined 15-neighbourhood. For example, the 11-neighbourhood combines a 5-neighbourhood in the XY plane (light grey), a 5-neighbourhood in the YZ plane (mid-grey) and a 6-neighbourhood in the XZ-plane (dark grey). Note that two grey levels are marked for voxels from overlapping neighbourhoods, while (i, j, k) is part of all three neighbourhoods.

chain runs for a sufficient time using the scanning scheme presented by Wu *et al.* (2006), in which the simulated structure from the chain is used as a basis to represent the features of the soil sample. More details of the method are given by Wu *et al.* (2004, 2006).

The model outlined above has been applied to simulate a number of soil and rock samples, ranging from highly porous sandstone to dense clay soil. For each sample, prior to cutting the sample in the three principal directions to obtain thin sections, a flow experiment was carried out first to measure the hydraulic conductivity and water retention curve, which are used as references to test the accuracy of the MCMC model. From the three thin sections shown in Fig. 2.3a, we can calculate the Markov transition probability and then use it to reconstruct the 3D structure using the 15-neighbourhood model. Depending on the resolution of the thin section and computer power, the size of the simulated 3D structure can range from centimetres to metres. For the soil images we investigated, the size of pixels in the thin section ranged from 0.1 μm (for fine media) to more than 10 μm (for coarse sandstone); the associated size of the numerically generated 3D sample therefore ranges from a few cubic micrometres to more than 10 cm^3. Simulation revealed that the MCMC chain constructed using the above method converges approximately after $200 \times 200 \times 200$ iterative steps, meaning that the minimum size of the constructed structure consists of at least $200 \times 200 \times 200$ voxels.

The target of the reconstructed structure is to reproduce the topological features of the real 3D soil using the information extracted from several 2D thin sections in such a way that the porosity, pore size distribution, pore connectivity, hydraulic conductivity and water retention characteristics of the reconstructed structure are close to those of the real samples. Table 2.1 shows four samples with contrasting structures generated using the MCMC model. Also shown in the table is a comparison of the experimentally measured hydraulic conductivity with that numerically calculated from a simulation of water flow in the reconstructed 3D structure using the lattice Boltzmann model. The novelty of the MCMC model is that its chain needs only one pass rather than the intensive iterations used in standard MCMC; the size of the chain depends on the heterogeneity revealed by the thin sections. Figure 2.4 and Table 2.1 compare the 2D images taken from the reconstructed 3D images with the 2D thin-section image on which the 3D structures are generated.

Multiple-component model: microbial distribution

The spatial organization of soil and bacterial cells can be visualized and quantified using a combination of thin sections, image processing and analysis procedures (Nunan *et al.*, 2002). Spatial associations of bacteria with transmission porosity have also been quantified (Nunan *et al.*, 2003). Incorporating microbial data into models for soil structure would allow interactions between bacteria and their microhabitat to be explored. The spatial distribution of microbes in soil is associated with pore geometry and is therefore also spatially correlated. Scales of micrometre, or lower, are needed to identify the bacteria in an image, whilst the modelling of soil structure operates at the centimetre scale in order to describe the pore morphology that controls fluid flow and nutrient transport. We therefore face a scale issue in relating the microbial data measured at a smaller spatial scale to their habitat at larger scales. We have therefore developed an upscaling method to bridge this gap, based on the Markov chain model.

Measurement of bacteria at 0.16 μm scale

Quantitative analysis of the spatial distribution of bacteria in soil requires the location of the bacterial cells to be measured accurately at different spatial scales. The number and *in situ* spatial distribution of bacteria at a scale of 0.16 μm is determined by an image-processing and analysis procedure (Nunan *et al.*, 2003). Composite (tessellated) images in which individual bacteria can be resolved within an area of

Table 2.1. Comparisons of various properties between experimental and simulation data.

Properties		Sample 1	Sample 2	Sample 3	Sample 4
2D thin section image (20 × 20 mm²) Horizontal					
Vertical					
3D simulation image Realization 1					
Realization 2					
Realization 3					
Porosity (%)	2D image average	11.94	9.27	11.86	15.86
	Simulation 1	10.97	8.36	14.54	19.25
	Simulation 2	9.35	12.94	13.12	14.53
	Simulation 3	8.32	12.20	12.28	15.23
Hydraulic conductivity K_{sat} (m/d)	Laboratory measurement	0.467	1.389	3.604	45.035
	Simulation 1	0.5492	2.0511	2.3072	36.5691
	Simulation 2	0.7828	1.5387	3.6156	54.9234
	Simulation 3	0.3284	0.8705	2.4003	40.5444

0.282 mm² were acquired by means of a motorized scanning microscope stage. To visualize the bacteria in soil, high magnification imaging has to be involved, and at these magnifications single fields of view through optical microscopes present only small image areas (typically ~0.01 mm²). One resultant microbial image under high magnification (×650) is shown in Fig. 2.5a. The microbial data are at such a small scale that there is an issue in matching the lower magnification scale of soil structure, because this has to be considered at a lower resolution image within computing limits

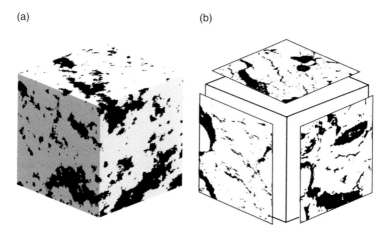

Fig. 2.4. (a) Simulated 3D heterogeneous soil based on (b) three perpendicular thin sections.

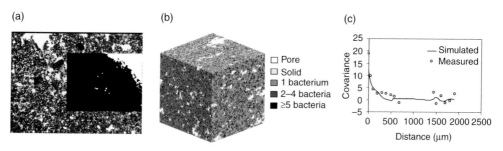

Fig. 2.5. Simulation of a 3D heterogeneous soil structure in association with microbial distribution. (a) Soil biological thin section; (b) 3D simulated bacterial distribution in relation to soil structure; (c) covariance of the simulated and measured microbial distributions.

to describe the macro-pores that control flow systems within the soil structure.

Determination of the impact of soil structure on microbial distribution: upscaling

The soil structure to be modelled is based on a low-resolution image with an area of 1.58×1.19 cm^2 (magnification ×18), equivalent to 670 high-resolution images (×650) each having an area of 0.28 mm^2. The microbial density maps were de-resolved by counting the bacterial numbers in a quadrate equivalent to one pixel of the low-resolution image. As an example, Fig. 2.5a shows bacterial density in a quadrat of 434 µm^2. It will be seen that most of the quadrate contains only one bacterium, which implies that the de-resolved maps still retain most of the individual's distribution information, sug-

gesting that the model at the low-resolution scale could successfully simulate the distribution of individual bacteria. We incorporated bacterial distribution into soil structure using upscaling from the micron to the centimetre scale, after de-resolving the high-resolution pore map and then combining them to provide information on the coexistence of bacteria and pore. These data are used to simulate microbial distribution in relation to soil structure.

Integration of microbial distribution model with soil structural model

The conditional probability distribution (probability of obtaining a specific state with a given neighbourhood) for bacteria can be obtained in a similar manner to that for soil structure, as outlined above. Once this is obtained we apply the MCMC to simulate bacterial distribution by the algorithm presented by Wu *et al.* (2004) to sample the posterior distribution. According to the histogram of the bacterial density map, five categories can be defined for each pixel: pore, solid, 1 bacterium, 2–4 bacteria and >4 bacteria. Because of the intensive computation of processing the images and counting the bacterial cells in thin sections, only small images (comprising 3840 × 2870 pixels per image) were randomly selected from each thin section to obtain bacterial density in relation to soil structure. We then measured these to obtain the transition matrix, similar to the parameterization for soil structure. Thus, in the model for bacterial distribution, we use five colours to represent the presence and number of bacteria in each pixel. Therefore, there are 15,625 parameters in the transition matrix for bacterial distribution and 86 parameters for soil structure.

We modelled bacterial distribution in conjunction with soil structure using a multi-dimensional Markov chain model. Unlike soil structure, in which the status of each voxel is binary, the status of the structural–microbial model has five possibilities. Figure 2.5b shows an example of the reconstructed 3D soil structure in association with spatial bacterial distribution based on the 2D thin section shown in Fig. 2.5a. The size of the 2D image is approximately 1.6×1.2 cm² (comprising 760 × 579 pixels, each of 22 μm). The simulation result shows that the total number of bacteria in the simulated image is close to that of the measurement. We also calculated the covariance of spatial microbial distribution in the simulated structure and in the thin section, with the comparison shown in Fig. 2.5c: they are very close.

5 Characterization of Pore Geometry and Topology

Soil structure is complex and difficult to quantify, yet such information is needed to adequately predict soil transport properties. Different approaches with various complexities have been proposed to quantify pore geometry. One purpose of characterization of pore geometry is to link the geometrical characteristics of soil to its transport properties (Bakke and Øren, 1997; Liang *et al.*, 2000; Lindquist *et al.*, 2000; Silin *et al.*, 2003). In the work presented here, we use the approach proposed by Jiang *et al.* (2007). There are two basic approaches in characterizing a 3D structure. The first is to fit pore elements into the void space of the structure so as to assign each pore voxel a pore size, and the other is to extract a backbone to make a network model with the pore bodies connected by throats. The method presented here belongs to the first approach; it is an 'element-fitting' method and the shape of the element is spherical.

The method consists of three main steps: (i) determination of pore size; (ii) partitioning and extraction of pore volume; and (iii) extraction of the pore network. After all pores have been measured and labelled, an invasion percolation is carried out to obtain the distributions of pore size and connectivity.

Pore size is defined as equivalence to the radius of the maximum ball that just fits within the pore. To measure pore size template spheres with different diameters, as shown in Fig. 2.6, are used. To improve the computational efficiency of the fitting algorithm, large pores are extracted first from the 3D structure. An estimation of the radius of the largest sphere is important in reducing computational time. Based on the relationship between porosity and pore size (Mulder, 1996), the likely largest pore radius can be estimated approximately from

$$M_r = \left(\frac{3v\phi}{4\pi}\right)^{\frac{1}{3}}, \qquad (2.7)$$

where M_r is the upper limit of the radius of the largest sphere, ϕ is the porosity and v is the volume of the structure. The radius of the largest pores in the image might be much smaller than that predicted from equation (2.7); therefore, in practice, a smaller radius ($M_{st} = M_r/3$) would be a reasonable approximation to the largest radius.

In a 3D structure comprising a number of voxels, a pore is defined as consisting of a

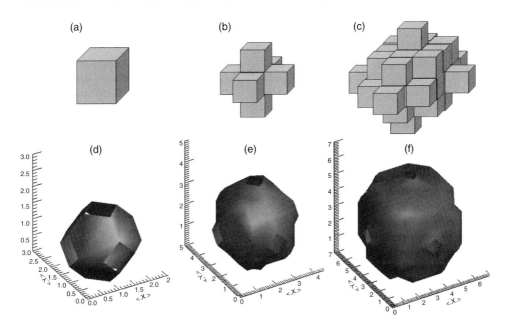

Fig. 2.6. Discrete template spheres (a, b and c with radius 0, 1 and 2, respectively) with corresponding real spheres (d, e, f).

26-/18-connected component of voxels. Most existing approaches for defining a pore rely on morphological operations (dilation or erosion) and cannot satisfy the requirements for detecting pore topology and geometry (Spanne *et al.*, 1994; Baldwin *et al.*, 1996). As a result, a new method is proposed here, combining geometrical transformation and morphological operation. In this approach, a pore is assumed to be equivalent in size to the radius r of the sphere. Apart from being large enough to contain an inscribed sphere, the pore should be able to preserve the irregularity of the void space in natural soils. A 26-/18-connectivity porous component X_r is defined as a pore of size r if (i) there exists a sphere $B(p, r-1)$ of radius $r-1$ such that $B(p, r-1) \subset X_r$; and (ii) the number of voxels in the set $\{p \mid p \in X_r, p \in B(p, r), p \notin B(p, r-1)\}^{\#}$ is larger than $N_{\min}(r)$, where $N_{\min}(r)$ is defined as follows:

$$N_{\min}(r) = N_{\min}(M_{sr})[\tfrac{1.0 - N_{\min}(1)}{M_{sr} - 1}(r-1) + N_{Min}(1)];$$

(2.8)

that is, a pore of size r must completely contain a sphere of radius $r-1$ and include

most of the voxels in the sphere with radius r. In practical terms, $N_{\min}(M_{sr})$ and $N_{\min}(1)$ are set to be 0.95 and 0.65, respectively, for a sandstone sample. Thus, the number of voxels in a pore of size r should be larger than the number of voxels in the current template sphere and the value of $N_{\min}(r)$ should be associated with the current radius r.

When a pore has been identified through the process described above, we use the cubic cutting method to remove it from the void space to avoid overlapping with pores that have not yet been extracted. After that, the pore is partitioned into an associated cubic volume with a side length r. Finally, the pore is combined with adjacent smaller pores under predefined criteria and then merged with all the remaining voxels without altering the pore connectivity, as shown in Fig. 2.7.

5.1 Pore connectivity: coordination and Euler numbers

Soil functions depend not only on pore size distribution but also on the connectivity of

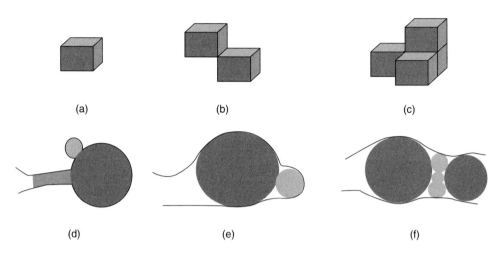

Fig. 2.7. Merging smaller pores and remaining voxels. (a–c) Remaining voxels with radius <1; (d) small pores contributing to pore throat (mid-grey) should remain; (e) smaller dead end-pores (light grey) with only one neighbour (grey) of larger size can be merged; (f) smaller pores (light grey) can be combined with larger pore (on left).

pores of different sizes. Here we use coordination and Euler numbers to describe pore connectivity at the local and global scales, respectively. Their calculation from a 3D structure is explained briefly below.

After all pores have been extracted from the structure, the centre of each pore is defined as the coordinate of the centre of its associated sphere. The contact area of the neighbouring pores is also recorded, and it is then possible to count the pore coordination number. The 6-connectivity and 26-connectivity neighbourhood models are commonly used in 3D images to define the coordination number, and we use this to quantify the connectivity. The 26 neighbours of each pore were checked to determine pore connectivity, which is subsequently used to calculate the coordinate number if the neighboured pores have different pore sizes. Figure 2.8a shows the distribution of coordination numbers of a 3D image directly measured from X-ray CT, in comparison with that calculated from a numerically generated structure using the MCMC method outlined above.

The Euler number is defined as $N - C + H$, where N is the number of isolated pores, C the number of tunnels and H the number of completely enclosed cavities. The smaller the Euler number, the better the connectivity

of the soil sample. There are a number of ways to estimate the Euler number – for details see Vogel (1997), Kong and Rosenfeld (1989) and Saha and Chaudhuri (1995). Here, we propose a novel method based on the concept of topological number and geodesic neighbourhood used by Kong and Rosenfeld (1989) to estimate the Euler number. A binary 3D image can be defined as $P = (V, 26, 6, B)$, where V is the set of all cubic grid points, B is the set of all pore voxels in V, and $V–B$ represents the solid voxels. Combined with the concept of topological number and using Kong's algorithm, the Euler number $\chi(\cdot)$ can be computed as follows.

$$\chi(V, 26, 6, B) = 0 \text{ if } B = \varnothing. \tag{2.9}$$

For any other point $x \in B$,

$$\chi(V, 26, 6, B) = \chi(V, 26, 6, B-\{x\}) + T_6(x, V-B) - T_{26}(x, B) \tag{2.10}$$

where $\chi(V, 26, 6, B)$ is the Euler number; $\chi(V, 26, 6, B-\{x\})$ is the Euler number excluding point x; $T_6(x, V-B)$ is the solid topological number of x; and $T_{26}(x, B)$ is the pore topological number of x. The distribution of pore connectivity against pore size can be defined as $G(r_0) = \chi\{ p \mid p \in P_r, r \leq r_0\}$, where P_r is all pores with radius r. Figure 2.8b shows the

(a)

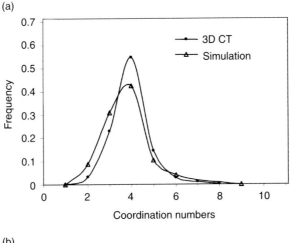

(b)

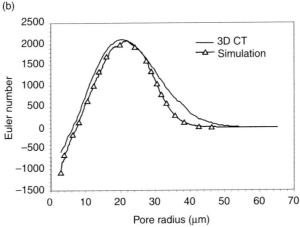

Fig. 2.8. (a) Coordination number distribution curve and (b) connectivity function (Euler number) curve derived from 3D CT of reconstructed sandstone.

relationship between Euler number and pore size calculated using this method for a sandstone sample. Obviously, the connectivity is high for small pores. However, there is a rapid increase in the specific Euler number for pore sizes in the range 50–60 mm, indicating that in this range the pores are likely to be more globally connected.

5.2 Water distribution in the void space

The above characterization of pore geometry assigns each pore voxel in the 3D structure a pore radius, and the void space can therefore be viewed as an assembly of the pore voxels of different sizes, as shown in Fig. 2.9. Hence, the ability of soil to hold and conduct water is entirely determined by the spatial connectedness of these voxels, and their values can be calculated by mimicking the ways used in the laboratory to measure water characteristic curves and hydraulic conductivity. For example, to calculate the water distribution in the void space under different pressures, one can apply a series of suctions at the base of the structure either to drain or wet it, depending on the antecedent water content. For an applied suction h at the base of the

(a)   (b)

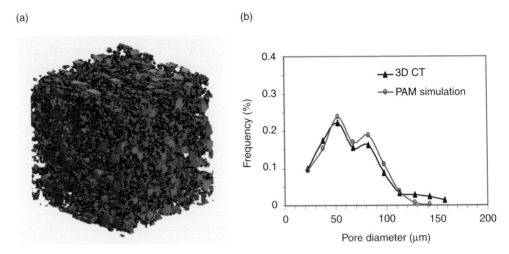

Fig. 2.9.   Illustration of pores of various sizes in a simulated 3D structure (represented by different colours) and pore size distribution curve (right) estimated from 3D CT and numerical structure.

structure, all pore voxels with radius $< r$ will be filled with water if they form one or a few clusters that extend to the base where the suction is applied. The value of r is given by $r = \gamma\cos\alpha/\rho gh$, where γ is the surface tension of water, α is the contact angle, ρ is the density of water and g is the acceleration of gravity. For calculating water distribution in the void space under each applied suction, we use a binary variable to describe the status of each pore voxel such that, if the pore voxel has already been filled with water, the value of its status is 1, otherwise it is 0. Therefore, during the wetting process, a pore voxel with diameter $< r$ can only be filled by water when at least one of its six face-to-face neighbours has already been filled by water. To be more specific, for an air-filled pore voxel located at (i, j, k) with radius $< r$, the value of its binary status is determined as follows:

$$X(ijk) = \begin{cases} 1 & \text{if at least one of the six} \\ & X(i\pm1, j\pm1, k\pm1) \text{ is equal to 1} \\ 0 & \text{otherwise} \end{cases}$$

(2.11)

For each applied suction h, the associated water distribution in the void space of the structure can be obtained by applying equation (2.11) to all the pore voxels iteratively until no voxels show status change during the iteration. Repeating the above procedure for different suctions gives the water characteristic curve. The water retention curve associated with the wetting and drying processes can be calculated by initially drying or saturating the structure, respectively. Figure 2.10 shows a combination of the use of the above model to reconstruct a 3D structure based on a 2D thin section, followed by characterization of pore geometry and, finally, estimation of the saturation–suction curves as shown in Fig. 2.10c.

6 Simulating Water Flow and Chemical Diffusion in Soil Structure

Because of the inherent complexity and opaqueness of soil structures, most soil and hydrology textbooks often claim that any attempts at modelling water flow and chemical transport processes through the void space in soils are redundant (e.g. Bear, 1972). The rapid development of computer power and computational physics over the past 20 years, however, has proved this thought premature. In fact, the advent of imaging technologies and computational fluid dynamics over the past two decades has made a direct simulation of water flow

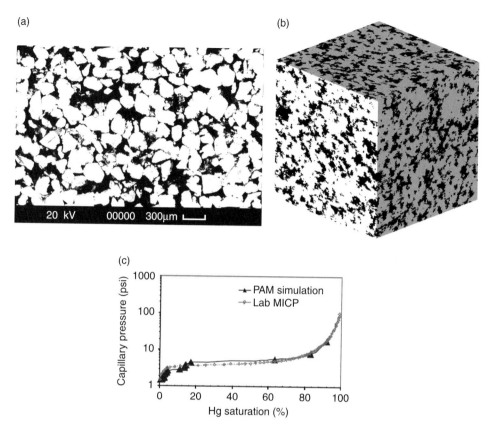

Fig. 2.10.   (a) 2D sandstone thin section image; (b) reconstructed 3D structure based on the thin section; (c) comparison between measured saturation–suction relationship with that estimated from numerical structure using invasion percolation.

and chemical transport in porous media at the pore scale feasible and renewed interest in many fields, including petroleum engineering and hydrology and soil science. The most commonly used model for simulating such flow and transport processes is the lattice Boltzmann (LB) method. The LB method is a numeral model developed over the last two decades based on kinetic theory to simulate fluid dynamics (Chen *et al.*, 1992). The concept of the LB method was introduced by McNamara and Zanetti (1988), aimed at overcoming some drawbacks of its predecessor, the lattice gas algorithm, proposed by Frisch *et al.* (1986) to simulate fluid dynamics. It was soon found that the LB method can also be derived independently from the continuous Boltzmann equation in kinetic theory through a discretization of both phase space and physical space (He and Luo, 1997).

Unlike conventional computational fluid dynamics, the LB method does not directly solve the partial differential equations; instead, it tracks the movement and collision of a number of fictitious particles in a lattice. Entities such as fluid density and velocity at each node of the lattice are calculated by summing up the corresponding moments of all the particles present in the node. The advantage of the LB method is its ease of use when dealing with complicated boundaries and various forces, and it is therefore efficient at simulating complex flow in porous media. It has been applied in hydrology and soil science to simulate a variety of processes such as multiple-phase flow (Pan *et al.*, 2004; Sukop and Or, 2004), biofilm growth (Knutson *et al.*, 2005) and single- and multiple-component reactive transport (Kang *et al.*, 2006; Zhang and Lv, 2007). In what follows, we will briefly explain the LB method and its application for modelling single-phase flow and solute movement in the void space of soils.

6.1 LB model for water flow

Several LB methods have been proposed, but the simplest and most widely used is the following single-relaxation time BGK model (Chen *et al.*, 1992; Qian *et al.*, 1992), in which the movement and collision of the fictitious particles are described by

$$f_i\left(x + \delta t\xi_i, t + \delta t\right) = f_i\left(x, t\right)$$
$$+ \frac{1}{\tau}\left[f_i^{eq}\left(x, t\right) - f_i\left(x, t\right)\right], \quad (2.12)$$

where $f_i(x,t)$ is the particle distribution function – the probability of finding a particle at position x at time t moving with velocity ξ_i, $f_i^{eq}(x,t)$ is the equilibrium distribution function – the value of $f_i(x,t)$ when it is in equilibrium state, τ is a dimensionless time that controls the rate of $f_i(x,t)$ approaching $f_i^{eq}(x,t)$ and δt is time step. We use a cubic lattice, as shown in Fig. 2.11, to simulate water movement in which the particles move in 19 directions with velocities $\xi_0 = (0, 0)$, $\xi_{1,2} = (\pm\delta x/\delta t, 0, 0)$, $\xi_{3,4} = (0, 0, \pm\delta x/\delta t)$, $\xi_{5,6} = (0, 0, \pm\delta x/\delta t)$, $\xi_{7-10} = (\pm\delta x/\delta t, \pm\delta x/\delta t, 0)$, $\xi_{11-14} = (\pm\delta x/\delta t, 0, \pm\delta x/\delta t)$ and $\xi_{15-18} = (0, \pm\delta x/\delta t, \pm\delta x/\delta t)$, where δx is the side length of the cubes.

The original equilibrium distribution functions given by Chen *et al.* (1992) and Qian *et al.* (1992) that have been widely used since then are

$$f_i^{eq}\left(\rho, \boldsymbol{u}\right) =$$
$$w_i\rho\left[1 + 3\xi_i \cdot \boldsymbol{u} + 4.5\left(\xi_i \cdot \boldsymbol{u}\right)^2 - 1.5u^2\right], \quad (2.13)$$

with the weighting factor $w_i = 1/3$, for $|\xi_i| = 0$, $w_i = 1/18$, for $|\xi_i| = \delta x/\delta t$ and $w_i = 1/36$ for $|\xi_i| = \sqrt{2}\delta x/\delta t$); the fluid density ρ

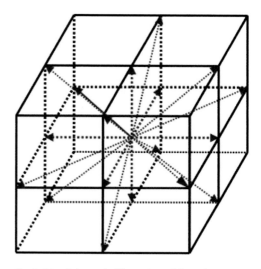

Fig. 2.11. Schematic illustration of the cubic 3D LB model for water flow and solute transport.

and velocity vector u in equation (2.13) are calculated from

$$\rho = \sum_{i=0}^{18} f_i = \sum_{i=0}^{18} f_i^{eq},$$
$$\rho u = \sum_{i=0}^{18} f_i \xi_i = \sum_{i=0}^{18} f_i^{eq} \xi_i. \quad (2.14)$$

The above LB method recovers the following Navier–Stokes equation in the limit for small Knudsen and Mach numbers in second-order accuracy:

$$\frac{\partial \rho}{\partial t} + \nabla \cdot \boldsymbol{u} = 0,$$
$$\rho\frac{\partial \boldsymbol{u}}{\partial t} + \rho\boldsymbol{u} \cdot \nabla \cdot \boldsymbol{u} = \mu\nabla^2\boldsymbol{u} - \nabla P, \quad (2.15)$$

with the kinematic viscosity given by $\mu = (\tau - 0.5)\delta x^2/3\delta t$ and the pressure by $P = \rho\delta x^2/3\delta t^2$. Because of the dependence of pressure on density, the above LB method actually simulates the uncompressible flow using a compressible approximation. As a result, it could give rise to significant compressibility errors when modelling flow driven by pressure gradient. To demonstrate this, we simulated the 2D Poiseuille flow driven by a pressure gradient; the pressure gradient was created by maintaining different pressure (i.e. fluid density) at the two ends of a pipe, as shown in Fig. 2.12a.

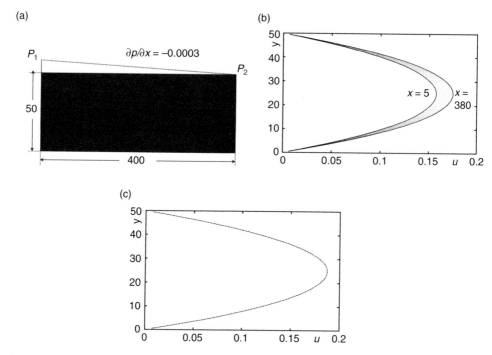

Fig. 2.12. Comparison of velocity profile simulated using original and modified LB models. The 2D flow domain (a); velocity profiles simulated by LB with the equilibrium distribution function defined by equation (2.13) (b), in comparison with that simulated with the modified equilibrium distribution function defined by equation (2.16) (c).

The simulated velocity profiles (along the vertical y direction) at different locations in the x direction away from the inlet are shown in Fig. 2.12b; it is obvious that, for a given y, the value of the velocity varies with x, which is incorrect because of the above compressibility errors. The errors induced by the above LB method increase with pressure gradient.

Several modifications have been proposed in attempts to reduce or eliminate such compressibility errors. One simple, yet very efficient, approach is the method proposed by Zou and He (1997) by modifying the equilibrium distribution functions as follows:

$$f_i^{eq}(\rho, \boldsymbol{u}) =$$
$$w_i\left[p + 3\xi_i \cdot \boldsymbol{u} + 4.5(\xi_i \cdot \boldsymbol{u})^2 - 1.5\boldsymbol{u}^2\right]. \quad (2.16)$$

The associated fluid density and velocity are calculated from

$$\rho = \sum_{i=0}^{18} f_i = \sum_{i=0}^{18} f_i^{eq},$$
$$\boldsymbol{u} = \sum_{i=0}^{18} f_i \xi_i = \sum_{i=0}^{18} f_i^{eq} \xi_i. \quad (2.17)$$

This modified LB model recovers the following Navier–Stokes equation in the limit:

$$\frac{\partial \rho}{\partial t} + \nabla \cdot \boldsymbol{u} = 0,$$
$$\frac{\partial \boldsymbol{u}}{\partial t} + \boldsymbol{u} \cdot \nabla \cdot \boldsymbol{u} = \nu \nabla^2 \boldsymbol{u} - \nabla P, \quad (2.18)$$

with the kinematic viscosity given as $\mu = (\tau - 0.5)\delta x^2/3\delta t$ and the pressure still by $P = \rho \delta x^2/3\delta t^2$. This modification significantly reduces the compressibility errors, as revealed in Fig. 2.12c, for the same problem as shown in Fig. 2.12 a and b, but simulated with the modified LB model.

The LB method involves two steps to advance one time step: a collision step and streaming step. The collision step involves calculating the term on the right-hand side of equation (2.12) as $f_i(x, t)^* = f_i(x, t) + [f_i^{eq}(x, t) - f_i(x, t)]/\tau$ for all fluid particle distribution functions, and the streaming step moves the collision result to a new position to become $f_i(x + \xi_i \delta t, t + \delta t) = f_i(x, t)^*$ at the end of each time step.

6.2 The impact of soil particles: the boundary treatment

The impact of soil particles on water flow comes into effect in the LB model when fluid particles hit a solid wall during the streaming step. How to handle collisions between solid wall and fluid particles is an issue that has not been well resolved in modelling of fluid flow in soils; the difficulties lie in the quantification of the impact of the solid wall on water flow. It is understood that most soil particles in natural and managed lands are coated with organic material such as biofilms, which may make soils hydrophobic and less affinitive to water. Such a coating is spatially heterogeneous and its effect on water movement is difficult to measure and quantify. For simplicity, most studies have treated the solid wall as a non-slip boundary where the water velocity is zero. Such a treatment is obviously an approximation because, strictly speaking, in the presence of organic coating water could slip over the solid wall and the water velocity on the wall is therefore unlikely to be zero. Quantification of the slip induced by organic coating on soil particles is an area still under development. Here we present only the methods of how to handle the non-slip boundary in the LB model. The non-slip boundary might not adequately describe the effect of soil particles on water movement, and it serves only as an approximation.

The simplest method of solving the non-slip boundary is the so-called bounce-back method, which is a direct inheritance of the lattice gas algorithm achieved by mimicking the movement of a ball when it hits a solid wall. To explain the method, we take a 2D problem as an example. The 2D lattice consists of a number of pixels, each fully occupied by either solid or water. The particle distribution functions are located at the centre of the pixels. The boundary treatment involves dealing with the collision between solid wall and fluid particles during the streaming step when the particles hit the wall. To explain this, we take the sample shown in Fig. 2.13 as an example, where the water pixel centred at 0 is bordered by two solid pixels. During the streaming step, the particle distribution functions f_2 and f_6 hit the wall of the two solid pixels. Because the particle distribution functions are located at the centre of the pixel, they hit the wall halfway in their one-step journey, i.e. $\delta t/2$ after collision. With the bounce-back method, the wall simply bounces the particles back to the centre of the fluid pixel at the end of each time step; that is, $f_1(x_0, t + \Delta t) = f_2(x_0, t)^*$ and $f_8(x_0, t + \Delta t) = f_6(x_0, t)^*$, where x_0 is the location of the centre of the water pixel. Application of the above bounce-back method to other scenarios and 3D is straightforward. From the definition of the velocity in equation (2.17), it is obvious that velocity at the solid wall is naturally zero at time $\delta t/2$ after each collision.

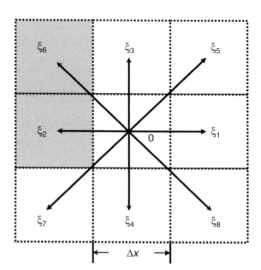

Fig. 2.13. Schematic representation of the bounce-back method for treating the water–solid boundary in the LB method (grey pixels represent solid).

6.3 LB model for solute transport

The LB model for solute transport is similar to that for water flow; it tracks the movement and collision of a number of theoretical solute particles, the evolution of which is described by

$$g_i\left(x + \delta t \xi_i, t + \delta t\right)$$
$$= g_i\left(x, t\right) + \frac{1}{\tau_c}\left[g_i^{eq}\left(x, t\right) - g_i\left(x, t\right)\right], \quad (2.19)$$

where $g_i(x, t)$ is the particle distribution function in the ith direction for the solute, $g_i^{eq}(x, t)$ is the value of $g_i(x, t)$ at the equilibrium state and τ_c is a dimensionless time that controls the rate of $g_i(x, t)$ approaching $g_i^{eq}(x, t)$. For illustration, solute transport is assumed to consist of convection and molecular diffusion only, and the diffusion is isotropic. Therefore, unlike water flow, the macroscopic variable to be recovered in the LB model for solute transport is concentration, which is a scalar variable. For computational efficiency, we use the same lattice as for water flow but allow the solute particles to move only in seven directions with velocities $\xi_0 = (0, 0)$, $\xi_{1, 2} = (\pm \delta x/\delta t, 0, 0)$, $\xi_{3, 4} = (0, \pm \delta x/\delta t, 0)$ and $\xi_{5, 6} = (0, 0, \pm \delta x/\delta t)$. Following Zhang and Lv (2007), the equilibrium distribution functions for the solute movement are defined as follows:

$$g_i^{eq}\left(c, \boldsymbol{u}\right) = w_i c\left[1 + 3.5 \xi_i \cdot \boldsymbol{u}\right], \quad (2.20)$$

where the weighting coefficient w_i is a constant ($w_i = 1/7$). The fluid velocity u in equation (2.20) can take the value simulated from the LB model for water flow, and the concentration c at position x and time t is calculated from $c = \sum_{i=0}^{6} g_i\left(x, t\right)_i = \sum_{i=0}^{6} g_i^{eq}\left(x, t\right)$. As shown by Zhang *et al.* (2002), the above LB model for solute transport recovers the following convection–diffusion equation

$$\frac{\partial c}{\partial t} + \nabla \cdot \left(\boldsymbol{u} c\right) = D_0 \Delta c, \quad (2.21)$$

where the molecular diffusion coefficient is given by $D_0 = (\tau_c - 0.5)\delta x/3.5\delta t$. Similarly, the implementation of the LB model

for solutes also includes a collision step and a streaming step to advance one time step. The collision step is to calculate the right-hand term in equation (2.19) as $g_i(x, t)^* = g_i(x, t) + [g_i^{eq}(x, t) - g_i(x, t)]/\tau_c$ and the streaming step is to move the collision result to a new position to become $g_i(x + \xi_i \delta t, t + \delta t) = g_i^*(x, t)$. The impact of the solid particles on solute movement is more complicated than for water movement, as most chemical and biological reactions of solutes in soils occur on the surface of soil particles. The simplest case is for inert solutes that do not react with soils, so that the impact of soil particles on solute movement can be solved in the same way as for water flow, i.e. by treating the wall as a non-slip boundary in that any solute particles hitting the wall during the streaming step are simply bounced back to where the particles come from at the beginning of each time step.

7 Applications of Pore-scale Modelling

Pore-scale modelling of fluid flow and chemical transport has progressed rapidly over the past decade and is now capable of simulating complicated multiple-phase flow and multiple-component reactive solute movement in 3D soil structures, obtained using imaging technology such as x-ray CT (Chen and Doolen, 1998). Apart from being used as a predictive tool to estimate transport parameters at the macroscopic scale (REV scale), the significance of pore-scale modelling is to provide a complement to investigate some fundamental transport processes in soils that otherwise remain invisible in experimentation. One of these processes is the velocity field in void space, which has a profound impact on chemical transport and microbial activity. Since these processes are immeasurable, pore-scale modelling can provide an alternative to bridging this gap and ultimately improves modelling of these processes at macroscopic scales. Here, we take chemical movement as an example to demonstrate how pore-scale modelling improved our understanding of fundamental transport processes.

(a)

(b)

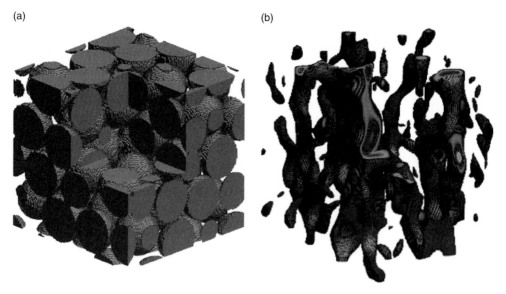

Fig. 2.14. X-ray tomography of glass beads (a), the simulated fluid velocity field with the magnitude of the velocity being represented by different colours (b).

(a)

(b)

Fig. 2.15. Idealized porous medium packed regularly by discs of diameter either 34 or 40 pixels (a); the simulated velocity field (b).

It has been generally accepted in reference to soil and hydrology that, in homogeneous media, the movement of inert solutes can be described by the classical advection–dispersion equation (ADE), with the dispersion coefficient increasing monotonically with the average pore water velocity. Strictly speaking, this perception could be wrong as recent experiments and pore-scale simulation both show that, in carefully packed uniform columns, the velocity field appears to be spatially erratic and water flows mainly along a number of channels. Even in a medium where all pores are well connected hydraulically, water could still bypass some large pores with water flowing along a few channels (Zhang *et al.*, 2005). As an example, we present the findings of simulated water flow in a column packed with glass beads, as shown in Fig. 2.14a. The diameter of the glass beads was 2.11 mm and the column is therefore very homogeneous. However, the velocity field in the pore space is erratic and water movement takes place along a few channels, as indicated in Fig. 2.14b. Such a channel-type flow has a significant impact on chemical movement, as well as its associated physical and biochemical processes in soil. For example, for microbial activity, this channel-type flow (hence nutrient movement) implies that the bioavailability of nutrients for microbes is likely to be in the vicinity of these channels. Such a flow-induced spatial heterogeneity of nutrients also renders the microbial distribution spatially heterogeneous, which has profound impact on microbial biodiversity and associated functions (Long and Or, 2005).

    Another consequence of the fast movement of water along few channels is that it could result in the spatial distribution of a pulse of tracers no longer Gaussian. As a result, the spatio-temporal evolution of the chemical plume cannot be described by classical ADE. Some recent experiments (Cortis et al., 2004) and re-examination of the benchmark experiments conducted in the 1960s, which underpinned ADE as a valid model to describe solute movement in homogeneous soils, do indicate that ADE gives rise to systematic errors (Cortis and Berkowitz, 2004). To demonstrate this further, we simulated water flow and solute movement in an idealized 2D uniform medium, as shown in Fig. 2.15a, which was regularly packed with discs of diameter either 34 or 40 pixels. The size of the column was 903 × 6170, and water flow was from left to right, driven by a pressure gradient (0.001 in the lattice unit). The top and bottom sides of the column were treated as periodic boundaries for both water and solute in which any particle moving out of the domain from one side re-entered the domain through its opposite side without changing its momentum. We began the water flow simulation first. Once flow reached steady state, a pulse of inert tracer was applied uniformly at the inlet boundary of the column,

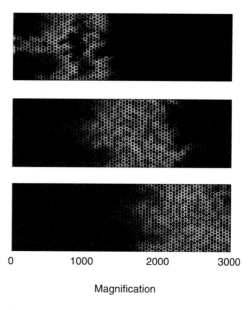

0   1000   2000   3000

Magnification

Fig. 2.16. Snapshots of concentration distribution in the void space.

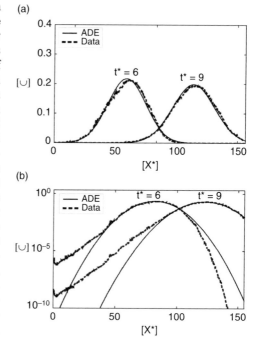

Fig. 2.17. Concentration profiles estimated from pore-scale simulation (broken lines) in comparison with best fitting ADE (solid lines), expressed in normal (a) and semi-log plots (b).

and then displaced by a tracer-free fluid from left to right to mimic the miscible displacement conducted in the laboratory. During the simulation, the tracer concentration in each water pixel was measured, from which we calculated the averaged breakthrough of the tracer at each cross-section of the column, as well the spatial distribution of the average resident concentration along the column. Water flow was simulated with $\tau = 1.0$ and solute transport using different values of τ_c.

Figure 2.15b shows the velocity field when flow reached steady state, and Fig. 2.16 shows snapshots of the concentration contour. To analyse how well ADE described the distribution of the averaged resident concentration along the column, Fig. 2.17 compares the measured results from the pore-scale simulation with those calculated from ADE using the average

pore-water velocity directly estimated from water flow simulation; the value of the dispersion coefficient in ADE was calibrated against the plume. Overall, ADE captured the main feature of the plume but systematically underestimated its tail. The discrepancy between the measurements and the best fitting of ADE can be seen more clearly in a semi-log plot, as shown in Fig. 2.17b. The measured concentration profile has a tail that decays with the distance from the peak in $c \propto \exp(-a|x|)$ rather than in $c \propto \exp(-ax^2)$ as predicted by ADE, while at the front the measured concentration decays with distance faster than predicted by ADE. Such anomalous dispersion is well described by the continuous time-random walk model (Zhang and Lv, 2007), a detailed description of which is beyond the scope of this book, and interested readers are referred to Berkowitz et al. (2006).

References

Adler, P.M., Jacquin, C.G. and Quiblier, J.A. (1990) Flow in simulated porous media. *International Journal of Multiphase Flow* 16, 691–712.

Adler, P.M., Jacquin, C.G. and Thovert, J.F. (1992) The formation factor of reconstructed porous media. *Water Resources Research* 28, 1571–1576.

Bakke, S. and Øren, P.E. (1997) 3-D pore-scale modeling of sandstone and flow simulations in pore networks. *SPE Journal* 2, 136.

Baldwin, C.A., Sederman, A.J., Mantle, M.D., Alexander, P. and Lynn, F. (1996) Determination and characterization of the structure of a pore space from 3D volume images. *Journal of Colloid and Interface Science* 181, 79–92.

Bear, J. (1972) *Dynamics of Fluids in Porous Media*. Dover, New York.

Berkowitz, B., Cortis, A., Dentz, M. and Scher, H. (2006) Modelling non-Fickian transport in geological formations as a continuous time random walk. *Reviews of Geophysics* 44, Art. No. RG2003.

Biswal, B., Manswarth, C., Hilfer, R., Bakke, S. and Øren, P.E. (1999) Quantitative analysis of experimental and synthetic microstructures for sedimentary rocks. *Physica A* 273, 452–475.

Chen, H., Chen, S. and Matthaeus, M. (1992) Recovery of the Navier–Stokes equations using a lattice-gas Boltzmann method. *Physical Review A* 45, R5339–R5342.

Chen, S. and Doolen, G.D. (1998) Lattice Boltzmann method for fluid flows. *Annual Review of Fluid Mechanics* 30, 329–364.

Childs, E.C. and Collis-George, N. (1950) The permeability of porous materials. *Proceedings of the Royal Society A* 201, 392–405.

Coker, D.A., Torquato, S. and Dunsmoir, J.H. (1996) Morphology and physical properties of Fontainebleau sandstone via tomographic analysis. *Journal of Geophysical Research* 101, 497–506.

Coles, M.E., Spanne, P., Muegge, E.L. and Jones, K.W. (1994) Computer microtomography of reservoir core samples. *Proceedings of the 1994 Annual SCA Meeting*, Stavanger, Norway, 12–14 September.

Cortis, A. and Berkowitz, B. (2004) Anomalous transport in 'classical' soil and sand columns. *Soil Science Society of America Journal* 68, 1539–1548.

Cortis, A., Chen, Y.J., Scher, H. and Berkowitz, B. (2004) Quantitative characterization of pore-scale disorder effects on transport in 'homogeneous' granular media. *Physical Review E* 70, Art. No. 041108.

Crawford, J.W. (1994) The relationship between structure and hydraulic conductivity of soil. *European Journal of Soil Science* 45, 493–502.

Crawford, J.W., Harris, J.A., Ritz, K. and Young, I.M. (2005) Towards an evolutionary ecology of life in soil. *Trends in Ecology and Evolution* 20, 81–87.

Currie, J.A. (1984) Gas diffusion through soil crumbs: the effects of compaction and wetting. *Journal of Soil Science* 34, 217–232.

Dexter, A.R. (1976) Internal structure of tilled soil. *European Journal of Soil Science* 27, 267–278.

Dunsmoir, J.H., Ferguson, S.R., D'Amico, K.L. and Stokes, J.P. (1991) X-ray microtomography: a new tool for the characterization of porous media. *Proceedings of the 1991 SPE Annual Technical Conference and Exhibition*, Dallas, Texas, 6–9 October.

Frisch, U., Hasslacher, B. and Pomeau, Y. (1986) Lattice-gas automata for the Navier–Stokes equation. *Physical Review Letters* 56, 1505–1508.

Geman, S. and Geman, D. (1984) Stochastic relaxation, Gibbs distributions, and the Bayesian restoration of images. *IEEE Transactions on Pattern Analysis and Machine Intelligence* PAMI-6(6), 721–741.

Gimenez, D., Perfect, E. and Rawls, W.J. (1997) Fractal models for predicting soil hydraulic properties: a review. *Engineering Geology* 48, 161–183.

Hazlett, R.D. (1995) Simulation of capillary dominated displacements in microtomographic images of reservoir rocks. *Transport in Porous Media* 20, 21–35.

Hazlett, R.D. (1997) Statistical characterization and stochastic modeling of pore networks in relation to fluid flow. *Mathematical Geology* 29, 801–822.

He, X. and Luo, L. (1997) A priori derivation of the lattice Boltzmann equation. *Physical Review E* 55, 6333–6336.

Jiang, Z., Wu, K., Couples, G.D., van Dijke, M.I.J. and Sorbie, K.S. (2007) Efficient extraction of networks from 3D porous media. *Water Resources Research* 43, W12S03.

Joshi, M. (1974) A class of stochastic models for porous media. PhD thesis, University of Kansas, Lawrence, Kansas.

Kang, Q.J., Lichtner, P.C. and Zhang, D.X. (2006) Lattice Boltzmann pore-scale model for multicomponent reactive transport in porous media. *Journal of Geophysical Research Solid Earth* 111 (B5), Art. No. B05203.

Knutson, C.E., Werth, C.J. and Valocchi, A.J. (2005) Pore-scale simulation of biomass growth along the transverse mixing zone of a model two-dimensional porous medium. *Water Resources Research* 41 (7), Art. No. W07007.

Kong, T.Y. and Rosenfeld, A. (1989) Digital topology: Introduction and survey. *Computer Vision, Graphics, Image Processing* 46, 357–393.

Liang, Z, Ioannidis, M.A. and Chatzis, I. (2000) Geometric and topological analysis of three-dimensional porous media: pore space partitioning based on morphological skeletonization. *Journal of Colloid Interface Science* 221, 13–24.

Lindquist, W.B., Venkatarangan, A., Dunsmuir, J. and Wong, T.F. (2000) Investigating 3D geometry of porous media from high resolution images. In: *Physics and Chemistry of the Earth, Part A—Solid Earth and Geodesy*. Elsevier, Amsterdam, pp. 593–599.

Long, T. and Or, D. (2005) Aquatic habitats and diffusion constraints affecting microbial coexistence in unsaturated porous media. *Water Resources Research* 41, Art. No. W08408.

Manswart, C. and Hilfer, R. (1998) Reconstruction of random media using Monte Carlo methods. *Physical Review E* 59, 5596–5599.

Manswart, C., Torquato, S. and Hilfer, R. (2000) Stochastic reconstruction of sandstones. *Physical Review E* 62, 893–899.

Matthews, G.P., Canonville, C.F. and Moss, A.K. (2006) Use of a void network model to correlate porosity, mercury porosimetry, thin section, absolute permeability, and NMR relaxation time data for sandstone rocks. *Physical Review E* 73, Art. No. 031307.

McNamara, G. and Zanetti, G. (1988) Use of the Boltzmann equation to simulate lattice gas automata. *Physical Review Letters* 61, 2332–2335.

Moran, C.J. and McBratney, A.B. (1997) A two-dimensional fuzzy random model of soil pore structure. *Mathematical Geology* 29, 755–777.

Mulder, M. (1996) Transport in porous membranes. Porous membranes for microfiltration and ultrafiltration. Lecture on School Membranes and Membrane Separation Techniques of Joint European Network in UMK, JEN-04720- PL-95, TorunÂ, pp. 23–29.

Nunan, N., Wu, K., Young, I.M., Crawford, J.W. and Ritz, K. (2002) In situ spatial patterns of soil bacterial populations, mapped at multiple scales, in an arable soil. *Microbial Ecology* 44, 296–305.

Nunan, N., Wu, K.J., Young, I.M., Crawford, J.W. and Ritz, K. (2003) Spatial distribution of bacterial communities and their relationships with the micro-architecture of soil. *FEMS Microbiology Ecology* 44, 203–215.

Okabe, H. and Blunt, J.M. (2004) Prediction of permeability for porous media reconstructed using multiple-point statistics. *Physical Review E* 70, 066135.

Øren, P.E. and Bakke, S. (2003) Reconstruction of Berea sandstone and pore-scale modelling of wettability effects. *Journal of Petroleum Science and Engineering* 39, 177–199.

Pan, C., Hilpert, M. and Miller, C.T. (2004) Lattice-Boltzmann simulation of two-phase flow in porous media. *Water Resources Research* 40, Art. No. W01501.

Qian, W. and Titterington, D.M. (1991a) Multidimensional markov-chain models for image-textures. *Journal of the Royal Statistical Society Series: B Methodological* 53, 661–674.

Qian, W. and Titterington, D.M. (1991b) Pixel labelling for three-dimensional sense based on markov mesh models. *Signal Processing* 22, 313–328.

Qian, Y.H., Dhumieres, D. and Lallemand, P. (1992) Lattice BGK models for Navier–Stokes equation. *Europhysics Letters* 17, 479–484.

Quiblier, J.A. (1984) A new three-dimensional modelling technique for studying porous media. *Journal of Colloid and Interface Science* 98, 84–102.

Rappoldt, C. and Crawford, J.W. (1999) The distribution of anoxic volume in a fractal model of soil. *Geoderma* 88, 329–347.

Roberts, A.P. (1997) Statistical reconstruction of three-dimensional porous media from two-dimensional images. *Physical Review E* 56, 3203–3212.

Saha, P.K. and Chaudhuri, B.B. (1995) A new approach to compute the Euler characteristic. *Pattern Recognition* 28, 1955–1963.

Silin, D.B., Jin, G. and Patzek, T.W. (2003) Robust determination of the pore space morphology in sedimentary rocks. SPE 84296, in *Proceedings of SPE Annual Technical Conference and Exhibition*, Denver, Colorado, October.

Spanne, P., Thovert, J.F., Jacquin, C.J., Lindquist, W.B., Jones, K.W. and Adler, P.M. (1994) Synchrotron computed microtomography of porous media: topology and transports. *Physical Review Letters* 73, 2001–2004.

Sukop, M.C. and Or, D. (2004) Lattice Boltzmann method for modeling liquid-vapor interface configurations in porous media. *Water Resources Research* 40, Art. No. W01509.

Vogel, H.J. (1997) Morphological determination of pore connectivity as a function of pore size using serial sections. *European Journal of Soil Science* 48, 365–377.

Vogel, H.J., Tolke, J., Schulz, V.P., Krafczyk, M. and Roth, K. (2005) Comparison of a Lattice-Boltzmann model, a full-morphology model, and a pore network model for determining capillary pressure-saturation relationships. *Vadose Zone Journal* 4, 380–388.

Wu, K., Nunan, N., Ritz, K., Young, I.M. and Crawford, J.W. (2004) An efficient Markov chain model for the simulation of heterogeneous soil structure. *Soil Science Society of America Journal* 68, 346–351.

Wu, K., van Dijke, M.I.J., Couples, G.D., Jiang, Z., Ma, J. and Sorbie, K.S. (2006) A new 3D stochastic model to characterize heterogeneous porous media, applications to reservoir rocks. *Transport in Porous Media* 65, 443–467.

Yeong, C.L.Y. and Torquato, S. (1998a) Reconstructing random media. *Physical Review E* 57, 495–506.

Yeong, C.L.Y. and Torquato, S. (1998b) Reconstructing random media: II. Three-dimensional media from two-dimensional cuts. *Physical Review E* 58, 224–233.

Young, I.M. and Crawford, J.W. (2004) Interactions and self-organisation in the soil–microbe complex. *Science* 304, 1634–1637.

Young, I.M. and Ritz, K. (2000) Tillage, habitat space and function of soil microbes. *Soil and Tillage Research* 53, 201–213.

Young, I.M., Crawford, J.W. and Rappoldt, C. (2001) New methods and models for characterizing structural heterogeneity of soil. *Soil and Tillage Research* 61, 1–13.

Zhang, X.X. and Lv, M.C. (2007) Persistence of anomalous dispersion in uniform porous media demonstrated by pore-scale simulations. *Water Resources Research* 43, Art. No. W07437.

Zhang, X.X., Bengough, A.G., Deeks. L.K., Crawford, J.W. and Young, I.M. (2002) A novel 3-D lattice Boltzmann model for advection and dispersion of solute in variably saturated soils. *Water Resources Research* 38(9), Art. No. 1167.

Zhang, X.X., Deeks, L.K., Bengough, A.G., Crawford, J.W. and Young, I.M. (2005) Determination of soil hydraulic conductivity using lattice Boltzmann method and thin-section technique. *Journal of Hydrology* 306, 59–70.

Zou, Q. and He, X. (1997) On pressure and velocity boundary conditions for the lattice Boltzmann BGK model. *Physics of Fluids* 9, 1591–1598.

3 Microbial Regulation of Soil Structural Dynamics

Claire Chenu[*] and Diego Cosentino

1 Introduction

Soil microorganisms live in a complex three-dimensional framework of pores and solid particles having variable sizes, shapes, continuity, and one that varies with time, depending on rain events, machinery traffic or events such as the burrowing of soil by earthworms. Besides being the habitat of soil organisms, soil structure, which is usually defined as 'the heterogeneous arrangement of solid and voids that exist in soil at a given time' (Kay, 1990), has a pivotal importance. It controls the spatial and temporal distribution of air, water and solute flows. It thus affects root respiration and penetration, seedling emergence, microbial respiration, defines the accessibility of nutrients and pollutants to the surfaces of soil constituents and the mechanical properties of soil.

Soil structure may be viewed either from the solid phase, as the clustering of soil particles into aggregates of various sizes and shapes, or from the void phase as a network of pores of different sizes, shapes and connectivity. Both views describe the same complexity, in a range of spatial scales that span the nanometre to the metre. However, these two models of soil structure involve a different array of methods usually mastered

by different scientists. The different areas of soil science tend to favour either one or the other, and therefore these views are seldom associated.

Kay (1990) describes soil structure by several terms. Structural form corresponds to the above definition. Structural stability is the ability of a given arrangement to retain its structural form when exposed to different stresses, such as the disruptive action of water, erosion or mechanical stresses due to heavy machinery. In the literature, the term structural stability most often refers to stability of soil structure towards water. The resiliency of soil structure is its ability to 'recover its initial form though natural processes when the applied stresses are reduced or removed' (Kay, 1990). Soil structure is then both a static and dynamic property. Changes in soil structure, i.e. the acquisition or loss of a given structural form, occur at diverse time scales and can be frequent.

Aggregates are subsets of soil structure, and were defined by Martin et al. (1955) as a 'naturally occurring cluster or group of particles in which the forces holding the particles together are much stronger than the forces between aggregates'. Aggregates may be easily observed in some situations

---

*Corresponding author: claire.chenu@grignon.inra.fr

© CAB International 2011. *The Architecture and Biology of Soils: Life in Inner Space*
(eds K. Ritz and I. Young)

37

(e.g. in the rhizosphere of a grass) or, more often, may be revealed only by the application of an external stress acting on the zones of low cohesion (e.g. in the ploughed layer of a cultivated soil). An aggregated structure is favourable, because it implies the coexistence of fine pores (within the aggregates) and coarse pores (between the aggregates or forming the failure zones). Some stability of such an arrangement is required to resist the frequent external stresses.

Aggregate formation, stabilization and destruction occur in a sequential manner and the three can be simultaneous. Separating the two phenomena of aggregate formation and stabilization may be difficult as, by definition, aggregates are clusters having some degree of cohesion and separating them implies the application of an external stress. For example, several experiments investigating the role of microorganisms in soil aggregation were based on the stimulation of microbial growth in disaggregated soil (Caesar-TonThat and Cochran, 2000; Bossuyt et al., 2001). After different incubation times, the material was gently broken by hand and wet-sieved. Stable aggregates were then isolated: both aggregate formation and aggregate stabilization had occurred. Other studies concentrate on the stabilization of pre-existing aggregates (Cosentino et al., 2006; Abiven et al., 2007). Aggregate dynamics take place at different scales. In the short term, wetting–drying events and the incorporation of fresh organic matter in soils influence aggregate dynamics. At a seasonal scale, variations of climate and biological activity control the aggregate dynamics rather than land use or cultivation practices (e.g. Perfect et al., 1990).

Microorganisms are major agents of soil structure formation, stabilization and destruction and, as such, they influence soil structural dynamics. Their role in aggregate stabilization was recognized very early (Martin and Waksman, 1940). Aggregate destruction, via the consumption of organic binding substances, has been investigated more recently.

The role of microorganisms in the temporal dynamics of aggregation is the subject of the present chapter. Bacteria, fungi, actinomycetes and unicellular algae have minute dimensions, yet they can affect soil structure not only at their own scale but also at much larger scales (i.e. millimetres to centimetres). The regulation of the temporal and spatial dynamics of aggregate formation, stabilization and destruction by microorganisms involves various feedbacks, because microorganisms may directly or indirectly benefit or may have a reduced activity, because of the structures they contributed to create.

2 Microbial Controls on Soil Structure

2.1 Structural form

A structured soil differentiates from a mere random packing of solid particles in that the three-dimensional (3D) arrangement of particles reveals preferential orientations, compacted zones or large voids and cracks created by the repositioning of particles. Wetting and drying events are the major processes creating structural form in soils, followed by the activity of roots and fauna. The influence of microorganisms on soil structural form is generally restricted to small spatial scales (< 20 µm). Experiments designed to assess the role of microorganisms in soil structure formation are generally based on suppressing or stimulating microorganisms in soils or on mineral particles in which the structure had been previously destroyed.

Changes in soil or clay fabric are observed in the vicinity of microorganisms. Clay particles are reoriented and aligned parallel to the cell wall of hyphae over a few microns (Fig. 3.1a, b; Dorioz et al., 1993). This was ascribed to local compaction and water uptake associated with hyphal growth, such as has been described for roots at a larger scale. Similar orientations are observed around bacteria, although more thinly (Fig. 3.1c, d). Extracellular polysaccharides often ensure continuity between the bacterial wall and mineral particles, and impregnate soil within a few microns of the microbe's surfaces (Dorioz et al., 1993). In

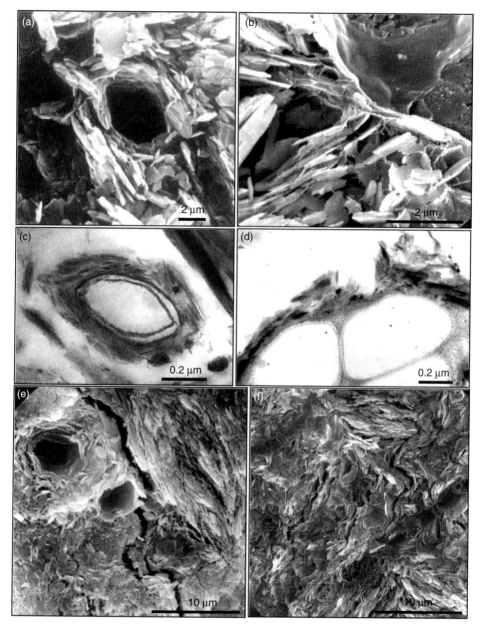

Fig. 3.1. Changes in soil structural form due to microorganisms. (a) *Chaetomium* sp. growing in kaolinite, by cryoSEM (Dorioz *et al.*, 1993); (b) detail of contact between *Fusarium oxysporum* cell and kaolinite, by cryoSEM (Dorioz *et al.*, 1993); (c) remnants of bacteria in a luvisol with oriented clay particles, by TEM (Chenu, unpublished); (d) remnants of bacterial colony in a luvisol with oriented clay particles, by TEM; extracellular polysaccharides are stained with silver and appear as black dots (Chenu, unpublished); (e) cracks in *Trichoderma viride* cultures in kaolinite after alternate 10/0.01 MPa, by cryoSEM (Dorioz *et al.*, 1993); (f) cracks in *Rhizoctonia solani* cultures in kaolinite after alternate pF2/pF5.8, by cryoSEM (from Dorioz, unpublished).

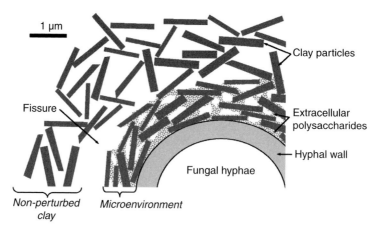

1 µm

Clay particles

Fissure

Extracellular
polysaccharides

Hyphal wall

Fungal hyphae

Non-perturbed   Microenvironment
clay

Fig. 3.2. Schematic representation of structural changes in the vicinity of a growing fungal hypha (adapted from Robert and Chenu, 1992).

these experiments, cracks appear following wetting–drying events, usually at the limit between the modified zone and bulk soil (Figs 3.1 and 3.2).

Preston *et al.* (1999) incubated soil slurries with different substrates and analysed the cracks generated during desiccation at a scale of millimetres to centimetres. The addition of easily utilizable organic substrates modified the cracking pattern (in terms of connectivity), density and heterogeneity. At various scales from microns (Dorioz *et al.*, 1993) to centimetres (Preston *et al.*, 1999), cracks are generated upon drying–re-wetting events where physical discontinuities exist, these being due to either the microbial bodies themselves or contrasted shrink–swell capacities of polysaccharide-rich zones compared with more mineralized zones (Fig. 3.2). Indeed Chenu (Chenu, 1989, 1993) and Czarnes *et al.* (2000) showed that microbial extracellular polysaccharides strongly increased the swelling capacity of clay.

Can fungi create pores like roots, penetrating the soil matrix and leaving channels behind after their death and decomposition? This was suggested by Emerson and McGarry (2003), who proposed that the relative increase in volume of pores < 30 µm with increasing organic matter in a soil was due to voids left by fungal hyphae, whose walls had been stabilized by extracellular polysaccharides. However, there is no direct evidence of the persistence of channels created by hyphae in soils, except in petrocalcic horizons, where the surroundings of fungal hyphae had been consolidated by calcium carbonate (Bruand and Duval, 1999). Very limited quantitative data are available regarding the impacts of soil microorganisms on total soil porosity, pore size distribution or pore connectivity. In one experiment, bacteria were shown to decrease hydraulic conductivity in sand columns because they clogged pores with their biomass and extracellular polysaccharides (Vandevivere and Baveye, 1992). Similar observations were reported in soil columns (Wu *et al.*, 1997).

In other experiments, substrate additions and associated microbial growth caused a 67% increase in the volume of mesopores (0.6–10.0 µm) at the expense of both smaller and larger pores (Strong *et al.*, 1998). Using computer-assisted X-ray tomography, Feeney *et al.* (2006) showed that the volumes of pores > 4.4 µm increased by 55% in a quantity of soil explored by rhizosphere fungal hyphae but from which roots were excluded, compared with an area explored by roots. Computed tomography also allowed De Gryze *et al.* (2006) to demonstrate an increase in the volume of pores of 12–50 µm in the vicinity of incubated straw residues. Hence, besides their small size, fungi and other microorganisms can alter soil pore size distribution.

2.2 Aggregate stabilization

When soil aggregates are exposed to rapid wetting, they may break down due to several mechanisms (Grant and Dexter, 1990; Le Bissonnais, 1996), which are summarized in Fig. 3.3: the kinetic energy of raindrops mechanically disperses the particles; differential swelling of clays within the aggregate creates internal stresses, which cause cracks and may disrupt the aggregate; upon rapid wetting air is entrapped in the porosity; the subsequent internal pressure developed breaks the aggregate if it overcomes the internal cohesion of the aggregate. This rupture mechanism is referred to as slaking, and colloidal particles, especially those saturated with monovalent ions, undergo physico-chemical dispersion due to interfacial forces.

Thus, any process increasing the cohesion between soil particles will increase the resistance of soil structure to any of the above breakdown mechanisms. Reducing the swelling of clay will also limit the extent of aggregate breakdown by differential swelling. Additionally, slowing the rate of entry of water gives time for air to escape, decreases the build-up of air pressure and distributes the mechanical tensions due to swelling or slaking over increased time; hence it stabilizes aggregates against slaking and differential swelling (Le Bissonnais, 1989; Grant and Dexter, 1990). The rate of entry of water into an aggregate depends on several characteristics: the aggregate pore size distribution, pore connectivity and tortuosity and surface characteristics of the solid particles. Solid surfaces may display contrasted affinity for water. Water spreads readily on hydrophilic surfaces with very low or nil contact angles, whereas water does not spread on hydrophobic or water-repellent surfaces and those exhibiting contact angles > 90°.

Aggregate stability is a complex integrative soil property, resulting from several basic properties: aggregate internal cohesion, pore volume, connectivity and tortuosity, and hydrophobicity of pore walls. Studies dealing with aggregate stability are generally based on the use of empirical tests, reproducing the actions of water on soil (see for example Le Bissonnais (1996) for a review). None the less, very few studies have directly addressed the basic properties underlying aggregate stability, although a clearer understanding of agents and factors may be expected should that situation change. It is not clear which property predominantly controls aggregate stability, and several may work synergistically (Grant and Dexter, 1990). According to Caron *et al.* (1996), increased water entry is the major mechanism for decreased aggregate stability. Aggregate cohesion and swelling would, therefore, be less important processes (Zaher *et al.*, 2005).

It is now many years since microorganisms were shown to enhance aggregate stability, such demonstrations usually being based on manipulating the microflora and

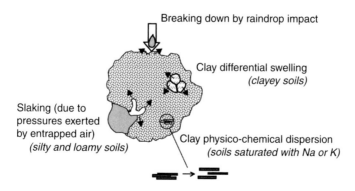

Breaking down by raindrop impact

Clay differential swelling
(clayey soils)

Slaking (due to pressures exerted by entrapped air)
(silty and loamy soils)

Clay physico-chemical dispersion
(soils saturated with Na or K)

Fig. 3.3. Mechanisms of aggregate destabilization by water.

investigating the impacts on aggregate stability. Studies reported in the literature involve (i) stimulating soil microflora via the addition of readily usable organic substrates to soil; (ii) inoculating sterile soil or sterile model minerals with microorganisms; or (iii) selectively inhibiting certain classes of microorganisms (bacteria or fungi) and comparing with untreated soil. Other series of experiments have been based on comparing soils characterized by different crops, agricultural practices or land uses and correlating their aggregate stability with their microbiological characteristics. In regard to the latter, many factors of aggregate stability other than microbial activity are involved. Extensive literature has been written on the subject, and reviews published (e.g. Robert and Chenu, 1992; Amezketa, 1999; Baldock, 2001). A still very relevant critical perspective was presented by Allison (1968) and a more recent one by Six et al. (2004). In this chapter we focus on the mechanisms by which microorganisms modify soil aggregate stability, i.e. increasing aggregate cohesion through binding substances, physical enmeshment or increasing aggregate hydrophobicity. Figure 3.4 illustrates the different mechanisms by which microorganisms influence aggregate stability.

2.3 Adhesion of microbial cells to solid particles

A large proportion of soil microorganisms are thought to adhere to solid particles, both mineral and organic. Direct quantification of cells adhering to solid particles in soils is very difficult to achieve in practice, and the limited extent of leaching of bacteria from soils and the difficulty in recovering these bacteria by direct separation techniques (e.g. Ramsay, 1984; Richaume et al., 1993) is mostly due to adhesion. Furthermore, bacteria, algae and fungi adhering to soil particles are easily observed (Chenu and Stotzky, 2002). The strength of adhesion of microorganisms to solid surfaces may be great (Chappell and Evangelou, 2002;

Kendall and Lower, 2004) and can easily explain the binding of several particles, in particular fine clay to microbial cells, and thus their aggregation (e.g. Fig. 3.1).

Production of binding substances: extracellular polysaccharides

Microorganisms frequently produce extracellular polymeric substances, most of which are extracellular polysaccharides. The abbreviation EPS is used for both in the literature, but will be used here to refer to extracellular polysaccharides. That microbial EPS are responsible for aggregate stabilization has long been suspected, and was demonstrated by (i) adding pure EPS to either soils or soil constituents (Martin et al., 1959, 1972; Martin, 1971; Chenu, 1989; Chenu et al., 1994; Czarnes et al., 2000; Palma et al., 2000); (ii) stimulating their production in situ by inoculating microorganisms known to produce them (e.g. Gouzou et al., 1993; Amellal et al., 1998; Maqubela et al., 2009); or (iii) correlating aggregate stability with hot water-extractible polysaccharides used as a surrogate for plant and microbial extracellular polysaccharides (see references in Table 3.1). The ability to produce EPS is a widely distributed trait among soil bacteria, fungi and unicellular algae, but occurs quite independently of taxonomy (Chenu, 1995).

Any estimation of in situ quantification of EPS production is seriously hampered by the fact that microbial extracellular polysaccharides cannot be selectively extracted and quantified: they have the same composition, in terms of individual sugar components, as cytoplasmic or wall microbial polysaccharides and partly plant polysaccharides (Chenu, 1995). Hence direct assessment of the abundance of microbial EPS in soils is not possible. A few workers have attempted to quantify microbial EPS. Using data from pure cultures in liquid or sand, Chenu (1995) assumed an EPS:microbial biomass production ratio of 10% in soils, and from that estimated that newly formed EPS would comprise 0.2–0.3% of soil organic carbon and 2–3% soil

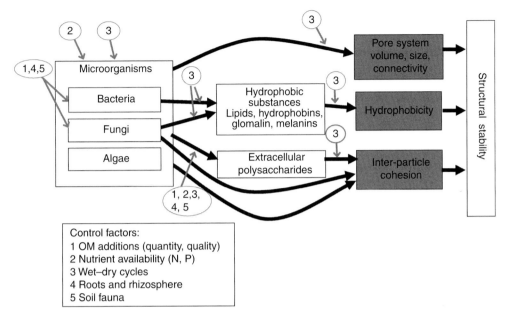

Fig. 3.4. Schematic representation of the main mechanisms by which microorganisms stabilize soil aggregates.

polysaccharides. To estimate the contents of EPS produced *in situ* by a soil bacterium, Cohen (2002) used measurements of the viscosity of soil pastes, calibrated by EPS additions in known amounts. She found that 2–8% of assimilated carbon was exuded as EPS, which corresponded to an EPS:biomass ratio of 5–19% and an abundance of newly formed EPS in a medium-textured soil of 0.5–2.0% of soil polysaccharides. Newly formed EPS thus represent a minor proportion of soil organic matter and soil carbohydrates; however, they may persist and accumulate in the soil.

Mechanisms

The effects of microbial EPS on aggregate stabilization are twofold: first, they have a strong tendency to adsorb to mineral surfaces because of their surface reactivity. Their adsorption involves mainly weak and electrostatic bonds, but, since microbial EPS are macromolecules of high molecular weight (10^5 to $> 10^6$ Daltons), their adsorption on clay minerals and silicates exhibits high affinity and low reversibility (Chenu

et al., 1987; Malik and Letey, 1991; Khandall *et al.*, 1992); and, secondly, they bridge mineral particles together. Having a linear conformation, a high molecular weight and the ability to form intermolecular linkages, EPS form strands of a few nanomicrons in width and > 100 nm length and plurimolecular networks (Morris and Norton, 1983; McIntire *et al.*, 1997). This ability is expressed in the presence of soil minerals, as demonstrated by the observation of organo-mineral networks of EPS with clay and soil particles (Chenu, 1989; Chenu and Jaunet, 1992; Chenu *et al.*, 1994). Microbial EPS thereby increase inter-particle cohesion (Chenu and Guérif, 1991).

Which EPS are effective?

The efficiency of EPS as an aggregating agent should therefore depend on both their molecular weight and their ability to form pluri-molecular networks. Indeed, Chenu *et al.* (1994) and Czarnes *et al.* (2000) found that polysaccharides exhibiting such structures (e.g. xanthan, scleroglucan, polygalacturonic acid) were more efficient than

Table 3.1. Correlations between aggregate stability and microbial characteristics (determination coefficients).

Reference	Soil type or texture	Total organic C	Total polysaccharides	Dilute acid-soluble polysaccharides	Hot water-soluble polysaccharides	Microbial biomass	Hyphal length	Other fractions measured
Angers and Mehuis (1989)	Clay	0.001 (ns)		0.63				
Angers et al. (1993)	Clay	0.6		0.28	0.43	0.33		
Angers (1992)	Loam					0.4 (ns)		
Angers et al. (1999)	Sandy loam, Podzol	0.77				0.57		
Ball et al. (1996)	Loam	0.83	0.97	0.8	0.98			
Bethlenfalvay et al. (1999)	Silt loam			ns			0.565	
Bissonnette et al. (2001)	Silty clay	0.95		0.89		0.9		
Capriel et al. (1990)	20% clay/63% silt					0.82		Lipids
Carter (1992)	Sandy loam, Podzol	0.942				0.947		
Carter et al. (1994)	Sandy loam	ns				ns		
Chan and Heenan (1999)	Oxic Paleustalf	ns		<0.25	<0.25	0.64		
Chantigny et al. (1997)	Silt clay loam and Clay loam			0.42/0.05		ns		Fungal glucosamine (r = 0.68); bacterial muramic acid (r = 0.48)
Degens et al. (1994)	Sandy loam	ns		ns	0.05	ns	ns	
Degens and Sparling (1996)	Sandy, Podzol			ns	ns	0.54–0.83		
Degens et al. (1996)	Sandy					0.26	0.41	
Drury et al. (1991)	Loam, illitic					0.59		
Denef and Six (2005)	Clayey, ferralsol					0.21		
Haynes and Swift (1990)	Silt loam	0.58	0.57 (ns)	0.56 (ns)	0.74			
	Sandy loam	0.66	0.67		0.83			
Haynes et al. (1991)	Silt loam	0.77	0.75	0.76	0.84	0.957[a]		
	Clay loam	0.72	0.76	0.6 (ns)	0.79			
Haynes (1999)	Silt loam	0.92			0.72 (ns)	0.8		
Haynes (2000)	Silt loam	0.61	0.66		0.99			Mineralizable C

Reference	Soil type							Notes
Jastrow et al. (1998)	Silt loam	0.43						
Kiem and Kandeler (1997)	6–30% clay	0.34			0.55	0.65	0.89	
Kinsbursky et al. (1989)	3–44% clay				0.58–0.85	0.59–0.87		
Lax and García-Orenes (1993)	Clay loam		0.88					
Metzger et al. (1987)	Sandy clay loam		0.5					
Perfect et al. (1990)					0.68		0.74	
Roberson et al. (1991)	Loam	ns			0.11	0.71		Heavy-fraction carbohydrates ($r = 0.9$); light-fraction carbohydrates ($r = $ ns)
Roberson et al. (1995)	Sandy loam	ns				ns in summer: 0.52		Heavy-fraction carbohydrates with slacking resistance ($r = 0.74/0.81$)
Roldan et al. (1994)	Loamy clay			0.66–0.28				
Sparling (1992)						0.88		
Tisdall and Oades (1980)	Red-brown earth	0.93						
Wright and Upadhyaya (1998)	37 soils of varying texture	0.65	0.84 (easily extractable glomalin)				0.77	

ns, not significant.

[a]Calculated.

those having random coil structures, such as dextran or amylopectin (Fig. 3.5).

Factors relating to regulation

Extracellular polysaccharide synthesis is typically stimulated when carbon is available in excess and when other nutrients limit the build-up of biomass (Sutherland, 1985) (Fig. 3.4). EPS production by soil bacteria is enhanced by nitrogen limitation (Linton *et al.*, 1987; Auer and Seviour, 1990; Roberson, 1991; Breedveld *et al.*, 1993; Lopez-Garcia *et al.*, 2001; Otero and Vincenzini, 2003), although this effect is less pronounced for algae (Kroen, 1984; Kroen and Rayburn, 1984). The presence of solid surfaces, such as sand grains, markedly increases the production of EPS by several soil and sediment bacteria (Bengtsson, 1991; Roberson, 1991; Vandevivere and Kirchman, 1993). Desiccation also stimulates EPS production (Roberson and Firestone, 1992). In soil, high carbon availability conditions are met in the rhizosphere in particular. Indeed, several rhizosphere bacteria of the genera *Rhizobium* or *Pantotea* were found to exhibit large EPS production, and they increased the stability of aggregates when

inoculated into soil in the presence of plants (Hebbar *et al.*, 1992; Amellal *et al.*, 1998, 1999; Alami *et al.*, 2000; Kaci *et al.*, 2005).

Scales

Aggregation by EPS occurs at the micro scale. This has been demonstrated by visual observation (Foster, 1988; Dormaar and Foster, 1991; Chenu, 1995; Chenu *et al.*, 2001) as well as by size fractionation techniques (Hu *et al.*, 1995). However, the effect of EPS can span larger scales when the microorganisms themselves are of greater dimensions. Fungal EPS that bind soil at the 20–2000 µm scale with their 'sticky string bags' (Oades and Waters, 1991) combine physical entanglement and binding.

Polysaccharides are not the only microbial extracellular polymers that have the ability to adsorb to soil mineral surfaces and to bind particles. Soil microorganisms may also exude extracellular proteins (Achouak *et al.*, 1994) or complex glycoproteins. The aggregating action of a class of extracellular proteins from mycorrhizae, operationally defined and named glomalin, has been extensively investigated (Driver *et al.*, 2005; Rillig and Mummey, 2006). Although the

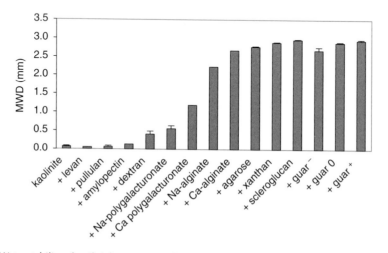

Fig. 3.5. Water stability of artificial aggregates of kaolinite, either pure or with added polysaccharide (1% w/w). Results are expressed as mean weight diameter (MWD), after an immersion test, of aggregates initially having a diameter of 2–4 mm. Among the polysaccharides, levan, pullulan, dextran, amylopectin and Na-polygalacturonate typically exhibit a random coil tertiary structure, whereas the other polysaccharides exhibit ordered helical structures with intermolecular linkages (from Chenu and Robert, unpublished).

correlation between glomalin fractions and stable aggregation is strong (Wright and Upadhyaya, 1998; Wright *et al.*, 1999; Franzluebbers *et al.*, 2000; Rillig, 2004) it is unlikely that these proteins can be efficient binding agents, due to their low molecular weight.

2.4 Enmeshment by fungi and actinomycetes

Many studies report that fungi observed the formation of hyphal networks on aggregate surfaces (e.g. Harris, 1972; Tisdall and Oades, 1982; Marinissen and Dexter, 1990; Guggenberger *et al.*, 1999). Microscopic observations revealed the physical entanglement of soil particles by fungal hyphae; Fig. 3.6). Hyphae are bound with sufficient strength to freely suspend sand grains (Degens *et al.*, 1996). These visual observations have been supported by numerous significant correlations between hyphal length,

as measured in soil smears, and aggregate stability (Table 3.1).

Mechanisms

Although many papers dealing with microbially mediated aggregation refer to physical entanglement as being a frequent and important mechanism, to date there has been no investigation, in soil science, of either its biophysics or biomechanics (Ritz and Young, 2004). An efficient aggregation of soil particles by a network of fungal hyphae is expected to depend on several factors (Figs 3.6, 3.7): the strength of adhesion of the fungal hyphae to soil particles, tensile strength of the hyphae and the architecture of the mycelial network in the soil matrix.

As seen previously, the strength of adhesion of the fungal hyphae to soil particles may be very pronounced, in particular in the presence of extracellular

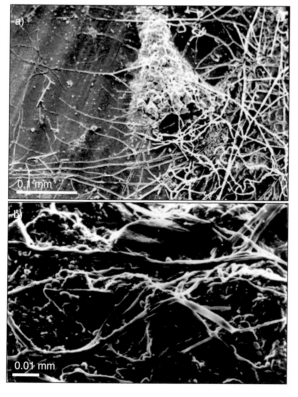

Fig. 3.6a, b. Physical entanglement of soil particles by fungal hyphae. Entanglement is related to the growth of saprophytic fungi decomposing wheat straw (from Chenu and Angers, unpublished).

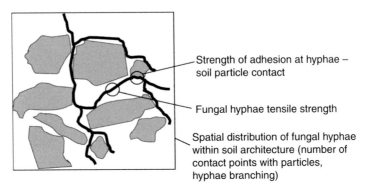

Strength of adhesion at hyphae –
soil particle contact

Fungal hyphae tensile strength

Spatial distribution of fungal hyphae
within soil architecture (number of
contact points with particles,
hyphae branching)

Fig. 3.7.   Properties involved in the efficiency of physical entanglement by fungal hyphae.

polysaccharides. Very few data are available concerning the mechanical properties of fungi, with those that are available coming from studies on fermentation technology (e.g. Thomas *et al.*, 2000; Li *et al.*, 2002). For example, the tensile strength of the actinomycete *Saccharopolyspora erythraea* was measured using a micromechanics device and was found to range between 20 and 40 MPa (Thomas *et al.*, 2000). This, however, varies with the age of the mycelia (Li *et al.*, 2002) and is expected to vary among species with regard to hyphal diameter and cell wall thickness. For example, the tensile strength of the mycelial networks of *Aspergillus niger*, *Aspergillus oryzae* and *Rhizopus oligosporus* growing between two pellets of ground wheat was much less, reaching only 0.003 MPa (Schutyser *et al.*, 2003). These data can be compared with the tensile strength of individual soil aggregates, which typically range from 0.05 to 0.20 MPa (Munkholm *et al.*, 2002). Hence, fungal bridges have the potential to increase soil tensile strength. Furthermore, the spatial distribution of hyphal growth within the soil architecture (e.g. the volume explored, the numbers of contact points with soil particles, the number of soil particles in contact with fungi, fungal branching, etc.) will probably modulate the efficiency of the bonds at a smaller scale (Fig. 3.7), although this has not been studied to date. Hadas (Hadas *et al.*, 1994) suggested that the observed increases in tensile strength after addition of cotton residues to soil

were due to fungal hyphae on the surface of aggregates externally enforcing their strength. However, no substantive evidence for this was presented.

Effective agents

The capacity of fungi to aggregate particles is species specific (Tisdall *et al.*, 1997). These authors found, for example, that saprophytic fungi (*Rhizoctonia solani* and *Hyalodendron* sp.) were more efficient at aggregating a clay soil than mycorrhizal species (*Hymenoscyphus ericae* and *Hebeloma* sp.), despite the latter exhibiting longer hyphal length; other mechanisms of aggregate stabilization were probably at play. It should be noted here that several cyanobacteria produce filamentous colonies (e.g. *Nostoc*) and can entangle soil particles, especially in microbiotic soil crusts (Malam Issa *et al.*, 2001; Maqubela *et al.*, 2009).

2.5 Decreasing the hydrophobicity of particles and aggregates

Hydrophobicity and aggregate stability

A reduction in the wetting rate of soils has been associated with increased aggregate stability (Concaret, 1967). Virgin soils, stable and rich in organic matter, have lower rates of water uptake (Caron *et al.*, 1996; Hallett *et al.*, 2001a) and lower water drop penetration times (WDPTs; Chenu *et al.*, 2000) than their arable counterparts. Strong correlations were found between hydrophobic components of soil organic matter, such as aliphatic fractions resembling those from microorganisms, humic acids and aggregate stability (Chaney and Swift, 1984; Capriel *et al.*, 1990). In some soils organic substances induce severe water repellency, impeding water infiltration and causing overflow erosion (Wallis and Horne, 1992), whereas subcritical water repellency is widespread. The repellency of soils can be assessed by several methods such as measuring WDPTs, measuring the sorptivity of water and ethanol, from which a repellency index is derived (Tillman *et al.*, 1989), or using capillary rise measurements, from which the solid–water contact angles can be calculated (see Wallis and Horne, 1992 for a review; Michel *et al.*, 2001). The contact angles of water on solid soil particles can be directly measured but only on flat and cohesive surfaces, and this technique is therefore restricted for soils with clay deposits (Chenu *et al.*, 2000).

Evidence that microorganisms increase soil hydrophobicity and thereby increase aggregate stability

When adding organic substrates to soil and thereby stimulating microbial growth and activity, water sorptivity decreases and the repellency index increases (Hallett and Young, 1999; De Gryze *et al*, 2006; Feeney *et al.*, 2006) and water drop penetration time increases (Cosentino *et al.*, 2006, 2010). However, increases in aggregate stability and repellency are not always associated, as shown by De Gryze *et al.* (2006) in a sandy loam and two loams, in contrast to a clay loam.

Which organisms are active?

Repellency in the sandy soils of Australia was associated with the presence of fungi (Bond and Harris, 1964; Savage *et al.*, 1969). Hallett *et al.* (2001b) demonstrated that fungi are the major group of organisms imparting subcritical soil repellency. These workers treated soil with organic substrates and biocides; soil amended and treated with a bacterial biocide showed a higher degree of water repellency than untreated soil or amended soil treated with a fungicide (Fig. 3.8). Inoculation of soil with white rot fungi (White *et al.*, 2000) and mycorrhizae (Feeney, 2004) increased repellency. Whereas the predominant contribution of fungi to repellency over bacteria seems well established, the effect of different fungal species is variable (McGhie and

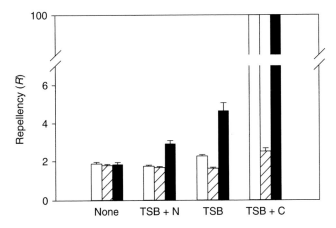

Fig. 3.8. Soil repellency induced by microorganisms. Microbial growth was stimulated by adding various combinations of tryptone soya broth (TSB), ammonium nitrate (N) and glucose carbon (C). Fungi were selectively inhibited using cycloheximide (black bars) and bacteria selectively inhibited with streptomycin (striped bars). Soil repellency was found to be predominantly due to fungi (from Hallett *et al.*, 2002).

Posner, 1980). The surface properties of filamentous fungi have been characterized, and a large variability has been found. Ascomycetes and basidiomycetes are often hydrophobic, with contact angles of water even exceeding 90° (Unestam, 1991; Smits *et al.*, 2003; Table 3.2). Hydrophobicity increased with mycelial age and is associated with aerial growth (Smits *et al.*, 2003). Cyanobacteria present in microbiotic crusts have been shown to induce repellency after drying of the crust (Kidron *et al.*, 1999), but this phenomenon has received little attention to date.

Which compounds are hydrophobic?

Being composed of carbohydrates, the fungal wall is hydrophilic; however, it may be impregnated with melanins. Melanins are dark-pigmented polymers, generated by the condensation of phenolic precursors present in the fungal wall or exuded in the media, and protect organisms from environmental stresses (Butler and Day, 1998). Melanins are hydrophobic and resistant to biodegradation (Martin *et al.*, 1959; Martin and Haider, 1979; Zavgorodnyaya *et al.*, 2002).

Fungi commonly exude a specific class of molecules in the medium, hydrophobins. Hydrophobins are small proteins, ubiquitous among the different fungal taxa, characterized by eight cysteine residues ordered in a defined sequence. When secreted in the media hydrophobins self-assemble to form highly insoluble amphiphilic membranes. These membranes typically have a water contact angle of 22–63° on their hydrophilic side and of > 100° on their hydrophobic side (Wosten and Vocht, 2000). After being exuded, hydrophobins self-assemble at the medium–air interface of an amphipathic membrane, causing a huge decrease in surface tension of the interface (Wosten, 2001). Being amphiphilic, hydrophobins can render hydrophilic surfaces hydrophobic and vice versa (e.g. Lumsdon *et al.*, 2005).

Table 3.2. Hydrophobic/hydrophilic fungi. Hydrophobicity may depend not only on species but also on culture conditions.

Hydrophobic forms		Hydrophilic forms	
Basidiomycetes			
Laccaria bicolor *Paxillus invollutus*	Smits *et al.* (2003)	*Thelephora terrestris, Cenococcum geophilum, Laccaria laccata*	Unestam (1995)
Hebeloma crustiliforme	Smits *et al.* (2003)	Hevbeloma crustuliniforme	Unestam (1995)
Tricholoma sp. *Cortinarius umbonatus Chlatrus gracilis Clavaria aurantia*	Bond and Harris (1964)		
Suillus bovinus Marasmius oreades	Smits *et al.* (2003) York and Canaway (2000)		
Coriolus versicolor Phanerochaete chrysosporium	White *et al.* (2000)		
Ascomycetes			
Cladosporium sp.	Smits *et al.* (2003)		
Deuteromycetes			
		Fusarium oxysporum Trichoderma harzianum	Smits *et al.* (2003)

The secretion and presence of hydrophobins in soil have never been studied. However, it is highly probable that they are present, since many soil-borne fungi are recognized for their ability to exude hydrophobins, and it may be hypothesized that they are responsible for fungally induced soil hydrophobicity.

As described previously, several mycorrhizae exude an insoluble class of proteins, glomalins, currently operationally isolated and quantified from soil as glomalin-related soil protein (GRSP; Rillig, 2004), which is correlated to aggregate stability (Wright and Upadhyaya, 1998; Rillig, 2004). GRSP was suggested to stabilize aggregates by rendering them more hydrophobic (Wright and Upadhyaya, 1996) and to belong to the hydrophobins (Rillig, 2005). Indeed, many proteins have an amphiphilic character. However, Feeney et al. (2004) conducted a pot experiment in which they grew pea and associated VAM (vesicular-arbuscular mycorrhizal) fungi; in samples of 1 cm they then measured glomalin content and water sorptivity; they found no correlation between glomalin content and soil repellency index. The mechanisms by which this class of molecules act on aggregate stability still remain to be investigated (Rillig and Mummey, 2006).

Surface or extracellular lipids may also be responsible for microbially induced soil repellency, as suggested by Capriel et al. (1990). Bacteria as well as yeasts exude lipids (Georgiou et al., 1992), which may also be involved.

Factors controlling microbially mediated hydrophobicity

Factors regulating the production of hydrophobic substances by microorganisms are far from understood. One key area is weather conditions, specifically the occurrence of wetting–drying events and the magnitude and duration of drying. Microbially derived hydrophobicity was found to be either increased by wetting–drying cycles or was unaffected (Hallett et al., 2002; Cosentino et al., 2006). Microbially derived hydrophobic substances are generally amphiphilic (e.g. hydrophobins, glomalin), hence they tend to expose their hydrophobic moieties towards the pores after drying (Michel et al., 2001).

2.6 Effective aggregate stabilizers: bacteria versus fungi

Which organisms are the most effective in stabilizing aggregates? A number of experiments have been designed to selectively inhibit either bacteria or fungi with antibiotics in order to assess their relative contributions to aggregation, and these showed that fungi played the dominant role in the stabilization of aggregates (Molope et al., 1987; Hu et al., 1995; Beare et al., 1997; Bossuyt et al., 2001; Denef et al., 2001). The pre-eminence of fungi can be explained by the variety of mechanisms by which they can aggregate soil (i.e. physical entanglement, production of EPS and production of hydrophobic substances). Furthermore, fungi are able to aggregate soil at spatial scales larger than bacteria are able to do, i.e. 20–2000 μm (Degens, 1997). Fungi have another advantage, which is that their action extends further than that of bacteria from favourable sites; VAM fungi can extend 10–30 mm into soil from the root surface; similarly, saprophytic hyphae were found to extend > 1 mm from decomposing plant residues (Gaillard et al., 1999). Another question relates to the efficacy of different fungi. A path analysis emphasized the importance of mycorrhizae compared with saprophytic fungi (Miller and Jastrow, 1990). However, evidence exists on the involvement of both groups in aggregate stabilization (saprophytic fungi – Kinsbursky et al., 1989; Tisdall et al., 1997; Caesar-TonThat and Cochran, 2000; Bossuyt et al., 2001; and mycorrhizal fungi – Miller and Jastrow, 1990; Tisdall, 1991; Thomas et al., 1993; Tisdall et al., 1997; Andrade et al., 1998; Bethlenfalvay et al., 1999; Bearden and Petersen, 2000; Feeney, 2004).

3 Factors Controlling Microbially Mediated Aggregation

3.1 Factors favouring bacteria versus fungi

Since fungi are more effective in changing soil structural form and stability, ecological factors influencing the fungal:bacterial ratio in soils should impact on soil structure. Nitrogen fertilization has been reported to decrease the fungal:bacterial ratio (Lavelle, 2002), as has amendment with organic matter of high C:N ratio (Henriksen and Breland, 1999; Vinten *et al.*, 2002; de Vries *et al.*, 2006). As a consequence, low-input cropping systems tend to favour fungi (de Vries *et al.*, 2006) although possible consequences for soil aggregation have not been explored to date. Zero tillage relatively increased the abundance of fungi, and was partly responsible for increased soil aggregation (Beare *et al.*, 1997).

Soil characteristics

Positive correlations between microbial biomass C and aggregate stability have been reported by most studies considering this variable (Table 3.1); however, several examples exhibit no significant correlation, these corresponding in particular to sandy (Carter *et al.*, 1994; Degens *et al.*, 1994) and ferrallitic soils (Denef and Six, 2005). One might suspect that in the former example sand grains are not favourable geometrically to be bound by microorganisms because they are too large and not suited to surface

interaction because of their low reactivity (Degens *et al.*, 1994); and in the latter that the dominant aggregating agent is iron oxide rather than microorganisms.

3.2 Aggregate destruction

Processes leading to aggregate destruction are climatic events (rain), compression or breaking by machinery, faunal burrowing or remoulding activity as well as decomposition processes. Microorganisms are directly involved in aggregate destruction by consuming organic binding agents, as well as hydrophobic agents and dead hyphae. The lifetime of aggregates depends thus on the lifetime of their aggregating agents. Data are scarce on the lifetime of fungal hyphae or molecular classes in soils, but we report a few examples in Table 3.3.

Feedbacks: creating a favourable habitat?

Soil architecture has profound influences on soil microorganisms as they live in the complex, tortuous, more or less interconnected pore network filled with either water or air, defined by the arrangement of soil particles (Young and Ritz, 2000).

By aggregating soil particles around them, microorganisms modify their immediate environment. Their structuring and aggregating action may have the 'purpose' of creating a more favourable habitat, or may be viewed as the side effect of the production of diverse molecules active on the

Table 3.3. Estimated residence/lifetime of selected aggregating agents.

Aggregating agent	Residence/lifetime (days)	Conditions	Reference(s)
Live hyphae	3–5		Staddon *et al.* (2003)
Dead hyphae	160	1-point data, calculation assuming first-order kinetics	Steinberg and Rillig (2003)
Extracellular polysaccharides	20	Bacterial polysaccharide luvisol, 24°C	Cohen (2002)
	5–10	Bacterial polysaccharide	Andreyuk *et al.* (1986)
Glomalin	240	1-point data, calculation assuming first-order kinetics	Steinberg and Rillig (2003)

soil structure, but there may be other major roles for microorganisms.

3.3 Extracellular polysaccharides: an intriguing case in microbial ecology

Extracellular polysaccharides represent an important expenditure of C and energy for the microbial cell. Their widespread occurrence among soil microorganisms must then be related to some competitive advantage, and they have been shown to protect microorganisms – especially bacteria and microalgae – from desiccation, to mediate attachment to surfaces and biofilm or microcolony formation and to act as ionic barriers (Chenu, 1995).

Soil is a drastic environment in terms of water potential fluctuations, which are frequent and large. Microorganisms being in equilibrium with the outside water potential, desiccation and rapid re-wetting can both be lethal. Microbial EPS exhibit properties that should allow them to protect microorganisms. They retain large amounts of water compared with minerals and remain water-saturated at low water potential (Chenu, 1993). Hence, they can provide an immediate environment for cells that remain hydrated and suitable for the diffusion of nutrients and substrates even in dry conditions (Chenu and Roberson, 1996). Furthermore, they decrease the drying rates of soils and minerals, thus acting as buffers during water potential fluctuations (Chenu, 1993). However, studies comparing the survival of EPS- and non-EPS-producing strains of bacteria show contrasting results, i.e. either positive or no effect on survival (see review in Chenu, 1995).

Extracellular polysaccharides also favour and strengthen the attachment of microorganisms to mineral surfaces. Being the outermost surface of many microbial cells, EPS have high adhesive properties and their secretion is stimulated by attachment to surfaces (Bengtsson, 1991; Vandevivere and Kirchman, 1993); hence, biofilm formation is stimulated. However, biofilms with no EPS have also been described (Chappell and Evangelou, 2002).

In the soil environment, ion and metal concentrations fluctuate widely. Microorganisms and, in particular bacteria, rely on diffusion for their livelihood as they cannot reach out and ingest a substrate or disperse a toxic substance. Extracellular polysaccharides, often having an anionic character and a strong cation-binding ability, would also act as ionic barriers protecting microorganisms from toxic substances (Geddie and Sutherland, 1993; Wolfaardt et al., 1994; Chen et al., 1995; Garcia-Meza et al., 2005).

Although it is still difficult to demonstrate in situ that EPS enhance microbial survival and activities in soils, it seems clear that EPS have many potential roles and are not (only) produced specifically to promote soil aggregation.

3.4 Why exude hydrophobic substances?

Soil is a triphasic environment, and one challenge for soil fungi is to extend through water–air interfaces, either to cross an unsaturated pore or to form aerial fruiting bodies. Hydrophobins are a class of small proteins (see section 2.2) exuded by filamentous fungi and having several functions, the major one being to help fungi to break these interfaces (Wosten et al., 1999). Submerged hyphal tips exude these molecules in solution. Hydrophobins self-assemble at the water–air interface over hyphae as an amphiphilic film, resulting in a huge decrease in the water surface tension from 72 J/m^2 to about 24 J/m^2 (Wosten et al., 1999). The hydrophobin film covers the surface of the hyphae as it emerges from solution, attached to the hyphal wall by its hydrophilic end and exposing the hydrophobic end. Aerial hyphae and spores thus become hydrophobic. Hydrophobins also enable the adhesion of fungi to hydrophobic surfaces, again by exposing the hydrophobic surface towards the outside. This allows the colonization of hydrophobic organic resources or hosts (plant cuticles, humic substances) by fungi. Hydrophobins are thus important molecules for fungal biology – habitat colonization, fruit body formation, sporulation and infection of hosts. Hydrophobins are widespread among basidiomycetes and ascomycetes and are also produced by zygomycetes (Wosten, 2001). To date they have not been identified and quantified in soils, but they have been isolated from fungi that inhabit soils. Their potential effect on soil aggregation is likely to be a side effect.

4 Are Aggregates Favourable Habitats?

A frequent assumption, or myth, is to consider that aggregates are favourable habitats for microorganisms. This requires careful examination: why would microbes form and stabilize aggregates? Do they derive any benefit from it? In other words, is a stable aggregated soil a more favourable medium than non-aggregated soil for microorganisms? A point that must be stressed is that microorganisms do not 'perceive' aggregates, but rather pores, in which they live. In a non-aggregated soil, pores are generated by the packing of solid particles, which has been termed textural porosity (Fiès, 1992) as opposed to structural porosity due to cracks, aggregation and faunal activity. The volume of textural pores is less than total porosity (Fiès, 1992). A non-aggregated soil has less pore volume and fewer coarse pores compared with aggregated soil (Dexter, 1988). Hence, in terms of habitat, aggregated soils offer a wider diversity of pore sizes and thus of physical habitats, and provide more unsaturated macropores and more micropores situated at a short distance from unsaturated pores (Parry et al., 1999).

4.1 Microbial activities and survival are affected by pore size

Hattori (1988) proposed that micropores are important for bacterial survival and that macropores, i.e. the outer surfaces of aggregates, are involved in aerobic microbial activities. Indeed, it was shown that microorganisms are protected from predation by microfauna when located in pores < 6 µm (Elliott et al., 1980; Hattori, 1988; Kuikman et al., 1990; Heijnen and van Veen, 1991).

Micrometre-size pores were also found to protect bacteria from desiccation (Chao and Alexander, 1982; Osa-Afiana and Alexander, 1982). However, aggregation of mineral particles in the immediate environment of bacteria – as observed for example in Fig. 3.1c and d – may be detrimental to bacterial activity. Bacteria located in micropores filled with EPS (i.e. 'bacterial micro-aggregates'; Chenu, 1995) depend on the diffusion of oxygen, enzymes and substrate between them and the nearest air-filled or substrate-containing pore for their metabolism. Diffusion of glucose is much slower in EPS than in water (Chenu and Roberson, 1996) and it is unlikely that enzymes can diffuse through an EPS–clay coating. Fungi coated with clay were found to be oxygen deprived (Stotzky, 1986). Hence, by aggregating mineral particles around its cell, the microorganism may change its environment towards one more buffered and protected in terms of desiccation and predation, but one that impedes activity. At this scale microbially induced aggregation can be considered to have negative feedbacks on microorganism activity.

On the other hand, C and N mineralization was significantly correlated with pores of diameter 15–60 µm and 0.6–30 µm, respectively (Strong et al., 1998, 2004). Added fructose decomposed faster in pores of 100–300 µm diameter than in 0.3–3.0 µm or 1–100 µm (Ruamps et al., 2011). These pore sizes would be very favourable to microbial aerobic activities because pores alternately filled with water and air would thus be provided with adequate water and oxygen resources. Fungi are aerobic organisms, and one of the major factors controlling their growth in soils is the availability of air-filled pore space (Glenn and Sivasithamparam, 1990; Otten et al., 1999). Accordingly, fungi have been located predominantly on the outer surfaces of aggregates measuring between 1 µm and 1 mm (Hattori, 1988) and in pores > 10 µm (Schack-Kirchner et al., 2000). Fungal spread is strongly dictated by large pores and tends to occur along their surfaces (Harris et al., 2003; Otten et al., 2004). Soil aggregation should thus facilitate fungal abundance and growth.

Relevant to microbial activities is also the distance to the nearest air-filled pore, which decreases with soil aggregation. Parry *et al.* (1999) found that denitrification was less in a well-aggregated prairie soil rich in OM (organic matter) than in a cultivated soil less aggregated and poorer in OM. In prairie soil, most micropores were found to be situated within a short distance of air-filled macropores and anoxic conditions were thus not reached. Aggregation favoured aerobic microbial activities compared with anaerobic.

Few studies have directly investigated the impact of soil aggregation on microorganisms. For example, Dabire *et al.* (2001) hypothesized that a soil well aggregated in the vicinity of roots would enhance the retention of spores from the bacterium *Pasteuria penetrans*, which is a parasite of nematodes and therefore used as a biocontrol agent. They inoculated soil with *Pseudomonas* bacteria to promote soil aggregation and observed no change in the abundance of water-stable aggregates after inoculation, but soil adherence to roots increased. *Pasteuria penetrans* spore retention was effectively increased and most spores were located in aggregates > 200 µm.

In summary, an aggregated soil offers a larger diversity of physical habitats to microorganisms, enabling survival and varied metabolic activities. A larger diversity of microorganisms can thus be expected in aggregated soil compared with non-aggregated. We are aware of no study on this issue, and the specific role of the diversity of habitats might be difficult to disentangle from the larger OM contents usually associated with aggregated soils.

Now, are stable aggregates better for microorganisms than unstable? For microorganisms the main consequence of an unstable structure is that the architecture of the soil will be disrupted frequently with drying and re-wetting events. For example, Strong *et al.* (1998) found that a drying–re-wetting event increased the volume of pores of < 0.6 µm at the expense of more favourable pores of diameter 0.6–30 µm. Unstable aggregation may cause the periodic disappearance of air-filled pores, favourable to

aerobic microbial activity and to fungi, unless the soil structure has a high resilience. On the other hand, the periodic disruption of soil aggregates in unstable soils causes the release of SOM (soil organic matter) previously protected in aggregates, which favours microbial decomposition.

Stable aggregation thus ensures permanence of the various habitats defined by the architecture of the aggregated soil. Unstable aggregation associated with high resilience ensures the periodic loss and return to the characteristic architecture of an aggregated soil, and unstable aggregation with low resilience of soil structure appears less favourable for both aerobic activities and fungi.

5 The Dynamics and Factors of Microorganism–Soil Structure Interaction

5.1 Microbial, organic and structural dynamics are coupled

A conceptual model linking soil organic matter dynamics with the formation, stabilization and destabilization of aggregates has progressively emerged from the work of several groups in the last 15 years, and has been discussed in several review papers (Angers and Chenu, 1998; Golchin *et al.*, 1998; Jastrow *et al.*, 1998; Balesdent *et al.*, 2000; Six *et al.*, 2002, 2004). Briefly, when fresh organic matter enters the soil (e.g. decaying roots, incorporation of crop residues) it acts as nuclei for aggregate formation and stabilization because it locally stimulates microbial activity, and thus microbial binding and stabilizing mechanisms. As decomposition proceeds, microbial activity diminishes and the newly formed aggregates eventually lose their stability. Microbial, organic and structural dynamics are thus coupled.

In terms of spatial scales, aggregate stability increases at the macro-aggregate scale (units > 200 µm) within a few weeks of the addition of OM (Schletcht-Pietsch *et al.*, 1994; Angers *et al.*, 1997; Gale *et al.*, 2000; Bossuyt *et al.*, 2001; Denef *et al.*, 2002). This is consistent with the fact that microbial abundances and enzymatic activities rapidly increase in the vicinity of incorporated fresh plant residues (distances 2 to 5 mm) (Ronn et al., 1996; Gaillard et al., 1999; Kandeler et al., 1999). Macro-aggregate stability increased in the detritusphere of wheat straw after 1 month of incubation (Gaillard, 2001). Carbon from residues is incorporated first in stable macro-aggregates (> 250 μm), then in micro-aggregates (50–250 μm) (Angers et al., 1997), and stable micro-aggregates are formed within macro-aggregates in relation to microbial activity (Six et al., 2000, 2004). In terms of temporal scales, methods of directly measuring the turnover rate of structural units have recently been developed (Plante et al., 1999, 2002; De Gryze et al., 2006). These studies showed that macro-aggregates turn over rapidly, having turnover times variously reported as 4–95 days (Plante et al., 2002) and 9–30 days (De Gryze et al., 2006), while micro-aggregates turn over more slowly (i.e. 17–88 days; De Gryze et al., 2006). Such studies are hard to find and need to be extended to different soils and to in situ conditions (for the measurement of micro-aggregates turnover), but they clearly emphasize how soil architecture is dynamic and changes frequently. Aggregates are formed and destroyed within days, which for microorganisms means that cracks are displaced or easily colonizable planes of weakness change, and that micropores may change to macropores and vice versa. One future question regards the turnover time of the immediate environment of bacteria located in the pore matrix, in pores filled with EPS that are presumably sites for survival rather than for activity (Chenu, 1995; Chenu and Stotzky, 2002).

5.2 Major controls on these interactions

Plants and their root systems

Plant roots influence the interactions between soil structure and soil microorganisms mainly through trophic relations:

- Rhizodeposition fuels the microbial biomass through the sloughing off of border cells, the secretion of polysaccharide mucilages and the exudation of a variety

of low-molecular weight organic molecules (Nguyen, 2003). The availability of C in the rhizosphere permits the growth of EPS-producing bacteria, which have a strong effect on soil aggregation (Amellal et al., 1993, 1998; Alami et al., 2000; Kaci et al., 2005).
- Mycorrhizae have a major role in soil aggregation (see previous sections).
- Plant roots are also responsible for dispersing microbial cells through soils, along their length, and thus influence the localization of aggregating microorganisms.

Fauna

Soil animals influence the interactions between microorganisms and soil structure by (i) displacing microorganisms within the soil structure, at the scale of the soil profile (e.g. earthworms); and (ii) at a much finer scale, by changing their immediate environment, rearranging soil mineral particles, organic matter, microbial cells (in their gut in the case of termites; Garnier Sillam et al., 1987) and earthworms (Shipitalo and Protz, 1989; Marinissen and Dexter, 1990; Barois et al., 1993; Martin and Marinissen, 1993). As in the rhizosphere, bacteria proliferate in the earthworm and termite gut where they produce EPS (Garnier Sillam et al., 1987; Barois et al., 1993). Thereby, soil fauna initiate the microbial aggregation of soil particles. Creating new aggregated structures, they also modify the physical environment of soil microorganisms and affect their activity. For example, the activity of microorganisms was presumably inhibited in casts of Milsonia anomala due to a cortex of compacted particles (Blanchart et al., 1993).

Climate (water potential, wetting–drying cycles)

Soil climatic conditions are constantly changing, with frequent successions of drying and re-wetting events. These directly and adversely influence soil structure, creating cracks following clay shrinkage and swelling, increasing clay cohesion following drying and entrapping air upon re-wetting and thereby causing the slaking of soil architecture (Le Bissonais, 1996; Fig. 3.3). In arable soils, the disruptive effects of wetting outweigh the stabilizing effects of drying. Drying–re-wetting events also influence microbially mediated aggregation, and contrasting effects are observed depending on the type of dry–wet cycle (fast re-wetting or slow re-wetting) and the type of test used to assess structural stability. Denef *et al.* (2001) applied dry–wet cycles characterized by rapid re-wetting and observed decreased macro-aggregate stability to slow re-wetting until the third cycle, after which this was restored to normality. Denef *et al.* (2002) and Cosentino *et al.* (2006) both applied dry–wet cycles characterized by slow re-wetting. Denef *et al.* (2002) found increased macro-aggregate stability in a slow-wetting test and no effect in a fast-wetting test; Cosentino *et al.* (2006) found decreased macro-aggregate stability in a slow-wetting test and similarly no effect in a fast-wetting test. Cosentino *et al.* (2006) also observed that aggregate cohesion was increased by dry–wet cycles.

These effects may be explained first by the direct impact of dry–wet cycles on microorganisms (Fig. 3.4), causing the death of some and a flush of activity on rehydration (Jager and Bruins, 1975). Fungi are generally more resistant than bacteria to low water potentials (Griffin, 1981), and Cosentino *et al.* (2006) found that fungi were little affected by dry–wet cycles. However, no consistent differential effects of drying–re-wetting cycles were found on bacteria and fungi by others (West *et al.*, 1987; Scheu and Parkinson, 1994; Denef *et al.*, 2001). Dry–wet cycles can also affect the production of aggregating substances (Fig. 3.4) such as EPS (Roberson and Firestone, 1992), and presumably that of hydrophobins since these molecules have a role in resistance to desiccation (Wosten, 2001). Finally, dry–wet cycles may influence the expression of aggregating substances (Fig. 3.4). Amphiphilic molecules have been shown to present their hydrophobic moieties towards the exterior upon drying, and several results showing increased aggregation of soil with drying may be interpreted as an increased adhesion of EPS due to the closer juxtaposition of particles on drying (Haynes and Swift, 1990; Caron *et al.*, 1992; Cosentino *et al.*, 2006).

6 Predicting and Managing Microorganism–Soil Structure Interactions

6.1 A hierarchy of agents?

One striking feature of the literature in this field is that the microbial agents of soil structure were identified very early, as well as the mechanisms involved (see reviews by Allison (1968) and Six *et al.* (2004)). However, we still lack the ability to predict microbially mediated aggregation, and to implement favourable management practices. Many studies have aimed to identify, among the variety of agents of aggregation, which were the more important; establishing correlations among biological variables and aggregation was the method used. As may be seen in Table 3.1, no particular fraction or type of organism has emerged from this approach when various data were compiled, and several reasons can be invoked: (i) methods of measuring aggregate stability are very diverse, more or less drastic and favour various mechanisms of disruption (Fig 3.3); (ii) several variables have seldom been compared in a single study; and (iii) the importance of soil microorganisms in aggregation is dependent on soil type. Furthermore, as summarized in Fig. 3.4, various mechanisms are simultaneously at play (increased cohesion, increased hydrophobicity, alteration of pore network ...), as well as various microbial agents (bacteria, fungi, algae ...) and biochemical agents (EPS, lipids, proteins, hydrophobins ...). Hence, establishing a strict hierarchy of agents and defining unique indicators may seem a hopeless task. One integrative approach, potentially promising when dealing with cultivation systems, might be to relate soil structure to carbon inputs, since most microorganisms having an effect on soil structure are heterotrophs.

6.2 Quantitative models describing microorganism–soil structure–OM interactions

To date few quantitative models have addressed the dynamics of soil structure. Even fewer take explicitly into account the relationships between soil organic matter, soil microorganisms and soil structure. Such models are truly needed to predict changes in soil structure associated with specific cropping practices, e.g. the incorporation of organic wastes in soil, or with changes in crops or cropping systems. Furthermore, as soil structure protects organic matter from decomposition, aggregate dynamics should be incorporated in predictive models of soil organic matter dynamics.

Kay *et al.* (1988) proposed a conceptual and quantitative model to examine the changes in soil structure with cropping systems that considers a given cropping system in comparison with a reference state, stable within the time course of the study. Relative changes in S, a soil structural form, attribute $S_i/S_0 = f$ (t), where i refers to the situation under study, 0 to the reference situation and t to time, were described as the addition of: a biological function, describing the influence of soil biota on structural form; and a physico-chemical function depending on both the relative stability of soil structure (R_i/R_0) and the magnitude of the forces exerted in the cropping system.

The relative stability of soil structure depends linearly on the relative abundance of aggregating agents $R_i/R_0 = \alpha$ $(C_i/C_0) + \beta$, where C_i and C_0 are the aggregating agents and α and β are constants (Perfect and Kay, 1990). These workers applied this model to five different cultivation systems over 15 years and to several soil structure attributes, but did not derive predictions due to the high number of missing variables (including the nature and abundance of the aggregating agents). This model had the potential for providing a framework for many field and laboratory studies, but was unfortunately not developed further.

Many studies have shown that the addition of OM to soil temporarily increases aggregation and that the magnitude of effects, as well as their duration, varies widely with (i) the nature and decomposability of the added organic matter; (ii) the amounts incorporated; and (iii) the soil type (e.g. Metzger *et al.*, 1987; Fortun and Fortun, 1996; Kiem and Kandeler, 1997; Chantigny *et al.*, 1999; Bossuyt *et al.*, 2001; Denef *et al.*, 2002; Abiven *et al.*, 2007; Annabi *et al.*. 2007; De Gryze *et al.*, 2005b; Cosentino *et al.*, 2006). Monnier (1965) proposed a conceptual model in which aggregate stability kinetics after addition of OM to the soil was related to the decomposability of the added OM (Fig. 3.9). The magnitude and duration of microbial effects both depend on the decomposability of the added OM (Abiven *et al.*, 2009) (Fig. 3.10). Monnier's scheme and Fig. 3.11 show the complexity of the temporal pattern that needs to be predicted if one wishes to assess the short-term effect of a single OM addition on soil structure.

Recent studies have focused on predicting stable aggregate formation following the addition of fresh organic matter to soil. De Gryze *et al.* (2005a) quantified the influence of plant residue addition to soil on aggregate stability and tested the performance of different mathematical models in predicting the observed changes after 21 days of incubation for three different soils (i.e. prediction of e_{max} in Fig. 3.11). Two models were purely deterministic (linear and sigmoidal models), whereas the other two were mechanistic in that they related the rates of formation and destruction of aggregates with the abundance of macro-aggregates/non-aggregated soil. They found that the four models performed equally well (Fig. 3.12). Model 4 was selected because it enabled the calculation of an aggregate turnover rate and allowed the aggregate formation rate to decrease with time during the incubation according to several experimental observations.

Abiven *et al.* (2008) aimed to predict the decrease in aggregation that occurs at a later stage (i.e. the complete curve shown on Fig. 3.11). They developed statistical relationships between the initial biochemical characteristics of the plant residues and the shape parameters for the aggregate stability curve with time. Modulator functions

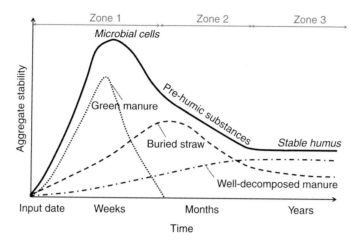

Fig. 3.9. Monnier's (1965) conceptual model: evolution of soil structural stability after the addition of organic matter characterized by either high decomposability (green manure), intermediate (buried straw) or low (well-decomposed manure).

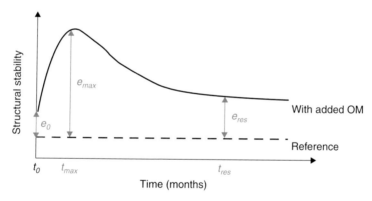

Fig. 3.10. Possible impacts of addition of organic matter to soil on structural stability and parameters that would allow the prediction of temporal patterns. t_0, time zero (addition of organic matter); e_0, direct effect of added organic matter at time 0; t_{max}, time at which effect on soil structure is maximal; e_{max}, maximal effect on soil structure; t_{res}, time at which a residual effect is observed (slope of aggregate stability versus time is low); e_{res}, residual effect.

temperature, C:N and humidity modifiers were used under field conditions. Cosentino (2006) proposed to couple a decomposition model to predict microbial biomass and activity (respiration) with a statistical relationship linking these variables to aggregate stability. This approach allows one to take into account not only the quality but also the amount of added organic matter. Validation and new developments are obviously needed in this field.

7 Conclusions

The interactions between soil microorganisms and soil structural dynamics are manifold and of great importance. Although the effects of microorganisms on structural stability have been studied for nearly 70 years, the ability to utilize the acquired knowledge as predictive tools of soil structural dynamics and related properties (e.g. C sequestration in soils), and into management options to

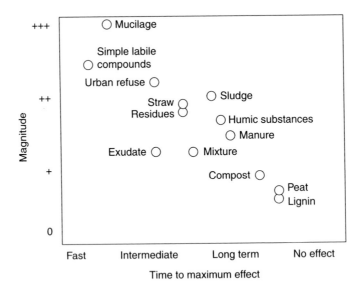

Fig. 3.11. Schematic representation of the relationships between the magnitude of the effect of an added organic matter category (emax as defined in Fig. 3.9) and time to maximum effect (t_{max}) (Abiven *et al.*, 2009). Categories of organic matter are defined in the cited paper.

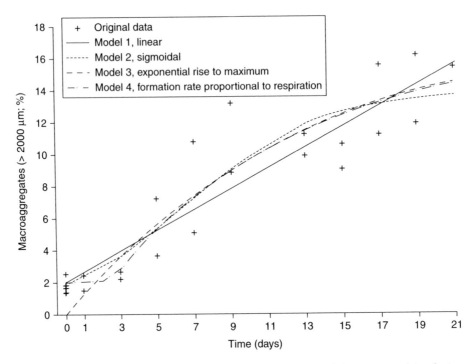

Fig. 3.12. Abundance of water-stable macro-aggregates in a silty loam soil during a 3-week incubation after addition of 1.5 g wheat residue/100 g of soil. The data were fitted using four different models (from De Gryze *et al.*, 2005a).

preserve or improve soil architecture, is still somewhat disappointing. Much effort should now be addressed to integrating existing knowledge into a comprehensive conceptual framework on the effects of soil biota on soil structure, including feedbacks on microorganisms, which can serve to test hypotheses and hierarchize processes and agents of soil structural dynamics in a given situation.

The effects of soil structure on microbial dynamics, activities and functions have received much less attention to date. The spatial heterogeneity of soil conditions at the microorganism scale must be taken into account if one wants to be able truly to describe and predict soil functioning, rather than considering average – and sometimes meaningless – microbial abundances and fluxes.

Acknowledgements

This chapter has been partly prepared under the auspices of the French Ministry of Ecology GESSOL programme 'MOST'.

References

Abiven, S., Menasseri, S., Angers, D.A. and Leterme, P. (2007) Dynamics of aggregate stability and biological binding agents during decomposition of organic materials. *European Journal of Soil Science* 58, 239–247.

Abiven, S., Menasseri, S., Angers, D.A. and Leterme, P. (2008) A model to predict soil aggregate stability dynamics following organic residue incorporation under field conditions. *Soil Science Society of America Journal* 72, 119–125.

Abiven, S., Menasseri, S. and Chenu, C. (2009) The effects of organic inputs over time on soil aggregate stability – a literature analysis. *Soil Biology and Biochemistry* 41, 1–12.

Achouak, W., Heulin, T., Villemin, G. and Balandreau, J. (1994) Root colonization by symplasmata-forming *Enterobacter agglomerans*. *FEMS Microbiology Ecology* 13, 287–294.

Alami, Y., Achouak, W., Marol, C. and Heulin, T. (2000) Rhizosphere soil aggregation and plant growth promotion of sunflowers by an exopolysaccharide-producing *Rhizobium* sp. strain isolated from sunflower roots. *Applied and Environmental Microbiology* 66, 3393–3398.

Allison, F.E. (1968) Soil aggregation. Some facts and fallacies as seen by a microbiologist. *Soil Science* 106, 136–143.

Amellal, N., Burtin, G., Bartoli, F. and Heulin, T. (1993) Soil aggregation in the rhizosphere of wheat: effect of inoculation with *Pantoea dispersa* on soil adhesion and structure. In: Pankhurst, C.E. (ed.) *Soil Biota. Management in Sustainable Farming Systems*. CSIRO, Victoria, Australia, pp. 7–8.

Amellal, N., Burtin, G., Bartoli, F. and Heulin, T. (1998) Colonization of wheat roots by an exopolysaccharide-producing *Pantoea agglomerans* strain and its effect on rhizosphere soil aggregation. *Applied and Environmental Microbiology* 64, 3740–3747.

Amellal, N., Bartoli, F., Villemin, G., Talouizte, A. and Heulin, T. (1999) Effects of inoculation of EPS-producing *Pantoea agglomerans* on wheat rhizosphere aggregation. *Plant and Soil* 211, 93–101.

Amezketa, E. (1999) Soil aggregate stability: a review. *Journal of Sustainable Agriculture* 14, 83–151.

Andrade, G., Mihara, K.L., Linderman, R.G. and Bethlenfalvay, G.J. (1998) Soil aggregation status and rhizobacteria in the mycorrhizosphere. *Plant and Soil* 202, 89–96.

Andreyuk, E.I., Lasik, J. and Iyatinskaya, G.A. (1986) Utilization of bacterial exopolysaccharides by soil microorganisms. *Zentralblatt für Mikrobiologie* 141, 83–88.

Angers, D.A. (1992) Changes in soil aggregation and organic carbon under corn and alfalfa. *Soil Science Society of America Journal* 56, 1244–1249.

Angers, D.A. and Chenu, C. (1998) Dynamics of soil aggregation and C sequestration. In: Lal, R., Kimble, J., Follet, R.F. and Stewart, B.A. (eds) *Soil Processes and the Carbon Cycle*, CRC Press, Boca Raton, Florida, pp. 199–206.

Angers, D.A. and Mehuis, G.R. (1989) Effects of cropping on carbohydrate content and water stable aggregation of a clay soil. *Canadian Journal of Soil Science* 69, 373–380.

Angers, D.A., Bissonnette, N., Légère, A. and Samson, N. (1993) Microbial and biochemical changes induced by rotation and tillage in a soil under barley production. *Canadian Journal of Soil Science* 73, 39–50.

Angers, D.A., Recous, S. and Aita, C. (1997) Fate of carbon and nitrogen in water-stable aggregates during decomposition of (CN)-C-13-N-15-labelled wheat straw in situ. *European Journal of Soil Science* 48, 295–300.

Angers, D.A., Edwards, L.M., Sanderson, J.B. and Bissonnette, N. (1999) Soil organic matter quality and aggregate stability under eight potato cropping sequences in a fine sandy loam of Prince Edward Island. *Canadian Journal of Soil Science* 79, 411–417.

Annabi, M., Houot, S., Francou, C., Poitrenaud, M. and Le Bissonnais, Y. (2007) Soil aggregate stability improvement with urban composts of different maturities. *Soil Science Society of America Journal* 71, 413–423.

Auer, D.P.F. and Seviour, R.J. (1990) Influence of varying N sources on polysaccharide production by *Aerobasidium pullulans* in batch culture. *Applied Microbiology and Biotechnology* 32, 637–644.

Baldock, J.A. (2001) Interactions of organic materials and microorganisms with minerals in the stabilization of soil structure. In: Huang, P.M., Bollag, J.M. and Senesi, N. (eds) *Interactions between Soil Particles and Microorganisms: Impact on the Terrestrial Ecosystem.* Wiley and Sons, New York, pp. 85–131.

Balesdent, J., Chenu, C. and Balabane, M. (2000) Relationship of soil organic matter dynamics to physical protection and tillage. *Soil and Tillage Research* 53, 215–230.

Ball, B.C., Cheshire, M.V., Robertson, E.A.G. and Hunter, E.A. (1996) Carbohydrate composition in relation to structural stability, compactibility and plasticity of two soils in a long-term experiment. *Soil and Tillage Research* 39, 143–160.

Barois, I., Villemin, G., Lavelle, P. and Toutain, F. (1993) Transformation of the soil structure through *Pontoscolex corethrurus* (Oligochaeta) intestinal tract. *Geoderma* 56, 57.

Bearden, B.N. and Petersen, L. (2000) Influence of arbuscular mycorrhizal fungi on soil structure and aggregate stability of a vertisol. *Plant and Soil* 218, 173–183.

Beare, M.H., Hu, H., Coleman, D.C. and Hendrix, P.F. (1997) Influences of mycelial fungi on soil aggregation and organic matter storage in conventional and no-tillage soils. *Applied Soil Ecology* 5, 211–219.

Bengtsson, G. (1991) Bacterial exopolymer and PHB production in fluctuating ground water habitats. *FEMS Microbiology Ecology* 86, 15–24.

Bethlenfalvay, G.J., Cantrell, I.C., Mihara, K.L. and Schreiner, R.P. (1999) Relationships between soil aggregation and mycorrhizae as influenced by soil biota and nitrogen nutrition. *Biology and Fertility of Soils* 28, 356–363.

Bissonnette, N., Angers, D.A., Simard, R.R. and Lafond, J. (2001) Interactive effects of management practices on water stable aggregation and organic matter of a humic gleysol. *Canadian Journal of Soil Science* 81, 545–551.

Blanchart, E., Bruand, A. and Lavelle, P. (1993) The physical structure of casts of *Millsonia anomala* (Oligochaeta: Megascolecidae) in shrub savanna soils (Côte d'Ivoire). *Geoderma* 56, 119–132.

Bond, R.D. and Harris, J.R. (1964) The influence of microflora on the physical properties of soils. I. Effects associated with filamentous algae and fungi. *Australian Journal of Soil Research* 2, 111–122.

Bossuyt, H., Denef, K., Six, J., Frey, S.D., Merckx, R. and Paustian, K. (2001) Influence of microbial populations and residue quality on aggregate stability. *Applied Soil Ecology* 16, 195–208.

Breedveld, M.W., Zevenhuizen, L.P.T.M., Cremers, H.C.J.C. and Zehnder, A.J.B. (1993) Influence of growth conditions on production of capsular and extracellular polysaccharide by *Rhizobium leguminosarum*. *Antonie van Leeuwenhoek* 64, 1–8.

Bruand, A. and Duval, O. (1999) Calcified fungal filaments in the petrocalcic horizon of Eutrochrepts in Beauce, France. *Soil Science Society of America Journal* 63, 164–169.

Butler, M.J. and Day, A.W. (1998) Fungal melanins: a review. *Canadian Journal of Microbiology* 44, 1115–1136.

Caesar-TonThat, T.C. and Cochran, V.L. (2000) Soil aggregate stabilization by a saprophytic lignin-decomposing basidiomycete fungus I. Microbiological aspects. *Biology and Fertility of Soils* 32, 374–380.

Capriel, P., Beck, T., Borchert, H. and Härter, P. (1990) Relationships between soil aliphatic fraction extracted with supercritical hexane, soil microbial biomass, and soil aggregate stability. *Soil Science Society of America Journal* 54, 415–420.

Caron, J., Kay, B.D. and Stone, J.A. (1992) Improvement of aggregate stability of a clay loam with drying. *Soil Science Society of America Journal* 56, 1583–1590.

Caron, J., Espindola, C.R. and Angers, D.A. (1996) Soil structural stability during rapid wetting: influence of land use on some aggregate properties. *Soil Science Society of America Journal* 60, 901–908.

Carter, M.R. (1992) Influence of reduced tillage systems on organic matter, microbial biomass, macro-aggregate distribution and structural stability of the soil in a humid climate. *Soil and Tillage Research* 23, 361–372.

Carter, M.R., Angers, D.A. and Kunelius, H.T. (1994) Soil structural form and stability, and organic matter under cool-season perennial grasses. *Soil Science Society of America Journal* 58, 1194–1199.

Chan, K.Y. and Heenan, D.P. (1999) Microbial-induced soil aggregate stability under different crop rotations. *Biology and Fertility of Soils* 30, 29–32.

Chaney, K. and Swift, R.S. (1984) The influence of organic matter on aggregate stability in some British soils. *Journal of Soil Science* 35, 223–230.

Chantigny, M.H., Angers, D.A., Prevost, D., Vezina, L.P. and Chalifour, F.P. (1997) Soil aggregation and fungal and bacterial biomass under annual and perennial cropping systems. *Soil Science Society of America Journal* 61, 262–267.

Chantigny, M.H., Angers, D.A. and Beauchamp, C.J. (1999) Aggregation and organic matter decomposition in soils amended with de-inking paper sludge. *Soil Science Society of America Journal* 63, 1214–1221.

Chao, W.L. and Alexander, M. (1982) Influence of soil characteristics on the survival of *Rhizobium* in soils undergoing drying. *Soil Science Society of America Journal* 46, 949–952.

Chappell, M.A. and Evangelou, V.P. (2002) Surface chemistry and function of microbial biofilms. *Advances in Agronomy*, 76, 163–199.

Chen, J.H., Czajka, D.R., Lion, L.W., Shuler, M.L. and Ghiorse, W.C. (1995) Trace metal mobilization in soil by bacterial polymers. *Environmental Health Perspectives* 103, 53–58.

Chenu, C. (1989) Influence of a fungal polysaccharide, scleroglucan, on clay microstructures. *Soil Biology and Biochemistry* 21, 299–305.

Chenu, C. (1993) Clay- or sand-polysaccharides associations as models for the interface between microorganisms and soil: water-related properties and microstructure. *Geoderma* 56, 143–156.

Chenu, C. (1995) Extracellular polysaccharides: An interface between microorganisms and soil constituents. In: Huang, P.M., Berthelin, J., Bollag, J.M., McGill, W.B. and Page, A.L. (eds) *Environmental Impact of Soil Component Interactions. Vol. I. Natural and Anthropogenic Organics.* Lewis, Boca Raton, Florida, pp. 217–233.

Chenu, C. and Guérif, J. (1991) Mechanical strength of clay minerals as influenced by an adsorbed polysaccharide. *Soil Science Society of America Journal* 55, 1076–1080.

Chenu, C. and Jaunet, A.M. (1992) Cryoscanning electron microscopy of microbial extracellular polysaccharides and their association with minerals. *Scanning* 14, 360–364.

Chenu, C. and Roberson, E.B. (1996) Diffusion of glucose in microbial extracellular polysaccharide as affected by water potential. *Soil Biology and Biochemistry* 28, 877–884.

Chenu, C. and Stotzky, G. (2002) Interactions between microorganisms and soil particles: An overview. In: Huang, P.M., Bollag, J.M. and Senesi, N. (eds) *Interactions between Soil Particles and Microorganisms.* Wiley and Sons, New York, pp. 3–40.

Chenu, C. Pons, C.H. and Robert, M. (1987) Interaction of kaolinite and montmorillonite with neutral polysaccharides. In: Schultz, L.G., Van Olphen, H. and Mumpton, F.A. (eds) *International Clay Conference Proceedings*, Denver, Colorado, 1985. The Clay Minerals Society, Bloomington, Indiana, pp. 375–381.

Chenu, C., Guérif, J. and Jaunet, A.M. (1994) Polymer bridging: a mechanism of clay and soil structure stabilization by polysaccharides. In: *XVth World Congress of Soil Science*, Acapulco, Mexico, pp. 403–410.

Chenu, C., Le Bissonnais, Y. and Arrouays, D. (2000) Organic matter influence on clay wettability and soil aggregate stability. *Soil Science Society of America Journal* 64, 1479–1486.

Chenu, C., Arias, M. and Besnard, E. (2001) The influence of cultivation on the composition and properties of clay-organic matter associations in soils. In: Rees, R.M., Ball, B.C., Campbell, C.D. and Watson, C.A. (eds) *Sustainable Management of Organic Matter.* CABI International, Wallingford, UK, pp. 207–213.

Cohen, V. (2002) Dynamique et rôle d'un exopolysaccharide bactérien produit par la souche *Rhizobium* sp. YAS34 dans le sol et la rhizosphère de Brassicacées. PhD dissertation, Université d'Aix Marseille, France, 230 pp.

Concaret, J. (1967) Etude des mécanismes de la destruction des agrégats de terre au contact de solutions aqueuses. *Annales Agronomiques* 18, 65–90.

Cosentino, D. (2006) Contribution des matières organiques à la stabilité de la structure des sols limoneux cultivés. Effet des apports organiques à court terme. PhD dissertation, Institut National Agronomique, Paris, 214 pp.

Cosentino, D., Le Bissonnais, Y. and Chenu, C. (2006) Aggregate stability and microbial community dynamics under drying-wetting cycles in a silt loam soil. *Soil Biology and Biochemistry* 38, 2053–2062.

Cosentino, D., Hallett, P.D. Michel, J.C. and Chenu, C. (2010) Do different methods for measuring the hydro-phobicity of soil aggregates give the same trends in soil amended with residue? *Geoderma* 159, 221–227.

Czarnes, S., Hallett, P.D., Bengough, A.G. and Young, I.M. (2000) Root- and microbial-derived mucilages affect soil structure and water transport. *European Journal of Soil Science* 51, 435–443.

Dabire, K.R., Duponnois, R. and Mateille, T. (2001) Indirect effects of the bacterial soil aggregation on the distribution of *Pasteuria penetrans*, an obligate bacterial parasite of plant-parasitic nematodes. *Geoderma* 102, 139.

Degens, B.P. (1997) Macro-aggregation of soils by biological bonding and binding mechanisms and the fac-tors affecting these: a review. *Australian Journal of Soil Research* 35, 431–459.

Degens, B. and Sparling, G. (1996) Changes in aggregation do not correspond with changes in labile organic C fractions in soil amended with C-14-glucose. *Soil Biology and Biochemistry* 28, 453–462.

Degens, B.P., Sparling, G.P. and Abbott, L.K. (1994) The contribution from hyphae, roots and organic carbon constituents to the aggregation of a sandy loam under long term clover-based and grass pastures. *European Journal of Soil Science* 45, 459–468.

Degens, B.P., Sparling, G.P. and Abbott, L.K. (1996) Increasing the length of hyphae in a sandy soil increases the amount of water-stable aggregates. *Applied Soil Ecology* 3, 149–159.

De Gryze, S., Six, J., Brits, C. and Merckx, R. (2005a) A quantification of short-term macro-aggregate dynam-ics: influences of wheat residue input and texture. *Soil Biology and Biochemistry* 37, 55–66.

De Gryze, S., Jassogne, L., Bossuyt, H., Six, J. and Merckx, R. (2005b) Water repellence and soil aggregate dynamics in a loamy grassland soil as affected by texture. *European Journal of Soil Science* 57, 235–246.

De Gryze, S., Six, J. and Merckx, R. (2006) Quantifying water-stable soil aggregate turnover and its implica-tion for soil organic matter dynamics in a model study. *European Journal of Soil Science* 57, 693–707.

Denef, K. and Six, J. (2005) Clay mineralogy determines the importance of biological versus abiotic processes for macro-aggregate formation and stabilization. *European Journal of Soil Science* 56, 469–479.

Denef, K., Six, J., Bossuyt, H., Frey, S.D., Elliott, E.T., Merckx, R. *et al.* (2001) Influence of dry-wet cycles on the interrelationship between aggregate, particulate organic matter, and microbial community dynam-ics. *Soil Biology and Biochemistry* 33, 1599–1611.

Denef, K., Six, J., Merckx, R. and Paustian, K. (2002) Short-term effects of biological and physical forces on aggregate formation in soils with different clay mineralogy. *Plant and Soil* 246, 185–200.

de Vries, F.T., Hoffland, E., van Eekeren, N., Brussaard, L. and Bloem, J. (2006) Fungal/bacterial ratios in grasslands with contrasting nitrogen management. *Soil Biology and Biochemistry* 38, 2092–2103.

Dexter, A.R. (1988) Advances in characterization of soil structure. *Soil and Tillage Research* 11, 199–238.

Dorioz, J.M., Robert, M. and Chenu, C. (1993) The role of roots, fungi and bacteria on clay particle organiza-tion. An experimental approach. *Geoderma* 56, 179–194.

Dormaar, J.F. and Foster, R.C. (1991) Nascent aggregates in the rhizosphere of perennial ryegrass (*Lolium perenne* L.). *Canadian Journal of Soil Science* 71, 465–474.

Driver, J.D., Holben, W.E. and Rillig, M.C. (2005) Characterization of glomalin as a hyphal wall component of arbuscular mycorrhizal fungi. *Soil Biology and Biochemistry* 37, 101–106.

Drury, C.F., Stone, J.A. and Filay, W.I. (1991) Microbial biomass and soil structure associated with corn, grasses and legumes. *Soil Science Society of America Journal* 55, 805–811.

Elliott, E.T., Anderson, R.V., Coleman, D.C. and Cole, C.V. (1980) Habitable pore space and microbial trophic interactions. *Oikos* 35, 327–335.

Emerson, W.W. and McGarry, D.M. (2003) Organic carbon and soil porosity. *Australian Journal of Soil Research* 41, 107–118.

Feeney, D.S. (2004) The influence of fungi upon soil structure and soil water relations. PhD thesis, University of Abertay, Dundee, UK.

Feeney, D.S., Daniell, T., Hallett, P.D., Illian, J., Ritz, K. and Young, I.M. (2004) Does the presence of glomalin relate to reduced water infiltration through hydrophobicity? *Canadian Journal of Soil Science* 84, 365–372.

Feeney, D.S., Hallett, P.D., Rodger, S., Bengough, A.G., White, N.A. and Young, I.M. (2006) Impact of fungal and bacterial biocides on microbial induced water repellency in arable soil. *Geoderma* 135, 72–80.

Fiès, J.C. (1992) Analysis of soil textural porosity relative to skeleton particle size using mercury porosimetry. *Soil Science Society of America Journal* 56, 1062–1067.

Fortun, A. and Fortun, C. (1996) Effects of two composted urban wastes on the aggregation and ion exchange processes in soils. *Agrochimica* 40, 153–165.

Foster, R.C. (1988) Microenvironments of soil microorganisms. *Biology and Fertility of Soils* 6, 189–203.

Franzluebbers, A.J., Wright, S.F. and Stuedemann, J.A. (2000) Soil aggregation and glomalin under pastures in the Southern Piedmont, USA. *Soil Science Society of America Journal* 64, 1018–1026.

Gaillard, V. (2001) Hétérogénéité spatiale de la biodégradation de résidus végétaux dans le sol: dynamique du carbone dans la résidusphère. PhD thesis, ENGREF, Paris.

Gaillard, V., Chenu, C., Recous, S. and Richard, G. (1999) Carbon, nitrogen and microbial gradients induced by plant residues decomposing in soil. *European Journal of Soil Science* 50, 567–578.

Gale, W.J., Cambardella, C.A. and Bailey, T.B. (2000) Root-derived carbon and the formation and stabilization of aggregates. *Soil Science Society of America Journal* 64, 201–207.

Garcia-Meza, J.V., Barrangue, C. and Admiraal, W. (2005) Biofilm formation by algae as a mechanism for surviving on mine tailings. *Environmental Toxicology and Chemistry* 24, 573–581.

Garnier Sillam, E., Villemin, G., Toutain, F. and Renoux, J. (1987) Contribution à l'étude du rôle des termites dans l'humification des sols forestiers tropicaux. In: Bresson, L.M., Coyrty, M.A. and Fedoroff, N. (eds) *Soil Micromorphology. Proceedings of the VIIth International Workshop Meeting on Soil Micromorphology.* AFES, Plaisir, France, pp. 331–335.

Geddie, J.L. and Sutherland, I.W. (1993) Uptake of metals by bacterial polysaccharides. *Journal of Applied Bacteriology* 74, 467–472.

Georgiou, G., Lin, S.-C. and Sharma, M.M. (1992) Surface-active compounds from microorganisms. *Natural Biotechnology* 10, 60.

Glenn, O.F. and Sivasithamparam, K. (1990) The effect of soil compaction on the saprophytic growth of *Rhizoctonia solani*. *Plant and Soil* 121, 282–286.

Golchin, A., Baldock, J.A. and Oades, J.M. (1998) A model linking organic matter decomposition, chemistry, and aggregate dynamics. In: *Soil Processes and the Carbon Cycle (Series: Advances in Soil Science).* CRC Press, Boca Raton, Florida, pp. 245–266.

Gouzou, L., Burtin, G., Philippy, R., Bartoli, F. and Heulin, T. (1993) Effect of inoculation with *Bacillus polymixa* on soil aggregation in the wheat rhizosphere: preliminary examination. *Geoderma* 56, 479–491.

Grant, C.D. and Dexter, A.R. (1990) Air entrapment and differential swelling as factors in the mellowing of moulded soil during rapid wetting. *Australian Journal of Soil Research* 28, 361–369.

Griffin, D.M. (1981) Water potential as a selective factor in microbial ecology of soils. In: Parr, J.F., Gardner, W.R. and Elliott, L.F. (eds) *Water Potential Relations in Soil Microbiology.* Soil Science Society of America, Madison, Wisconsin, pp. 141–151.

Guggenberger, G., Elliott, E.T., Frey, S.D., Six, J. and Paustian, K. (1999) Microbial contributions to the aggregation of a cultivated grassland soil amended with starch. *Soil Biology and Biochemistry* 31, 407–419.

Hadas, A., Rawitz, E., Etkin, H. and Margolin, M. (1994) Short-term variations of soil physical properties as a function of the amount and C/N ratio of decomposing cotton residues. I. Soil aggregation and aggregate tensile strength. *Soil and Tillage Research* 32, 199–212.

Hallett, P.D. and Young, I.M. (1999) Changes in water repellence of soil aggregates caused by substrate-induced microbial activity. *European Journal of Soil Science* 50, 35–40.

Hallett, P.D., Baumgartl, T. and Young, I.M. (2001a) Subcritical water repellency of aggregates from a range of soil management practices. *Soil Science Society of America Journal* 65, 184–190.

Hallett, P.D., Ritz, K. and Wheatley, R.E. (2001b) Microbial derived water repellencey in golf course soil. *International Turfgrass Society Research Journal* 9, 518–524.

Hallett, P.D., Feeney, D.S., Baumgartl, T., Ritz, K., Wheatley, R.E. and Young, I.M. (2002) Can biological activity dictate soil wetting, stability and preferential flow? In: *17th World Congress of Soil Science*, Bangkok, Thailand, 14–20 August. Soil and Fertilizer Society of Thailand, Bangkok, Thailand.

Harris, K., Young, I.M., Gilligan, C.A., Otten, W. and Ritz, K. (2003) Effect of bulk density on the spatial organisation of the fungus *Rhizoctonia solani* in soil. *FEMS Microbiology Ecology* 44, 45–56.

Harris, P.J. (1972) Microorganisms in the surface films from soil crumbs. *Soil Biology and Biochemistry* 4, 105–106.

Hattori, T. (1988) Soil aggregates as microhabitats for microorganisms. *Reports of the Institute of Agricultural Research Tohoku University* 37, 23–26.

Haynes, R.J. (1999) Labile organic matter fractions and aggregate stability under short-term, grass-based leys. *Soil Biology and Biochemistry* 31, 1821–1830.

Haynes, R.J. (2000) Labile organic matter as an indicator of organic matter quality in arable and pastoral soils in New Zealand. *Soil Biology and Biochemistry* 32, 211–219.

Haynes, R.J. and Swift, R.S. (1990) Stability of soil aggregates in relation to organic constituents and soil water content. *Journal of Soil Science* 41, 73–83.

Haynes, R.J., Swift, R.S. and Stephen, R.C. (1991) Influence of mixed cropping rotations (pasture–arable) on organic matter content, stable aggregation and clod porosity in a group of soils. *Soil and Tillage Research* 19, 77–87.

Hebbar, K.P., Guéniot, B., Heyraud, A., Colin Morel, P., Heulin, T., Balandreau, J. *et al.* (1992) Characterization of exopolysaccharides produced by rhizobacteria. *Applied Microbiology and Biotechnology* 38, 248–253.

Heijnen, C.E. and van Veen, J.A. (1991) A determination of protective microhabitats for bacteria introduced into soil. *FEMS Microbiology Ecology* 85, 73–80.

Henriksen, T.M. and Breland, T.A. (1999) Nitrogen availability effects on carbon mineralization, fungal and bacterial growth, and enzyme activities during decomposition of wheat straw in soil. *Soil Biology and Biochemistry* 31, 1121–1134.

Hu, H., Coleman, D.C., Beare, M.H. and Hendrix, P.F. (1995) Soil carbohydrates in aggrading and degrading agroecosystems – influences of fungi and aggregates. *Agriculture Ecosystems and Environment* 54, 77–88.

Jager, G. and Bruins, E.H. (1975) Effect of repeated drying at different temperatures on soil organic matter decomposition and characteristics and on the soil microflora. *Soil Biology and Biochemistry* 7, 153–159.

Jastrow, J.D., Miller, R.M. and Lussenhop, J. (1998) Contributions of interacting biological mechanisms to soil aggregate stabilization in restored prairie. *Soil Biology and Biochemistry* 30, 905–916.

Kaci, Y., Heyraud, A., Mohamed, B. and Heulin, T. (2005) Isolation and identification of an EPS-producing *Rhizobium* strain from arid soil (Algeria): characterization of its EPS and the effect of inoculation on wheat rhizosphere soil structure. *Research in Microbiology* 156, 341–347.

Kandeler, E., Luxhoi, J., Tscherko, D. and Magid, J. (1999) Xylanase, invertase and protease at the soil-litter interface of a loamy sand. *Soil Biology and Biochemistry* 31, 1171–1179.

Kay, B.D. (1990) Rates of change in soil structure under different cropping systems. *Advances in Soil Science* 12, 1.

Kay, B.D., Angers, D.A., Groenevelt, P.H. and Baldock, J.A. (1988) Quantifying the influence of cropping history on soil structure. *Canadian Journal of Soil Science* 68, 359–368.

Kendall, T.A. and Lower, S.K. (2004) Forces between minerals and biological surfaces in aqueous solutions. *Advances in Agronomy* 82, 1–54.

Khandall, R.K., Chenu, C., Lamy, I. and Tercé, M. (1992) Adsorption of different polymers on kaolinite and their effect on flumequine adsorption. *Applied Clay Science* 6, 343–357.

Kidron, G.J., Yaalon, D. and Vonshak, A. (1999) Causes for runoff initiation on microbiotic crusts: hydrophobicity and pore clogging. *Soil Science* 164, 18–27.

Kiem, R. and Kandeler, E. (1997) Stabilization of aggregates by the microbial biomass as affected by soil texture and type. *Applied Soil Ecology* 5, 221–230.

Kinsbursky, R.S., Levanon, D. and Yaron, B. (1989) Role of fungi in stabilizing aggregates of sewage sludge amended soils. *Soil Science Society of America Journal* 53, 1086–1091.

Kroen, W.K. (1984) Growth and polysaccharide production by the green alga *Chlamydomonas Mexicana* (Chlorophycae) on soil. *Journal of Phycology* 20, 253–257.

Kroen, W.K. and Rayburn, W.R. (1984) Influence of growth status and nutrients on extracellular polysaccharide synthesis by the soil alga *Chlamydomonas Mexicana* (Chlorophycae). *Journal of Phycology* 20, 253–257.

Kuikman, K., Van Elsass, J.D., Jansen, A.G., Burgers, S.L.G. and Van Veen, J.A. (1990) Population dynamics and activity of bacteria in relation to their spatial distribution. *Soil Biology and Biochemistry* 22, 1063–1073.

Lavelle, P. (2002) Functional domains in soils. *Ecological Research* 17, 441–450.

Lax A. and Garcia-Orenes, F. (1993) Carbohydrates of municipal solid-wastes as aggregation factor of soils. *Soil Technology* 6, 157–162.

Le Bissonnais, Y. (1989) Analyse des processus de microfissuration des agrégats à l'humectation. *Science du Sol* 27, 187–199.

Le Bissonnais, Y. (1996) Aggregate stability and measurement of soil crustability and erodibility: I. Theory and methodology. *European Journal of Soil Science* 47, 425–437.

Li, Z.J., Shluka, V., Wenger, K., Fordyce, A., Gade Pedersen, A. and Marten, M. (2002) Estimation of hyphal tensile strength in production scale *Aspergillus oryzae* fungal fermentations. *Biotechnology and Bioengineering* 77, 601–613.

Linton, J.D., Jones, D.S. and Woodward, S. (1987) Factors that control the rate of exopolysaccharide production by *Agrobacterium radiobacter* NICB 11883. *Journal of General Microbiology* 133, 2979–2987.

Lopez-Garcia, S.L., Vazquez, T.E.E., Favelukes, G. and Lodeiro, A.R. (2001) Improved soybean root association of N-starved *Bradyrhizobium japonicum*. *Journal of Bacteriology* 183, 7241–7252.

Lumsdon, S.O., Green, J. and Stieglitz, B. (2005) Adsorption of hydrophobin proteins at hydrophobic and hydrophilic interfaces. *Colloids and Surfaces B: Biointerfaces* 44, 172.

Malam Issa, O., Bissonnais, Y.l., Defarge, C. and Trichet, J., (2001) Role of a cyanobacterial cover on structural stability of sandy soils in the Sahelian part of western Niger. *Geoderma* 101, 15–30.

Malik, M. and Letey, J. (1991) Adsorption of polyacrylamide and polysaccharide polymers on soil materials. *Soil Science Society of America Journal* 55, 380–383.

Maqubela, M.P., Mnkeni, P.N.S., Malam Issa, O., Pardo, M.T. and D'Acqui, L.P. (2009) Nostoc cyanobacterial inoculation in South African agricultural soils enhances soil structure, fertility, and maize growth. *Plant and Soil* 315, 79–92.

Marinissen, J.C.Y. and Dexter, A.R. (1990) Mechanisms of stabilization of earthworm casts and artificial casts. *Biology and Fertility of Soils* 9, 163–167.

Martin, A. and Marinissen, J.C.Y. (1993) Biological and physico-chemical processes in excrement of soil animals. *Geoderma* 56, 331–347.

Martin, J.P. (1971) Decomposition and binding action of polysaccharides in soils. *Soil Biology and Biochemistry* 3, 33–41.

Martin, J.P. and Haider, K. (1979) Biodegradation of 14C-labeled model and cornstalk lignins, phenols, model phenolase humic polymers, and fungal melanins as influenced by a readily available carbon source and soil. *Applied Environmental Microbiology* 38, 283–289.

Martin, J.P. and Waksman, S.A. (1940) Influence of microorganisms on soil aggregation and erosion. *Soil Science* 50, 29–47.

Martin, J.P., Martin, W.P., Page, J.B., Raney, W.A. and De Ment, J.D. (1955) Soil aggregation. *Advances in Agronomy* 7, 1–37.

Martin, J.P., Ervin, J.O. and Shepherd, R.A. (1959) Decomposition and aggregating effect of fungus cell material in soil. *Soil Science Society of America Proceedings* 33, 717–720.

Martin, J.P., Ervin, J.O. and Richards, S.J. (1972) Decomposition and binding action of some mannose containing microbial polysaccharides and their FE, Al, and Cu complexes. *Soil Science* 113, 322–327.

McGhie, D.A. and Posner, A.M. (1980) Water repellence of a heavy textured Western Australian surface soil. *Australian Journal of Soil Research* 18, 309–323.

McIntire, T.M., David, A. and Brant, F. (1997) Imaging of individual biopolymers and supramolecular assemblies using noncontact atomic force microscopy. *Biopolymers* 42, 133–146.

Metzger, L., Levanon, D. and Mingelgrin, U. (1987) The effect of sewage sludge on soil structural stability: microbiological aspects. *Soil Science Society of America Journal* 51, 346–351.

Michel, J.C., Riviere, L.M. and Bellon-Fontaine, M.N. (2001) Measurement of the wettability of organic materials in relation to water content by the capillary rise method. *European Journal of Soil Science* 52, 459–467.

Miller, R.M. and Jastrow, J.D. (1990) Hierarchy of root and mycorrhizal fungal interactions with soil aggregation. *Soil Biology and Biochemistry* 22, 579–584.

Molope, M.B., Grieve, I.C. and Page, E.R. (1987) Contributions by fungi and bacteria to aggregate stability of cultivated soils. *Journal of Soil Science* 38, 71–77.

Monnier, G. (1965) Action des matières organiques sur la stabilité structurale des sols. *Annales Agronomiques* 16, 327–400.

Morris, E. and Norton, I.T. (1983) Polysaccharides aggregation in solutions and gels. In: *Aggregation Processes in Solution. Studies in Physical and Theoretical Chemistry, vol. 17.* Elsevier, Amsterdam.

Munkholm, L.J., Schjonning, P., Debosz, K., Jensen, H.E. and Christensen, B.T. (2002) Aggregate strength and mechanical behaviour of a sandy loam soil under long-term fertilization treatments. *European Journal of Soil Science* 53, 129–137.

Nguyen, C. (2003) Rhizodeposition of organic C by plants: mechanisms and controls. *Agronomie* 23, 375–396.

Oades, J.M. and Waters, A. (1991) Aggregate hierarchy in soils. *Australian Journal of Soil Research* 29, 815–828.

Osa-Afiana, L.O. and Alexander, M. (1982) Clays and the survival of Rhizobium in soil during desiccation. *Soil Science Society of America Journal* 46, 285–288.

Otero, A. and Vincenzini, M. (2003) Extracellular polysaccharide synthesis by Nostoc strains as affected by N source and light intensity. *Journal of Biotechnology* 102, 143.

Otten, W., Gilligan, C.A., Watts, C.W., Dexter, A.R. and Hall, D. (1999) Continuity of air-filled pores and invasion thresholds for a soil-borne fungal plant pathogen, *Rhizoctonia solani*. *Soil Biology and Biochemistry* 31, 1803–1810.

Otten, W., Harris, K., Young, I.M., Ritz, K. and Gilligan, C.A. (2004) Preferential spread of the pathogenic fungus *Rhizoctonia solani* through structured soil. *Soil Biology and Biochemistry* 36, 203–210.

Palma, R.M., DeCaire, G.Z., Zaccaro, M.C. and DeCano, M.S. (2000) Influence *of Nostoc muscorum* on microbial activity and soil aggregation. *Indian Journal of Agricultural Sciences* 70, 593–595.

Parry, S., Renault, P., Chenu, C. and Lensi, R. (1999) Denitrification in pasture and cropped soil clods as affected by pore space structure. *Soil Biology and Biochemistry* 31, 493–501.

Perfect, E. and Kay, B.D. (1990) Relations between aggregate stability and organic components in a silt loam soil. *Canadian Journal of Soil Science* 70, 731–735.

Perfect, E., Kay, B.D., Loon, W.K.P.v., Sheard, R.W. and Pojasok, T. (1990) Factors influencing soil structural stability within a growing season. *Soil Science Society of America Journal* 54, 173–179.

Plante, A.F., Duke, M.J.M. and McGill, W.B. (1999) A tracer sphere detectable by neutron activation for soil aggregation and translocation studies. *Soil Science Society of America Journal* 63, 1284–1290.

Plante, A.F., Feng, Y. and McGill, W.B. (2002) A modelling approach to quantifying soil macro-aggregate dynamics. *Canadian Journal of Soil Science* 82, 181–190.

Preston, S., Griffiths, B.S. and Young, I.M. (1999) Links between substrate additions, native microbes, and the structural complexity and stability of soils. *Soil Biology and Biochemistry* 31, 1541–1547.

Ramsay, A. (1984) Extraction of bacteria from soil: efficiency of shaking and ultrasonication as indicated by direct counts and autoradiography. *Soil Biology and Biochemistry* 16, 475–481.

Richaume, A., Steinberg, C., Jocteur-Monrozier, L. and Faurie, G. (1993) Differences between direct and indirect enumeration of soil bacteria: the influence of soil structure and cell location. *Soil Biology and Biochemistry* 25, 641–643.

Rillig, M.C. (2004) Arbuscular mycorrhizae, glomalin, and soil aggregation. *Canadian Journal of Soil Science* 84, 355–363.

Rillig, M.C. (2005) A connection between fungal hydrophobins and soil water repellency? *Pedobiologia* 49, 395–399.

Rillig, M.C. and Mummey, D.L. (2006) Mycorrhizas and soil structure. *New Phytologist* 171, 41–53.

Ritz, K. and Young, I. (2004) Interactions between soil structure and fungi. *Mycologist* 18, 52–59.

Roberson, E.B. (1991) Extracellular polysaccharide production by soil bacteria: environmental control and significance in agricltural soils. PhD thesis, University of California, Berkeley, California.

Roberson, E.B. and Firestone, M.K. (1992) The relationship between desiccation and exopolysaccharide production by a soil *Pseudomonas. Applied and Environmental Microbiology* 1284–1291.

Roberson, E.B., Sarig, S. and Firestone, M.K. (1991) Cover crop management of polysaccharide-mediated aggregation in an orchard soil. *Soil Science Society of America Journal* 55, 734–739.

Roberson, E.B., Sarig, S., Shennan, C. and Firestone, M.K. (1995) Nutritional management of microbial polysaccharide production and aggregation in an agricultural soil. *Soil Science Society of America Journal* 59, 1587–1594.

Robert, M. and Chenu, C. (1992) Interactions between microorganisms and soil minerals. In: Stotzky, G. and Bollag, J.M. (eds) *Soil Biochemistry, Vol. 7*. Marcel Dekker, New York, pp. 307–404.

Roldan, A., Garcia-Orenes, F. and Lax, A. (1994) An incubation experiment to determine factors involving soil aggregation changes in an arid soil receiving urban refuse. *Soil Biology and Biochemistry* 12, 1699–1707.

Ronn, R., Griffiths, B., Ekelund, F. and Christensen, S. (1996) Spatial distribution and successional pattern of microbial activity and micro-faunal populations on decomposing barley roots. *Journal of Applied Ecology* 33, 662–672.

Ruamps, L.S., Nunan, N. and Chenu, C. (2011) Microbial biogeography at the soil pore scale. *Soil Biology and Biochemistry* 43, 280–286.

Savage, S.M., Martin, J.P. and Letey, J. (1969) Contribution of some soil fungi to natural and heat-induced water repellency in sand. *Soil Science Society of America Proceedings* 33, 405–409.

Schack-Kirchner, H., Wilpert, K.V. and Hildebrand, E.E. (2000) The spatial distribution of soil hyphae in structured spruce-forest soils. *Plant and Soil* 224, 195.

Scheu, S. and Parkinson, D. (1994) Changes in bacterial and fungal biomass C, bacterial and fungal biovolume and ergosterol content after drying, remoistening and incubation of different layers of cool temperate forest soils. *Soil Biology and Biochemistry* 26, 1515.

Schletcht-Pietsch, S., Wagner, U. and Anderson, T.H. (1994) Changes in composition of soil polysaccharides and aggregate stability after carbon amendments to different textured soils. *Applied Soil Ecology* 1, 145–154.

Schutyser, M.A.I., Pagter, P.D., Weber, F.J., Briels, W.J., Boom, R.M. and Rinzema, A. (2003) Substrate aggregation due to aerial hyphae during discontinuously mixed solid-state fermentation with *Aspergillus oryzae*: experiments and modeling. *Biotechnology and Bioengineering* 83, 503–513.

Shipitalo, M.J. and Protz, R. (1989) Chemistry and micromorphology of aggregation in earthworm casts. *Geoderma* 45, 357–374.

Six, J., Elliott, E.T. and Paustian, K. (2000) Soil macro-aggregate turnover and micro-aggregate formation: a mechanism for C sequestration under no-tillage agriculture. *Soil Biology and Biochemistry* 32, 2099–2103.

Six, J., Connant, R.T., Paul, E.A. and Paustian, J. (2002) Stabilisation mechanisms of soil organic matter: implications for C-saturation of soils. *Plant and Soil* 241, 155–176.

Six, J., Bossuyt, H., De Gryze, S. and Denef, K. (2004) A history of research on the link between (micro)aggregates, soil biota, and soil organic matter dynamics. *Soil and Tillage Research* 79, 7–31.

Smits, T.H.M., Wick, L.Y., Harms, H. and Keel, C. (2003) Characterization of the surface hydrophobicity of filamentous fungi. *Environmental Microbiology* 5, 85–91.

Sparling, G.P. (1992) Ratio of microbial biomass carbon to soil organic C as a sensitive indicator of changes in soil organic matter. *Australian Journal of Soil Research* 30, 195–207.

Staddon, P.L., Ramsey, C.B., Ostle, N., Ineson, P. and Fitter, A.H. (2003) Rapid turnover of hyphae of mycorrhizal fungi determined by AMS microanalysis of 14C. *Science (Washington)* 300, 1138–1140.

Steinberg, P.D. and Rillig, M.C. (2003) Differential decomposition of arbuscular mycorrhizal fungal hyphae and glomalin. *Soil Biology and Biochemistry* 35, 191–194.

Stotzky, G. (1986) Influence of soil mineral colloids on metabolic processes, growth, adhesion, and ecology of microbes and viruses. In: Huang, P.M. and Schnitzer, M. (eds) *Interactions of Soil Minerals with Natural Organics and Microbes, Vol. 17.* Soil Science Society of America, Madison, Wisconsin, pp. 305–412.

Strong, D.T., Sale, P.W.G. and Helyar, K.R. (1998) The influence of the soil matrix on nitrogen mineralisation and nitrification. II. The pore system as a framework for mapping the organisation of the soil matrix. *Australian Journal of Soil Research* 36, 855–872.

Strong, D.T., Wever, H.D., Merckx, R. and Recous, S. (2004) Spatial location of carbon decomposition in the soil pore system. *European Journal of Soil Science* 55, 739–750.

Sutherland, I.W. (1985) Biosynthesis and composition of gram-negative bacterial extracellular and wall polysaccharides. *Annual Review of Microbiology* 39, 243–270.

Thomas, C.R., Zhang, Z. and Cowen, C. (2000) Micromanipulation measurements of biological materials. *Biotechnology Letters* 22, 531.

Thomas, R.S., Franson, R.L. and Bethlenfalvay, G.J. (1993) Separation of vesicular-arbuscular mycorrhizal fungus and root effects on soil aggregation. *Soil Science Society of America Journal* 57, 77–81.

Tillman, R.W., Scotter, D.R., Wallis, M.G. and Clothier, B.E. (1989) Water repellency and its measurement by using intrinsic sorptivity. *Australian Journal of Soil Research* 27, 637–644.

Tisdall, J.M. (1991) Fungal hyphae and structural stability of soil. *Australian Journal of Soil Research* 29, 729–743.

Tisdall, J.M. and Oades, J.M. (1980) The management of ryegrass to stabilise aggregates of a red brown earth. *Australian Journal of Soil Research* 18, 415–422.

Tisdall, J.M. and Oades, J.M. (1982) Organic matter and water-stable aggregates. *Journal of Soil Science* 33, 141–163.

Tisdall, J.M., Smith, S.E. and Rengasamy, P. (1997) Aggregation of soil by fungal hyphae. *Australian Journal of Soil Research* 35, 55–60.

Unestam, T. (1991) Water repellency, mat formation and leaf-stimulated growth of some ectomycorrhizal fungi. *Mycorrhiza* 1, 13–20.

Unestam, T. (1995) Extramatrical structures of hydrophobic and hydrophilic ectomycorrhizal fungi. *Mycorrhiza* 5, 301–311.

Vandevivere, P. and Baveye, P. (1992) Effect of bacterial extracellular polymers on the saturated hydraulic conductivity of sand columns. *Applied and Environmental Microbiology* 58, 1690–1698.

Vandevivere, P. and Kirchman, D.L. (1993) Attachment stimulates exopolysaccharide synthesis by a bacterium. *Applied and Environmental Microbiology* 59, 3280–3286.

Vinten, A., Whitmore, A., Bloem, J., Howard, R. and Wright, F. (2002) Factors affecting N immobilisation/mineralisation kinetics for cellulose-, glucose- and straw-amended sandy soils. *Biology and Fertility of Soils* 36, 190–199.

Wallis, M.G. and Horne, D.J. (eds) (1992) *Soil Water Repellency, Vol. 20.* Springer-Verlag, New York, pp. 92–146.

West, A.W., Sparling, G.P. and Grant, W.D. (1987) Relationships between mycelial and bacterial populations in stored, air-dried and glucose-amended arable and grassland soils. *Soil Biology and Biochemistry* 19, 599.

White, N.A., Hallett, P.D., Feeney, D., Palfreyman, J.W. and Ritz, K. (2000) Changes to water repellence of soil caused by the growth of white-rot fungi: studies using a novel microcosm system. *FEMS Microbiology Letters* 184, 88–95.

Wolfaardt, G.M., Lawrence, J.R., Headley, J.V., Robarts, R.D. and Caldwell, D.E. (1994) Microbial exopolymers provide a mechanism for bioaccumulation of contaminants. *Microbial Ecology* 27, 279–291.

Wosten, H.A.B. (2001) Hydrophobins: multipurpose proteins. *Annual Review of Microbiology* 55, 625–646.

Wosten, H.A.B. and Vocht, M.L.d. (2000) Hydrophobins, the fungal coat unravelled. *Biochimica et Biophysica Acta, Reviews on Biomembranes* 1469, 79–86.

Wosten, H.A.B., Wetter, M.A.v., Lugones, L.G., Mei, H.C.v.d., Busscher, H.J. and Wessels, J.G.H. (1999) How a fungus escapes the water to grow into the air. *Current Biology* 9, 85–88.

Wright, S.F. and Upadhyaya, A. (1996) Extraction of an abundant and unusual protein from soil and comparison with hyphal protein of arbuscular mycorrhizal fungi. *Soil Science* 161, 575–586.

Wright, S.F. and Upadhyaya, A. (1998) A survey of soil for aggregate stability and glomalin, a glycoprotein produced by hyphae of arbuscular mycorrhizal fungi. *Plant and Soil* 198, 97–107.

Wright, S.F., Starr, J.L. and Paltineanu, I.C. (1999) Changes in aggregate stability and concentration of glomalin during tillage management transition. *Soil Science Society of America Journal* 63, 1825–1829.

Wu, J., Gui, S., Stahl, P. and Zhang, R. (1997) Experimental study on the reduction of soil hydraulic conductivity by enhanced microbial growth. *Soil Science* 162, 741–748.

York, C.A. and Canaway, P.M. (2000) Water repellent soils as they occur on UK golf greens. *Journal of Hydrology* 126, 231–232.

Young, I.M. and Ritz, K. (2000) Tillage, habitat space and function of soil microbes. *Soil and Tillage Research* 53, 201–213.

Zaher, H., Caron, J. and Ouaki, B. (2005) Modelling aggregate internal pressure evolution following immersion to quantify mechanisms of structural stability. *Soil Science Society of America Journal* 69, 1–12.

Zavgorodnyaya, Y.A., Demin, V.V. and Kurakov, V.A. (2002) Biochemical degradation of soil humic acids and fungal melanins. *Organic Geochemistry* 33, 347–355.

4   The Zoological Generation of Soil Structure

Mark Bartlett[*] and Karl Ritz

1 Introduction

Whilst the soil fauna typically comprises a small proportion of the total biomass in soils, the animals that reside below ground have a large impact upon soil structure across a range of spatial and temporal scales. Such effects tend to be related to the body size of the organisms under consideration, and are predominantly founded upon physical movement of the soil – hence the concept that the soil fauna can be considered as principal 'engineers' of the soil system (Lavelle *et al.*, 1997; Wright and Jones, 2006). The movement of soil material by organisms is termed bioturbation, and such processes play important roles in pedogenesis and the structural and functioning of the soil system, with impacts ultimately manifest at the landscape scale (Wilkinson *et al.*, 2009). In this chapter we briefly review the effects of fauna on soil structure, restricting consideration to animals that engage in predominantly subterranean activity, hence excluding more surface-related phenomena such as impacts of trampling and surface foraging by animals on soil structure. To complement the more process-based recent review by Wilkinson *et al.* (2009), we focus on the range of faunal

types and the nature of the way they move and restructure soils.

All fauna that lives in or spends the majority of its life in the soil affects void ratio, and can potentially change the particle size distribution of the soil and alter the aggregate structure (FitzPatrick, 1980). The principal modes of animal-induced soil movement are a mixing, mounding and burial of soil fabric material, and formation of either open- or back-filled voids (Hole, 1981; Wilkinson *et al.*, 2009). Fauna can also restructure soil materials by processes of compression during passage through guts, and extrusion during voiding and defecation. Burrowing activities strongly affect the nature of the soil pore network, across a range of spatial scales related to the body size of the particular organism. Such body sizes range across several orders of magnitude, from a few microns for protozoan cells to almost one metre for some subterranean mammals. The different scales of soil organisms therefore have different roles and influences in and on the structure of the soil matrix. Organisms such as nematodes and protozoa are small in size (micro scale), but often have large populations typically confined to the top 50 mm. Soil macrofauna, such as oligochaetes, molluscs and

[*]Corresponding author: mark.bartlett@scotts.com

arachnids with body sizes from millimetres to centimetres, has more significant effects on soil structure and a greater role in bio-turbation processes. These organisms have influence at a wide range of depths through the soil, but dominate between the surface and up to 0.5 m. Although some macro-scale species will interact with the soil at a greater depth than this, the largest influ-ences and greatest depth of interactions come from larger-scale mammals, e.g. the Talpidae family. The increased body size of such species of soil fauna means that while they interact with the soil at greater depths they have a much lower abundance in the soil system as a whole.

2 Micro-scale Fauna

The smaller soil fauna, i.e. protozoa and nematodes, does not generally affect soil structure directly since the physical forces it imparts are relatively small compared with the larger fauna. Its organisms are also constrained to water films to effect any form of movement, which restricts its potential physically to move soil material. However, these organisms do have indirect effects on soil structural dynamics via their interac-tions with microbes. These are principally mediated by the role they play in grazing such microbes and thus regulating the bio-mass and activity of such organisms, with knock-on consequences for soil structure (see Chapter 3, this volume).

3 Macro-scale Organisms

Some of the smallest macro-scale organisms have only a limited influence on the physical structure of the soil system. Typically these organisms live opportunistically in the void spaces of the soil, rather than creating or modifying the structure of the matrix. These soil-dwelling fauna are affected by changes to soil structure because of their scale rela-tive to modification to the habitat space. Soil micro-arthropods and other mites, for exam-ple, can interact with the soil matrix but

have little influence on it. Increased abun-dance of mites has been observed when there are soil voids in the range of 60–300 µm in organic soils (Nielsen et al., 2008). Similar effects have also been observed in mineral soils (Ducarme et al., 2004). Relationships have been suggested between mite abun-dance and the structural complexity within the soil matrix. As soils become more com-pacted, with smaller void spaces and less pore connectivity, the abundance of micro-arthropods tends to decline (Borcard and Matthey, 1995; Heisler and Kaiser, 1995).

4 Soft-bodied Invertebrates

Terrestrial oligochaete worms account for approximately 25% of the whole of the phy-lum Annelida. Fossil evidence has shown that these globally distributed organisms began to colonize terrestrial environments about 500 million years ago, during the Cambrian explosion (Brusca and Brusca, 1990). They are now the most predominant species in the soil macrofauna of temperate soils (Lavelle, 2001). Their abundance within the soil system means they are also of considerable importance in pedogenesis and the development of distinguishable soil profiles (Lee, 1985; Brady and Weil, 1999). The physical size of the annelid has a sig-nificant implication for its ability to affect the structure of the soil. The largest reported free-living soil annelids are from the family Megascolideae, such as *Amynthas mekon-gianus*, native to South-east Asia, where individual worms can grow up to 2900 mm in length and consequently form extensive burrow networks and cycle large masses of soil to the surface (Van Praagh, 1992; Blake-more et al., 2007).

At the other extreme of the scale enchy-traeidea worms that live in the soil are typi-cally not more than 5 mm in length (Bardgett, 2005). Enchytraeidae-based soil casting in laboratory microcosms increases surface roughness at the micrometre scale (Schrader et al., 1997). The effects of enchytraeid worms are seen in the fine structure of the soil and surface texture. They can have a significant impact across a whole field because of high population numbers – up to 70,000 individuals/m^2 in agricultural fields (Dawod and FitzPatrick, 1993). Enchytraeid worms are positively correlated with aggregate stability (Bullinger-Weber et al., 2007). The physical structure of the soil has a bigger impact on these organisms than they do on it. Their activity and behaviour is considerably affected by changes to structure. Enchytraeid worms have also been highlighted as playing a significant role in the early stages of soil formation over granite outcrops in tropical environments such as French Guiana. The accumulation of Enchytraeidea casting material as an organic matrix for plant establishment is the first step in pedogensis. Enchytraeids therefore play a significant role in the development of a soil habitat and structure for other, larger soil fauna over slowly eroded igneous bedrock (Vaçulik et al., 2004).

The actions of earthworms in soils, and their significance for soil structure and bioturbation, have been considered for over 120 years. The first published work, by Darwin (1883), was a basic overview of the ecology and habitats of earthworms found in temperate climates, and understanding of the role of earthworms in the terrestrial environment has been considerably advanced since Darwin's time. A single earthworm can ingest between 2 and 30 times its body weight in soil per day, resulting in considerable impact on the architecture and other biotic activities within the soil (Lavelle et al., 1987; Curry, 1998; Curry and Schmidt, 2007). The physical activity of earthworms as they move though the soil therefore has significant impact on the structure of their environment and space within the soil system.

The concept of the drilosphere was introduced by Lavelle et al. (1987). This term encompasses the area of influence of the earthworm (in the same way that the rhizosphere describes the area affected by the roots of plants). In this way the whole soil system can be considered with the earthworm, including localized interactions with bacteria and fungi. The drilosphere can be defined as the soil within a 2 mm radius of an earthworm, accounting for roughly 3% of the soil volume; it can contain between 5 and 25% of soil microflora by volume (Lavelle et al., 1987). The drilosphere is primarily formed as the earthworm eats and pushes its way through the soil profile. The strength of the walls of these burrows is dependent on the hydrostatic pressure that the worm can generate in its coelomic cavity (Brusca and Brusca, 1990). Earthworms are capable of surviving in a wide range of soil conditions, with some species having a plastic life history, allowing them to adapt to life in a range of soil environments (Holmstrup, 2001; Bartlett et al., 2008).

Earthworms can be ordered into three ecological guilds with respect to the habitat-based niche that they occupy (Lee, 1985; Lavelle et al., 1987). These are (i) epigeic earthworms that live on and feed in leaf litter; (ii) anecic earthworms that maintain permanent or semi-permanent burrows; and (iii) endogeic earthworms that form burrows, but which are not permanent and are generally not open to the soil surface. Each ecological guild has a different impact in restructuring the soil matrix, as it either moves through it or feeds upon it.

Epigeic earthworms are typically found in woodland soils and have very shallow burrows in the soil system (Sims and Gerard, 1999). The action of epigeic earthworms such as Lumbricus rubellus have been shown to increase the storage of soil moisture in void spaces near the soil surface (Ernst et al., 2009). Other research has also implicated epigeic earthworms as an important factor in the formation of water-stable aggregates within alluvial soils (Bullinger-Weber et al., 2007), and Eisenia veneta has been shown to accelerate mineral weathering of clay particles in the soil (Carpenter et al., 2007). The action of these earthworms results in changes to both the soil's chemical and physical properties, both of which affect the physical structure and biological function of the soil voids. Endogeic earthworms have a similar effect on the soil structure, resulting in a high level of bioturbation within 200 mm of the soil surface (Lee, 1985). Species such as Aporrectodea rosea have been shown significantly to effect the structure of clayey soils, where processing of soil particles through their gut results in a platy structure of the soil at the field scale (Ester and Van Rozen, 2002).

Anecic earthworms such as *Lumbricus terrestris* and *Apporrectodea longa* form burrows that are typically open to the soil surface and represent a permanent home for them, from which they only travel a short distance to forage (Nuutinen and Butt, 2005). The burrow networks associated with these earthworms are considerably larger than those of the other two ecological guilds and their impact on the structure of the soils is correspondingly larger (Butt and Grigoropoulou, 2010). These burrow networks have

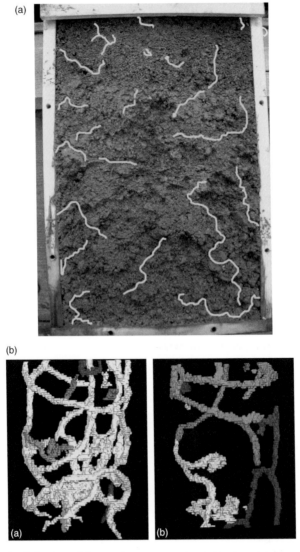

Fig. 4.1. The burrowing behaviour of the anecic earthworm *Lumbricus terrestris* has a strong influence on the architecture of the soil, creating semi-permanent burrows within the soil matrix. (a) Soil cross-section profile of an experimental Evans box showing the extent of earthworm-burrowing activity for foraging by *L. terrestris* in a sandy loam soil after 30 days. Width of box = 300 mm (photo courtesy of M. Bartlett); (b) 3D images generated using X-ray computed tomography of void spaces associated with *L. terrestris* activity within a soil monolith (from Langmaack *et al.*, 1999).

a considerable impact on the air-filled porosity of the bulk soil and associated water percolation rates (Fig. 4.1). Simple correlation between soil porosity and earthworm biomass is not apparent (Lee, 1985), and it has recently been suggested that it is the interaction between earthworms and the whole plant–soil system that causes the greatest effects in increasing void space (Milleret *et al.*, 2009).

Anecic earthworms also form surface casts and middens, thus playing a major role in the biotubation of a soil profile. Some of the first studies of earthworm ecology indicated that surface casting caused a considerable turnover of the soil surface, in the region of 15 t/ha/year, resulting in the burial of artefacts placed on the soil surface (Darwin, 1883), and contemporary research confirms these findings (Butt *et al.*, 2008; Hanson *et al.*, 2008). In the humid tropical soil systems of Nigeria, anecic earthworms (e.g. *Hyperiodrillus africanus*) will deposit as much as 36 t/ha/year (Hole, 1981), the impact of this behaviour on the structure of the soil being considerable. The feeding habit of anecic earthworms means that they can redistribute seeds within the soil seed bank. They can transport seeds to the surface via casting, where subsequently they can be stimulated to germinate; casts are a good environment for seed germination. Soil particles are loosely packed, allowing new roots to penetrate with greater ease (Decaens *et al.*, 2003). The combination of these properties means that earthworm casting activity can have a precipitous effect on gross soil structure, allowing the colonization of other flora and fauna that can have additional impacts on the whole soil system. Changes to the soil matrix caused by casting of earthworms create habitat spaces for some soil micro-arthropods, such as Collembola and prostigmatid mites (Hamilton and Sillman, 1989). Collembola are typically less that 6 mm long (Brusca and Brusca, 1990) but they can have a significant effect on increasing the surface roughness of soil (Schrader *et al.*, 1997). The consequence of this is that their activity accelerates the redistribution of the soil material that is deposited on the surface by earthworms.

All earthworm populations are affected by compaction and bulk density of the soil: the more compacted the soil the smaller the earthworm population (Langmaack *et al.*, 1999). When earthworms are removed from the soil system it tends to become more compact, as soil pore spaces are not altered by the burrowing action of the worms (Boag *et al.*, 1997). This created a more adverse environment for the earthworms, thus reducing their abundance. Earthworms play an important part in maintaining percolation and infiltration in soils, and thus their absence has a major impact on the rate of drainage because of the lack of biopores through the soil (Smettem, 1992). Quantitatively it has been shown that in soils processed by earthworm activity, gas volume increased by between 8 and 30%, and the bulked mass will drain between four and ten times faster than non-earthworm-worked soil (Edwards, 1996).

The texture of the soil also has an effect on the distribution of earthworms. The factor governing this is primarily the proportion of clay and sand particles. Experimentally it has been shown that a greater number of *Aporrectodea trapezoides* are found in light clay soils than sandy and clay loams (Baker *et al.*, 1998). The feeding of the earthworms is also selective relative to soil particle size. Where possible, earthworms will avoid larger sand particles; this can be seen in a size particle analysis of earthworm casts, which will typically contain less sand that the surrounding soil (Curry, 1989). The earthworm's preference is for organic matter, but the soil particles will be graded through the gut as a by-product. This action mixings the soil, forming new structural aggregates. The formation of micro- and macro-aggregates, and so to some extent soil structure and carbon cycling, is also dependent on earthworms (Bossuyt *et al.*, 2004). In an experiment where *Aporrectodea caliginosa* was introduced into soils, the formation of micro-aggregates in soil that had previously been homogenized to <250 µm was enhanced within 12 days. Earthworm gut mucus, microbial exudates and fungal hyphae will bind particles together and whether gut egestion takes place into or on to the soil profile has a significant effect on pedogenesis.

Slugs and snails are a significant group of agricultural pests because of their feeding behaviour, which is typically on the fleshy stems of young plants (Stephenson, 1968). Mollusca that burrow in the soil do not process the mineral matrix through their gut; they are more reliant on the structure of the soil that already exists rather than having the ability to modify it for their own purposes. Molluscs use the soil mineral matrix as a habitat, forming burrows and chambers to escape predation, and for over-winter protection (Barker, 1991). Unlike earthworms, however, slugs are poorly adapted to burrowing into the soil, and favour loosely compacted soils (Davis, 1989). In agricultural fields, species such as *Arion hortensis* burrow to feed on wheat grains, this activity resulting in bioturbation within the top 100 mm of the soil profile and increased availability of labile organic carbon as a substrate for other organisms (Glen *et al.*, 1988). The exudates that they produce also help cementation of particles within the soil matrix, which can help the formation of peds within the soil profile (Hole, 1981).

5 Hard-bodied Invertebrates

While not all species of ants live in the soil, where they do they are one of the most prominent groups of fauna that have influence on soil structure and have colonized all but the harshest environments (Paton *et al.*, 1995). Their tendency to form polymorphic social communities results in an extensive bioturbation of the soil. Even within the species of ants that do live in soil, the capacity of ant colonies to disturb and restructure the soil varies. Ants generally have a geographical preference for dry or loosely packed soils because of their burrowing behaviour and activity (Richards, 2009). Soil-dwelling species from the genus *Aphaenogaster*, which are found on the European, Asian, Australasian and American continents, have a geographical preference for sandy soils, but have also been reported in heavy clays and loams (Headley,

1949). Environmental parameters, such as aspect and topographic position, and factors that influence the rate of drainage play a major role in where nests are located within the landscape. Soils that have a high percolation rate are generally preferred (Saunders, 1967). Ant activity within the soil has two distinct modes of action: the deposit of material on the surface and the movement of soil within the profile.

Ants that produce surface mounts such as *Aphaenogaster barbigula* and *Aphaenogaster longiceps* typically start by excavation of narrow channels into the soil around 2 mm wide and, as the nest develops over time, the entrance is widened, with the spoil from excavation being deposited at the surface (Fig. 4.2). The surface mounds at the entrance to a fully established colony of *Camponotus intrepidus* can exceed 2300 cm^3 (Cowan *et al.*, 1985). This soil material is loosely packed, and very susceptible to erosion, from both wind and rain. The bulk soil matrix that has been worked by ants has, however, more water-stable aggregates than similar unworked soil material (Hole, 1981). Ant mounds are also dominated by mineral material that is finer than the surrounding soil matrix, and typically aggregate and grain sizes are smaller than 2.5 mm.

Behavioural studies on *Aphaenogaster* spp. have shown that their surface mounds are integral to below-ground nest structure; following storm events ants re-excavate their burrows of the finely particled soil materials that have been washed into them; this can result in distinct layering within the entrance mounds (Sloane and Sloane, 1964). *Formica negogagates* builds 55,000 mounds per hectare, each measuring *c.* 10 mm in height and 70 mm in diameter, mixing the upper 350 mm of the soil (Hole, 1981). A single nest may have multiple entrances and mounds associated with it, and the subterranean network may extend over several hundred square metres (Paton *et al.*, 1995). These structures within the soil have a considerable effect on increasing the rate of flow of water through the soil profile. The below-ground activity of ants can also have significant implications on the particle size distribution of the soil horizons in which

(a)

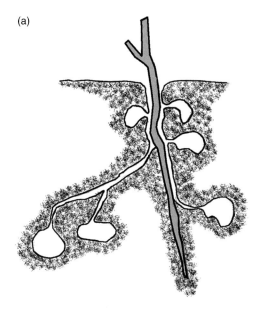

(b)

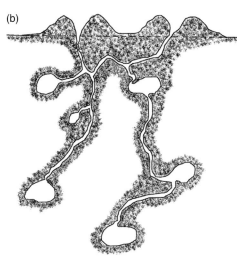

Fig. 4.2. The extent to which ants restructure the soil matrix to form their colony can have significant effects on the total void space of the soil. (a) Colonies may be formed around existing root macropores; (b) soil may be excavated to the surface and modify topographic surface features (from M. Bartlett).

they form their burrows. The selective movement of finer and smaller soil particles means that fine material is depleted from within the matrix, the result being that greater contrasts in texture between the

A and B horizons develop (Bishop *et al.*, 1980). Subsequently, this has implications for abiotic pedogenic soil processes such as the development of suspended water tables or the increase in the redox potential within the soil. Both of these changes significantly alter the capacity and species of vegetation that the soil is capable of supporting (Brady and Weil, 1999). The extent of the structural modifications that ants make to the soil matrix means that even once a colony has abandoned the nest the structure can persist. The geography of fossilized ant nests is extensive, with the longevity of the trace structures being closely related to both size- and species-specific excavation strategies (Tschinkel, 2003).

Subterranean termites creating tunnels and galleries underground, which are a compromise between foraging efficiency and similar environmental constraints to those of ants, such as soil hardness and moisture content (Lee and Su, 2010), have a significant impact on the structure of the soils in which they live. Termite activity has been shown to increase the macroporosity of the upper 100 mm of the soil profile by over 60%, by increasing the mean size of soil voids (Mando and Miedema, 1997). This increased void ratio results in more rapid percolation rates of rainfall in soils that have termite activity (Dawes, 2010). Their behaviour is frequently encouraged in soils where environmental or land management has resulted in hydrophobic crusts forming as their activity can break down these structures, increasing the abiotic functionality of the soil (Mando *et al.*, 1996). Research in Niger showed that approximately 30 foraging holes (predominantly from *Macrotermes subhyalinus*) were needed per square metre to significantly reduce surface runoff following rainfall events. At this level of termite activity in the soil infiltration rates are doubled compared with termite-free control soils (Léonard and Rajor, 2001). Deeper into the soil the effects of termites on structure are less easily quantified. While termite galleries are found up to 400 mm deep within the profile, the voids are typically backfilled with soil material. The activity of termites below 100 mm also has the effect of aggregating soil particles within the profile, and it is these biomediated peds that fill the void spaces (Mando and Miedema, 1997). As with ants, some of the material is taken to the surface by the colony, and this can have a significant influence on particle size distribution through the soil profile: smaller, finer soil particles are lifted to the surface.

Other species of termite form colonies that build structures on the soil surface, using materials taken from the soil matrix. These termites cement soil peds into a pedolithic mass (Hole, 1981). Typically, termite colonies of this type select clay particles from within the soil matrix as building materials, and within the mound this leads to a reduction in both pore size and rate of water diffusion. Soil organic matter that normally plays a key role in structuring soils has little or no effect on the stability of these termite mounds to processes such as rainfall erosion (Jouquet et al., 2004). Electrostatic forces between clay particles, especially in 2:1 clays, predominate in maintaining the structure. However, the high expansion rate of these clay materials on wetting also means that they are responsible for the breakdown of the colony's mound when water does penetrate.

Tunneller dung beetles (Scarabaeinae) play a role in the modification of soil structure, moving large quantities of soil to the surface during nesting. The nesting style of each beetle differs, but most construct underground tunnels with branching brood chambers. The whole nesting structure is backfilled with soil material of low bulk density to protect the developing larvae (Nichols et al., 2010). Minotaur beetles (*Typhaeus typhoeus*), for example, are found in sandy soils where they make backfilled burrows and produce soil mounds at the surface. Approximately 140 g of this loosely packed soil material is brought to the surface each year per burrow (Brussaard and Runia, 1984) and, despite this being relatively little soil movement in comparison with that of other soil-dwelling fauna, these beetles have considerable impact on their habitat space through the soil profile, often burrowing to 1 m in depth (Fig. 4.3).

Paracoprid dung beetles such as *Copris ochus* feed on herbage dung, and inhabit the soil matrix below or close to dung heaps. Research has shown that this species of beetle has a significant effect on the void space at the surface of the soil. The air permeability was found to be 1.5 times greater 100 mm from the surface of soil profiles where *C. ochus* was present compared with a beetle-free control, where the rate of diffusion was only 38 mm/h (Bang et al., 2005). Beetle activity was also confined to near the soil surface. No significant differences in air permeability at 200 mm were detected for treatments with and without beetles being present. Further quantitative evidence of the physical effects that beetle activity has on the soil structure is limited. Some soil-dwelling beetles are also capable of processing the mineral fraction of the soil, resulting in physical modification to the mineral matrix. The larvae of the beetle *Protaetia lugubris insperata* live within the soil matrix and feed on bacteria and fungi by direct ingestion of the mineral matrix. In an experiment that exposed these beetle larvae to grains of K-feldspar mineral, after only one day a significant change in the frequency distribution of grain sizes and the roundness of individual particles had occurred. Mineral particles that were analysed were both smaller and less angular as a result of being processed through the larval gut (Suzuki et al., 2003).

6 Winged Insects and Larvae

A wide range of winged insects and larvae burrow to make nests within the soil profile, and so increase the void space in the soil; bumblebees (*Bombus* spp.) and cicada (Homoptera: Cicadidae) are two common examples. Species of bumblebee such as *Bombus hortorum* that prefer to form nests at or just below the soil surface have a relatively low impact on the soil structure, often using abandoned burrows of larger soil fauna rather than creating their own burrows (Kells and Goulson, 2003). Species such as *Bombus terrestris* actively burrow in the soil to form their nesting sites and so modify the

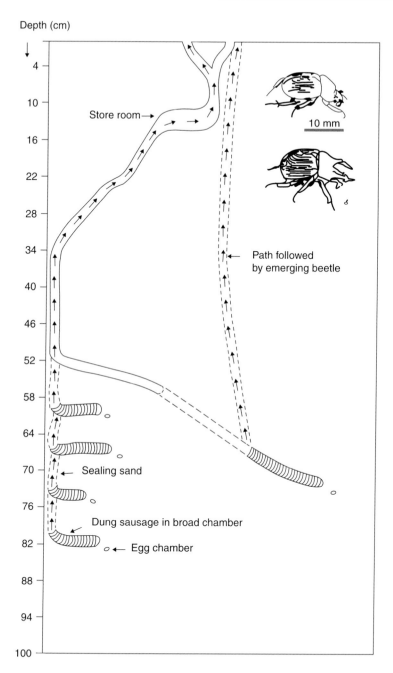

Fig. 4.3. Schematic diagram of the activity and nest structure of a tunnelling beetle *Typhaeus typhoeus* (Minotaur beetle) in a sandy soil (from Brussaard and Runia, 1984).

structure of the soil to suit their needs (Goulson *et al.*, 2002). An example of such a change to the structure of the soil profile is the action of the bee *Andrena prostimias* found in Japan. This bee species' life cycle lasts 12 months, and it nests for only a limited period during May and June, coinciding with the flowering of certain food plants

(Watanabe, 1998). In this two-month period these bees moved approximately twice as much soil to the surface as did the *L. terrestris* that Darwin (1883) studied in England in a whole year. The average nest size of this species is approximately 650 cm^3, consisting of chambers and galleries in which the bees lay larvae. The evacuated material is deposited on the surface as mounds of loose, low-bulk density soil that are quickly destroyed by rains after the bees vacate their nests in mid-June (Watanabe, 1998).

The decline in bee populations across Europe as a consequence of changing land patterns has been widely reported (e.g. Osborne and Williams, 2001). Research on *Bombus* spp. in the field has focused on their activity and geographical distribution. Typically this has then been related to above-ground ecosystem characteristics or functions (Goulson *et al.*, 2002; Kells and Goulson, 2003; McFrederick and LeBuhn, 2006). However, the close relationship that the bumblebee has with the soil as a habitat is frequently overlooked; only a very basic understanding of the relationships between *Bombus* spp. and the soil environment exists. Many species of soil invertebrates are significantly influenced by soil physico-chemical properties because they are in constant contact with the mineral matrix; however, these physico-chemical properties also significantly influence the physical structure and strength of the soil. These parameters should be of importance to burrowing *Bombus* species. Where soil particles are cemented by chemical processes or have particle size distributions dominated by clays, the ability to form voids and habitat spaces within the soil matrix will be difficult for these insects. The soil chemistry is also likely to play an indirect role, with the availability of nutrients from the soil influencing plant species composition and therefore availability as a food source. The use of spatially explicit mapping techniques of species composition and distributions accompanied by soil physical and chemical data may be useful in predicting changes to bumblebee populations and to provide insight into conservation strategies for these species, similar to those of other burrowing winged species such as cicadas (O'Geen *et al.*, 2002).

Cicadas are winged insects, the adults typically *c.* 50 mm in length, although *Pomponia imperatoria* found in Malaysia can grow up to 150 mm in length. Typically found in warm climates, they are poorly adapted to cooler temperatures (Brusca and Brusca, 1990). They feed on plants and can cause substantial damage to agricultural crops. Cicada nymphs live in the soil, and create extensive burrow systems which they occupy for up to 9 years prior to hatching. Their activity significantly modifies the void spaces within the soil, and some species also form chimneys of aggregates on the surface. Once the adult has left the burrow, structures called krotovina form, where the burrow fills with organic or mineral material from another soil horizon. These features are detectable through a wide range of soil types (O'Geen *et al.*, 2002). In soils where these insects are found, their activity is confined to the top 500 mm of the soil profile. However, the structural modifications that they make to the soil are so extensive and long lasting that the infill material of krotovina in the Pacific Northwest have been shown to have pollen and plant material that can be dated to 15,000 years BP (O'Geen and Busacca, 2001).

7 Mega-scale Organisms

Some vertebrates also have a subterranean habit, building nests, burrows and dens. The energy cost of living underground is significantly greater than living on the surface: between 340 and 3500 times more energy is required to dig 1 m underground than to walk 1 m on the surface (Reichman and Seabloom, 2002). Therefore the adaptive benefits of living in this way must be considerable. Commonly, the soil structures that these vertebrates engineer are used to protect themselves or their young from predation or unfavourable environments. The differences in physical size of these animals, however, mean that the changes to soil structure and above- and below-ground impacts result in changes in soil properties at both field and landscape scale.

8 Social and Gregarious Vertebrates

The European rabbit (*Oryctolagus cuniculus*) is a social animal, living in large underground, communal burrow networks. Rabbits use the same burrow networks for many generations, and their activity frequently depauperates the biological and physical soil system. Burrow construction is an ongoing process and can cover several hundred square metres on the surface. When the moisture content of the soil is favourable, to allow easy digging the rabbit colony will commonly expand its burrow network (Eldridge and Myers, 2001). During burrow construction large volumes of nutrient-poor soil are moved to the surface where they are deposited over the existing vegetation, normally resulting in it being totally smothered (Eldridge *et al.*, 2006). The digging behaviour of rabbits causes major changes to the organization of particle size distribution of soil horizons. Poorly developed pedological material from deeper in the soil profile is moved to the surface; this material typically has a higher clay content, which can form a slowly permeable barrier at the soil surface, leading to surface sealing and low infiltration rates. Where this material dries out it can also form physical crusts and barriers, impeding normal soil functions (Walker and Koen, 1995). Rabbits can have a destructive effect on both surface soils and vegetation in semi-arid soils. Their consumption of succulent vegetation and rampant breeding patterns mean that the stability of the soil ecosystem is frequently compromised when population numbers reach substantial levels. In one study, 5 years after extermination from a semi-arid woodland in Australia the effects of the European rabbit remained evident (Eldridge *et al.*, 2006).

In North American grasslands prairie dogs (*Cynomys* spp.) are a native species that lives in large colonies, building extensive underground towns with multiple entrances. The architecture of these towns is well structured and colonies have been shown to survive for up to 30 years. The prairie dog tunnels help to channel rainwater into the water table, reducing surface runoff and associated erosion (Lomolino *et al.*, 2004). As with the burrowing activity of rabbits the surrouningd soil–plant system is affected by these animals, with lower plant species diversity adjacent to the town and a significantly greater proportion of bare ground (Bangert and Slobodchikoff, 2000).

Gregarious breeding colony birds, such as the puffin (*Fratercula* spp.) and European bee-eater (*Merops apiaster*), form burrows in soil when they make breeding nests. Their activity is closely linked to soil strength. Where breeding colonies are found, banks that have significantly higher resistance to point penetration are left uncolonized (Heneberg, 2009). The nests of *M. apiaster* are built in sandy soils and are occupied for approximately 60 days during breeding, hatching and early chick rearing. They typically comprise a corridor approximately 1.5 m long, leading to a nest chamber approximately 100 mm in diameter (Ar and Piontkewitz, 1992). This activity plays a pivotal role in the arid environments in which these birds are most frequently found. The scale of the soil mixing, and the changes to the physical structure after the birds have vacated their nests, means that other fauna and flora can more efficiently colonize these sites. The changes that they make to the structure of the soil system are therefore important in maintaining the biodiversity of these arid environments (Casas-Crivillé and Valera, 2005).

9 Solitary Animals

Moles (family Talpidae) are highly adapted for digging in soil, with their forelimbs and sensory system being appropriate to this lifestyle (Sánchez-Villagra and Menke, 2005). Moles can bring up to 55 t/ha/year to the surface (Hole, 1981), with the soil in such mounds ('molehills') being highly disturbed and friable, and typically depauperated of fauna by the eating of prey by the moles. The European mole (*Talpa europaea*) creates extensive burrow systems in the surface soil horizons, where free passageway is maintained by periodic and regular passage to forage ('gleaning') for invertebrates that have entered such tunnels (Macdonald et al., 1997). The impact of such burrows upon soil macro- (arguably mega-) porosity and the implications for drainage are recognized by the agricultural engineering adoption of the term 'mole-drain' for certain types of connected drainage networks created by the use of tractor-borne implements that create tunnels in the soil.

Pocket gophers are described in several different families, and are found in North and South America, Africa and Europe, moving up to 57 m³ soil/ha/year to the surface (Reichman and Seabloom, 2002). They have a profound effect on community disturbance-dependent flora and feed on root material. Pocket gophers are territorial, with a population density of 20–40/ha being common (Smallwood and Morrison, 1999). The depths of the den that they dig vary with species, but they are typically within the plant root zone, between 60 and 200 mm from the soil surface, sometimes extending up to 1.5 m deep. A pocket gopher's den may have branched chambers that are backfilled by soil material. The proximity to growing roots means that this activity and modification of the soil structure stimulates root growth in surrounding plants (Reichman and Seabloom, 2002). Pocket gopher dens can also produce nutrient hotspots that can stimulate floral biodiversity in the loosely packed burrowing spoil; these materials are also susceptible to wind and water erosion. In some extreme cases where pocket gophers have built extensive dens on slopes, during extreme rainfall events flooded burrows can function as pipes through the soil.

The net effect (as well as the gophers' demise) is the acceleration of erosion processes. If the den totally collapses this can lead to the accelerated formation of a gully (Swanson et al., 1989).

10 Conclusions

Bioturbation by soil fauna has a very significant impact upon the structural organization of soils across a range of scales that transcend all other biotically mediated processes. The creation of biopores and the reduced bulk density that is typically a consequence of bioturbation means that the activity of animals within the soil has major consequences for the hydraulic characteristics of the whole soil. Their activity also plays a role in the net gas exchange between the atmosphere above and below ground. All living organisms within the soil matrix must take in O_2 and excrete CO_2; the activity of soil biota in modulating the rate of exchange is vitally important to the maintenance of a productive soil ecosystem (Stêpniewski and Gliñski, 1985). In natural systems such bioturbation underpins many of the pedogenic and successional series. In managed agricultural systems such zoologically mediated soil movement is largely substituted by the use of machinery, predominantly using fossil fuel-based energy. In such systems, the macro- and megafauna are largely excluded, but the role of invertebrates in 'biologically tilling' soils, for example under conservation tillage, is important and increasingly recognized.

References

Ar, A. and Piontkewitz, Y. (1992) Nest ventilation explains gas composition in the nest-chamber of European bee-eater. Respiration Physiology 87, 407–418.

Baker, G.H., Carter, P.J., Curry, J.P., Cultreri, O. and Beck, A. (1998) Clay content of soil and its influence on the abundance of Aporrectodea trapezoides (Duges). Applied Soil Ecology 9, 333–337.

Bang, H.S., Lee, J., Kwon, O.S., Na, Y.E., Jang, Y.S. and Kim, W.H. (2005) Effects of paracoprid dung beetles (Coleoptera: Scarabaeidae) on the growth of pasture herbage and on the underlying soil. Applied Soil Ecology 29, 165–171.

Bangert, R.K. and Slobodchikoff, C.N. (2000) The Gunnison's prairie dog structures a high desert grassland landscape as a keystone engineer. Journal of Arid Environments 46, 357–369.

Bardgett, R.D. (2005) The Biology of Soil. Oxford University Press, Oxford, UK.

Barker, G.M. (1991) Biology of slugs (Agriolimacidae and Arionidae: Mollusca) in New Zealand hill country pastures. *Oecologia* 85, 581–595.

Bartlett, M.D., James, I.T., Harris, J.A. and Ritz, K. (2008) Estimating species richness of earthworms on golf courses and implications for innovating environmentally benign control methods. *Acta Horticulturae* 783, 475–480.

Bishop, P.M., Mitchell, P.B. and Paton, T.R. (1980) The formation of Duplex soils on hillslopes in the Sydney Basin, Australia. *Geoderma* 23, 175–189.

Blakemore, R.J., Csuzdi, C., Ito, M.T., Kaneko, N., Paoletti, M.G., Spriridonov, S.E. *et al.* (2007) *Megascolex* (*Promegascolex*) *mekongianus* Cognetti, 1922 – its extent, ecology and allocation to *Amynthas* (Clitellata/Oligoghaeta: Megasacolecidae). *Opuscula Zoologica Budapest* 36, 19–30.

Boag, B., Palmer, L.F., Neilson, R., Legg, R. and Chambers, S.J. (1997) Distribution, prevalence and intensity of earthworm populations in arable land and grassland in Scotland. *Annals of Applied Biology* 130, 153–165.

Borcard, D. and Matthey, M. (1995) Effect of a controlled trampling of sphagnum mosses on their oribatid mite assemblages (Acari, Oribatei). *Pedobiologia* 39, 219–230.

Bossuyt, H., Six, J. and Hendrix, P.F. (2004) Rapid incorporation of carbon from fresh residues into newly formed stable microaggregates with earthworm casts. *European Journal of Soil Science* 55, 393–399.

Brady, N.C. and Weil, R.R. (1999) *Nature and Properties of Soil.* Chapman and Hall, Upper Saddle River, New Jersey.

Brusca, R. and Brusca, G. (1990) *Invertebrates.* Sinauer & Associates, Sunderland, Massachusetts.

Brussaard, L. and Runia, L.T. (1984) Recent and ancient traces of scarab beetle activity in sandy soils of The Netherlands. *Geoderma* 34, 229–250.

Bullinger-Weber, G., Le Bayon, R.-C., Guenat, C. and Gobat, J.-M. (2007) Influence of some physicochemcial and biological parameters on soil structure formation in alluvial soils. *European Journal of Soil Biology* 43, 57–70.

Butt, K.R. and Grigoropoulou, N. (2010) Basic research tools for earthworm ecology. *Applied and Environmental Soil Science* 2010, 1–12.

Butt, K.R., Lowe, C.N., Beasley, T., Hanson, I. and Keynes, R. (2008) Darwin's earthworms revisited. *European Journal of Soil Biology* 44, 255–259.

Carpenter, D., Hodson, M.E., Eggleton, P. and Kirk, C. (2007) Earthworm induced mineral weathering: Preliminary results. *European Journal of Soil Biology* 43, S176–S183.

Casas-Crivillé, A. and Valera, F. (2005) The European bee-eater (*Merops apiaster*) as an ecosystem engineer in arid environments. *Journal of Arid Environments* 60, 227–238.

Cowan, J.A., Humphreys, G.S., Mitchell, P.B. and Murphy, C.L. (1985) An assesment of pedoturbation by two species of mound-building ants, *Camponotus intrepidus* (Kirby) and *Iridomyrmex purpureus* (F. Smith). *Australian Journal of Soil Research* 23, 95–107.

Curry, J.P. (1989) The influence of invertebrates on soil fertility and plant growth in temperate grasslands. In: Clarholm, M. and Bergstrom, L. (eds) *Ecology of Arable Land: Perspectives and Challenges.* Kluwer Academic, Boston, Massachusetts.

Curry, J.P. (1998) *Earthworm Ecology.* Chapman and Hall, London.

Curry, J.P. and Schmidt, O. (2007) The feeding ecology of earthworms – a review. *Pedobiologia* 50, 463–477.

Darwin, C. (1883) *The Formation of Vegetable Mould, through the Action of Worms.* John Murray, London.

Davis, A.J. (1989) Effects of soil compaction on damage to wheat seeds by three pest species of slug. *Crop Protection* 8, 118–121.

Dawes, T.Z. (2010) Impacts of habitat disturbance on termites and soil water storage in a tropical Australian savannah. *Pedobiologia* 53, 241–246.

Dawod, V. and FitzPatrick, E.A. (1993) Some population sizes and effects of the Enchytraeidae (Oligochaeta) on soil structure in a selection of Scottish soils. *Geoderma* 56, 173–178.

Decaens, T., Mariani, L., Betancourt, N. and Jose Jaimenez, J. (2003) Seed dispersal by surface casting activities of earthworms in Colombian grasslands. *Acta Oecologica* 24, 175–185.

Ducarme, X.D., Andre, H.M., Wauthy, G. and Lebrun, P. (2004) Are there real endogeic species in temperate forest mites? *Pedobiologia* 48, 139–147.

Edwards, C.A. (1996) *Biology and Ecology of Earthworms.* Chapman and Hall, London.

Eldridge, D.J. and Myers, C.A. (2001) The impact of warrens of the European rabbit (*Oryctolagus cuniculus* L.) on soil and ecological processes in a semi-arid Australian woodland. *Journal of Arid Environments* 47, 325–337.

Eldridge, D.J., Costantinides, C. and Vine, A. (2006) Short-term vegetation and soil responses to mechanical destruction of rabbit (*Oryctolagus cuniculus* L.) warrens in an Australian box woodland. *Restoration Ecology* 14, 50–59.

Ernst, G., Felten, D., Vohland, M. and Emmerling, C. (2009) Impact of ecologically different earthworm species on soil water characteristics. *European Journal of Soil Biology* 45, 207–213.

Ester, A. and Van Rozen, K. (2002) Earthworms (Aporrectodea spp.; Lumbricidae) cause soil structure problems in young Dutch polders. *European Journal of Soil Biology* 38, 181–185.

FitzPatrick, E.A. (1980) *Soils: Their Formation, Classification and Distribution.* Longman, London.

Glen, D.M., Milsom, N.F. and Whiltshire, C.W. (1988) Effects of seed-bed conditions on slug numbers and damage to winter wheat in a clay soil. *Annals of Applied Biology* 115, 177–190.

Goulson, D., Hughes, W.O.H., Derwent, L.C. and Stout, J.C. (2002) Colony growth of the bumblebee, *Bombus terrestris*, in improved and conventional agricultural and suburban habitats. *Oecologia* 130, 267–273.

Hamilton, W.E. and Sillman, D.Y. (1989) Influence of earthworm middens on the distribution of soil micro-arthropods. *Biology and Fertility of Soils* 8, 279–284.

Hanson, I., Djohari, J., Furphy, P., Hodgson, C., Cox, G. and Broadbridge, G. (2008) New observations on the interactions between evidence and the upper horizons of the soil. In: Ritz, K., Dawson, L.A. and Miller, D. (eds) *Criminal and Environmental Soil Forensics.* Springer, Dordrecht, The Netherlands.

Headley, A.E. (1949) A population study of the ant *Aphaenogaster fulva* ssp. *aquia* Buckley. *Annals of the Entomological Society of America* 45, 435–442.

Heisler, C. and Kaiser, E.-A. (1995) Influence of agricultural traffic and crop management on collembolan and microbial biomass in arable soil. *Biology and Fertility of Soils* 19, 159–165.

Heneberg, P. (2009) Soil penetrability as a key factor affecting the nesting of burrowing birds. *Ecological Research* 24, 453–459.

Hole, F.D. (1981) Effects of animals on soil. *Geoderma* 25, 75–112.

Holmstrup, M. (2001) Sensitivity of life history parameters in earthworm *Aporrectodea caliginosa* to small changes in soil water potential. *Soil Biology and Biochemistry* 33, 1217–1223.

Jouquet, P., Tessier, D. and Lepage, M. (2004) The soil structural stability of termite nests: role of clays in *Macrotermes bellicosus* (Isoptera, Macrotermitinae) mound soils. *European Journal of Soil Biology* 40, 23–29.

Kells, A.R. and Goulson, D. (2003) Preferred nesting sites of bumblebee queens (Hymenoptera: Apidae) in agroecosystems in the UK. *Biological Conservation* 109, 165–174.

Langmaack, M., Schrader, S., Rapp-Bernhardt, U. and Kotzke, K. (1999) Quantitative analysis of earthworm burrow systems with respect to biological soil-structure regeneration after soil compaction. *Biology and Fertility of Soils* 28, 219–229.

Lavelle, P. (2001) *Soil Ecology.* Kluwer Academic Publishers, Dordrecht, The Netherlands.

Lavelle, P., Barois, I., Martin, A., Zaidi, Z. and Schaefer, R. (1987) Management of earthworm populations in agro-ecosystems: A possible way to maintain soil quality? In: Clarholm, M. and Bergstrom, L. (eds) *Ecology of Arable Land: Perspectives and Challenges.* Kluwer Academic, Boston, Massachusetts.

Lavelle, P., Bignell, D.E., Lepage, M., Volters, V., Roger, P., Ineson, P. *et al.* (1997) Soil function in a changing world: The role of invertebrate ecosystem engineers. *European Journal of Soil Biology* 33, 159–193.

Lee, K.E. (1985) *Earthworms: Their Ecology and Relationships with Soils and Land Use.* Academic Press, Sydney, Australia.

Lee, S. and Su, N. (2010) Simulation study on the tunnel networks of subterranean termites and their foraging behaviour. *Journal of Asia-Pacific Entomology* 13, 83–90.

Léonard, J. and Rajor, J.L. (2001) Influence of termites on runoff and infiltration: quantification and analysis. *Geoderma* 104, 17–40.

Lomolino, M.V., Smith, G.A. and Vidal, V. (2004) Long-term persistence of prairie dog towns: insights for designing networks of prairie reserves. *Biological Conservation* 115, 111–120.

Macdonald, D.W., Atkinson, R.P.D. and Blanchard, G. (1997) Spatial and temporal patterns in the activity of European moles. *Oecologia* 109, 88–97.

Mando, A. and Miedema, R. (1997) Termite-induced change in soil structure after mulching degraded (crust) soil in the Sahel. *Applied Soil Ecology* 6, 241–249.

Mando, A., Stroosnijder, L. and Brussaard, L. (1996) Effects of termites on infiltration into crusted soil. *Geoderma* 74, 107–113.

McFrederick, Q.S. and LeBuhn, G. (2006) Are urban parks refuges for bumble bees *Bombus* spp. (Hymenoptera: Apidae)? *Biological Conservation* 129, 372–382.

Milleret, R., Le Bayon, R.-C., Lamy, F., Gobat, J.-M. and Boivin, P. (2009) Impact of roots, mycorrhizas and earthworms on soil physical properties as assessed by shrinkage analysis. *Journal of Hydrology* 373, 499–507.

Nichols, E., Spector, S., Louzada, J., Larsen, T., Amezquita, S. and Favila, M.E. (2010) Ecological functions and ecosystem services provided by Scarabaeinae dung beetles. *Biological Conservation* 141, 1461–1474.

Nielsen, U.N., Osler, G.H.R., van der Wal, R., Campbell, C.D. and Burslem, D. (2008) Soil pore volume and the abundance of soil mites in two contrasting habitats. *Soil Biology and Biochemistry* 40, 1538–1541.

Nuutinen, V. and Butt, K.R. (2005) Homing ability widens the sphere of influence of the earthworm *Lumbricus terrestris* L. *Soil Biology and Biochemistry* 37, 805–807.

O'Geen, A.T. and Busacca, A.J. (2001) Faunal burrows as indicators of paleo-vegetation in eastern Washington, USA. *Palaeogeography, Palaeoclimatology, Palaeoecology* 169, 23–27.

O'Geen, A.T., McDaniel, P.A. and Busacca, A.J. (2002) Cicada burrows as indicators of paleosols in the inland Pacific Northwest. *Soil Science Society of America Journal* 66, 1584–1586.

Osborne, J.L. and Williams, I.H. (2001) Site constancy of bumble bees in an experimentally patchy habitat. *Agriculture, Ecosystems and Environment* 83, 129–141.

Paton, T.R., Humphreys, G.S. and Mitchell, P.B. (1995) *Soils: A New Global View.* UCL Press, London.

Reichman, O.J. and Seabloom, E.W. (2002) The role of pocket gophers as subterranean ecosystem engineers. *Trends in Ecology and Evolution* 17, 44–49.

Richards, P.J. (2009) Aphaenogaster ants as bioturbators: Impacts on soil and slope processes. *Earth-Science Reviews* 96, 92–106.

Sánchez-Villagra, M.R. and Menke, P.R. (2005) The mole's thumb – evolution of the hand skeleton in talpids (Mammalia). *Zoology* 108, 3–12.

Saunders, G.W. (1967) Funnel ants (*Aphaenogaster* spp., Formicidae) as pasture pests in Northern Queensland: I Ecological background, status and distribution. *Bulletin of Entomological Research* 57, 419–432.

Schrader, S., Langmacck, M. and Helming, K. (1997) Impact of Collembola and Enchytraeidae on soil surface roughness and properties. *Biology and Fertility of Soils* 25, 396–400.

Sims, R.W. and Gerard, B.M. (1999) *Synopses of the British Fauna (New Series): Earthworms.* Dorset Press, Dorchester, UK.

Sloane, H. and Sloane, A.I. (1964) The trap-nest or sand ant. *The Victorian Naturalist* 81, 165–167.

Smallwood, K.S. and Morrison, M.L. (1999) Spatial scaling of pocket gopher (Geomyidae) density. *The Southwestern Naturalist* 44, 73–82.

Smettem, K. (1992) The relation of earthworms to soil hydraulic properties. *Soil Biology and Biochemistry* 24, 1539–1543.

Stephenson, J. (1968) A review of the biology and ecology of slugs of agricultural importance. *Proceedings of the Malacological Society of London* 38, 169–179.

Stêpniewski, W. and Gliński, J. (1985) *Soil Aeration and its Role for Plants.* CRC Press Inc., Boca Raton, Florida.

Suzuki, Y., Matsubara, T. and Hoshino, M. (2003) Breakdown of mineral grains by earthworms and beetle larvae. *Geoderma* 112, 131–142.

Swanson, M.L., Kondolf, G.M. and Boison, P.J. (1989) An example of rapid gully initiation and extension by subsurface erosion: coastal San Mateo County, California. *Geomorphology* 2, 393–403.

Tschinkel, W.R. (2003) Subterranean ant nests: trace fossils past and future? *Palaeogeography, Palaeoclimatology, Palaeoecology* 192, 321–333.

Vaçulik, A., Kounda-Kiki, C., Sarthou, C. and Ponge, J.F. (2004) Soil invertebrates' activity in biological crusts on tropical inselbergs. *European Journal of Soil Science* 55, 539–549.

Van Praagh, B. (1992) The biology and conservation of the giant gippsland earthworm *Megascolides australis* McCoy, 1878. *Soil Biology and Biochemistry* 24, 1363–1367.

Walker, P.J. and Koen, T.B. (1995) Natural regeneration of groundstorey vegetation in a semi-arid woodland following mechanical disturbance and burning. 1. Ground cover levels and composition. *The Rangeland Journal* 17, 46–58.

Watanabe, H. (1998) Soil excavation by the deutzia andrenid bee (*Andrena prostimias*) in a temple garden in Hyogo Prefecture, Japan. *Applied Soil Ecology* 9, 283–287.

Wilkinson, M.T., Richards, P.J. and Humphreys, G.S. (2009) Breaking ground: Pedological, geological, and ecological implications of soil bioturbation. *Earth-Science Reviews* 97, 257–272.

Wright, J.P. and Jones, C.G. (2006) The concept of organisms as ecosystem engineers ten years on: progress, limitations, and challenges. *BioScience* 56, 203–209.

5 Biotic Regulation: Plants

Alain Pierret*, Christian Hartmann, Jean-Luc Maeght and Loïc Pagès

1 Root Growth and Root System Architecture

1.1 Introduction to root system architecture

The term 'architecture' of the root system refers to its shape and structure, both of which are of functional importance. The *shape* is the three-dimensional arrangement of roots that can be described using simple geometrical indicators such as lateral root extension or root depth or, more accurately, using the root length density distribution throughout the soil. The *structure* refers to the diversity of root types (components) and their relationships. According to their age and position within the root system, roots have specific physiological characteristics and functions (Waisel and Eshel, 2002). The way they are distributed and connected to each other (i.e. the topology of the system) is usually specific and important for the functioning of the whole system (Fitter, 1987). Root architecture dynamics is also important to consider, since the plant is a dynamic organism that develops in a changing environment. Interactions between roots and other organisms (e.g. competition, symbiosis) usually depend on dynamic

aspects and rates of development of the root system.

1.2 Overview of the root system architecture: classification and diversity

A distinction is commonly made between (i) fibrous (or diffuse, or fasciculate) root systems on the one hand, which are formed by a large number of equivalent roots, directly connected to the stem, and (ii) tap-rooted (or central, or conical) root systems on the other hand, which include a central, vertical main root (often called the tap root) along which a number of lateral roots originate at various depths. In addition to morphology at a given time, it is also important to consider the developmental sequence (or 'growth strategy'), as suggested by Cannon (1949). Cannon classified root system architecture according to two fundamental categories: (i) the primary root system, which originates from the radicle, grows downwards and branches acropetally (i.e. from its base to its tip); and (ii) the secondary root system (also called adventitious root system), which, in contrast to the primary system, emerges continuously from the shoot as it grows. Additional sub-types

*Corresponding author: apierret@gmail.com

are defined based on the distribution and relative importance of different root types.

This classification has been applied and completed by Krasilnikov (1968) and Kahn (1978) on trees. These authors have defined a number of morphological sub-types to refine the two basic categories defined by Cannon, and have shown that many root systems correspond to a combination of both strategies (mixed root systems).

It is noteworthy that root system architecture can be deeply altered by special adaptations to environmental conditions. Among these adaptations some roots, such as buttress and stilt roots, play a particular role for anchorage or support. Regarding nutrition, roots may exhibit several specializations: aerating roots (or pneumatophores) rise above the ground and provide oxygen, whilst tuberous roots work as storage organs and haustorial roots can absorb water and nutrients from other plants.

1.3 Elaboration of the root system architecture: components and developmental processes

To better understand root system architecture and its dynamics, it is worth analysing the system through its main components and developmental processes. Root systems develop according to basic morphogenetic production rules: (i) emergence of main axes (radicle, seminal or adventitious roots); (ii) growth of axes (including elongation and orientation – gravitropism); (iii) branching (new lateral axes); (iv) decay and abscission; and (v) radial root growth.

A large part of the architectural diversity of roots derives from the distribution and diversity of root apical meristems that are the key sub-structures for root elongation. These meristems provide the new cells in the apical part of the root, allowing apical elongation and maintenance of the root cap. In many plants, apical meristems exhibit large variations in diameter, and these variations are correlated to their morphogenetic capacities (Coutts, 1987; Pagès, 1995). Large meristems generally produce fast-growing roots with a long lifespan that are able to orient themselves through

various tropisms. These roots (macrorhizae) extend over the whole volume colonized by the root system. Conversely, small meristems make short and fine roots (brachyrhizae) that develop into soil pores, but stop growing after only a few days and subsequently become senescent and die. Thus, brachyrhizae exploit soil resources locally and temporarily. In addition to their transient role for the plant, their decomposition also contributes to the carbon efflux from the plant to the soil.

Branching is the major process through which root numbers, length and surface are increased. It is spatially and temporally organized, since lateral roots appear acropetally on their mother root. Lateral meristems are initiated as primordia very close to the meristem in the mother root, and emerge some days later as new lateral roots. Because of this process the most distal lateral roots are also the youngest.

Dicotyledonous plants have secondary meristems, which add layers of vessels and stiff tissues to the existing roots, thus increasing the conducting and support capacities of roots. Thanks to this secondary meristem functioning, roots become thicker, highly conductive for sap and better protected against adverse organisms and mechanical constraints. In monocotyledonous plants, original tissues are gradually complemented by stiff and protective substances, like lignin and suberin, but without any significant radial growth, due to the lack of secondary meristems producing additional tissues. These contrasted strategies induce large differences in the dynamics of these two categories of plants. Lateral root development is much more significant in plants that exhibit radial growth. In plants with radial growth there is a strong relationship, at each branching point, between the sectional area of a given mother root and the sum of the sectional areas of its daughter roots (Vercambre et al., 2003; Collet et al., 2005).

1.4 Root growth, tropisms and hormonal control

Root growth is influenced by a range of tropisms, among which gravitropism, thermotropism and phototropism, i.e. roots' response to gravity, temperature and light, respectively, are of central importance. Gravity causes localized expression of auxin in plant roots (Ottenschläger *et al.*, 2003). The distribution of soil temperatures modifies root developmental kinetics (Pregitzer *et al.*, 2000). Root growth is also partly driven by gradients of soil water potential, or hydrotropism (Tsutsumi *et al.*, 2004) and electrotropism, the sensitivity to electrical patterns that surrounds growing roots in the soil (Wolverton *et al.*, 2000).

The physical deformation of tissue resulting from root growth induces chemical changes leading to new organ formation: when a roots bends as it grows, concentrations of the plant hormone auxin increase along the outside of the bend and this determines how further roots branch from it, explaining both the inner/outer spacing, lateral inhibition and dynamics of lateral root initiation (Laskowski *et al.*, 2008; Richter *et al.*, 2009). A review of signalling that regulates root growth and root system architecture in response to physical cues can be found in Monshausen and Gilroy (2009).

Root growth rates are typically of the order of a few millimetres to a few centimetres per day. It has been known for a long time empirically that white and long root tips correspond to fast-growing roots and, conversely, that short and brown tips correspond to slow-growing roots. Remarkable root growth rates of 5–6 cm/day over 3 to 4 weeks have been reported for main axes in maize (Weaver and Clements, 1938). Intense root elongation can be sustained over long periods of time, as for example in *Eucalyptus camaldulensis*, for which root extension can exceed 2.5 m per year (Calder *et al.*, 1997). Morphological features of root apical regions, such as the length of the apical unbranched zone or root diameter, are closely correlated to root elongation rates and can therefore be reliably used to estimate root growth rates (Pagès *et al.*, 2009).

1.5 Modelling the root system architecture

Early models of the three-dimensional architecture of root systems (Diggle, 1988;

Pagès and Aries, 1988) have shown the potential of simulating the combination of several developmental processes in order to acquire a dynamic and quantitative representation of the complex root system architecture. Based on such models, linkages can be established between studies at both the individual root and whole root system level.

Later architectural models have integrated several root–soil interactions (Pagès *et al.*, 2004; Fig. 5.1) and have been merged with models of water and mineral transport in the soil (Clausnitzer and Hopmans, 1994). This approach allows a much more realistic simulation of plant development in a transient environment, and the quantitative study of the soil–plant system. Increasing the quality of predictions regarding the soil–plant–atmosphere fluxes is an important challenge, in which architectural models play a determinant role.

Other important integrative steps have been completed with the modelling of water (Doussan *et al.*, 1998) and sugar (Bidel *et al.*, 2000) transport within root system architecture. Such approaches allow assessment, at the whole root system level, of the functional importance of developmental and transport processes that can only be measured at the level of root components (meristems and root segments), bridging the two scales based on the distribution of local properties and associated processes throughout root system morphology.

Models allow the simulation of virtually an infinite number of combinations, by crossing plant and environmental situations. In this way, architectural diversity and functional consequences can be explored further.

2 Interactions between Plant Roots and the Soil Environment

2.1 Root growth in response to soil environmental conditions

The question of interactions between plants' roots and soil is a 'chicken and egg' situation: root growth is influenced, and some-

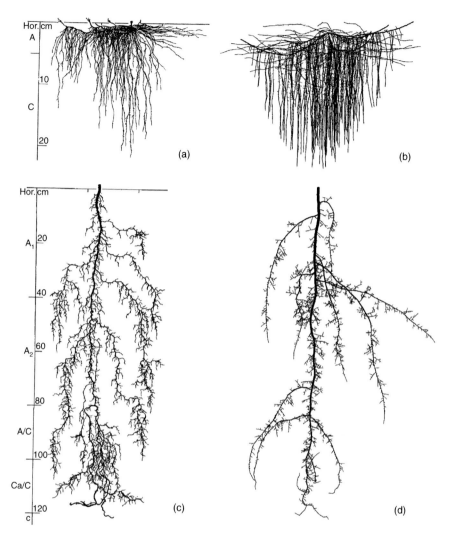

Fig. 5.1. Examples of drawn and simulated root systems, after Pagès *et al.* (2004). Drawn root systems (a, c) were observed in the field; simulated root systems (b, d) derive from a model combining several developmental processes, and take into account interactions with the soil.

times constrained, by the physico-chemical characteristics of the environment but, in the process, these also modify the growth medium through water and nutrient extraction, application of mechanical pressure and release of cells, ions and organic compounds into the soil (Hinsinger *et al.*, 2009).

Influence of chemical fertility

Soil nutrients most commonly occur non-uniformly, in a patchy manner, a situation to which root systems are known to adapt by means of morphological (Drew, 1975) and/or physiological (Jackson *et al.*, 1990) plasticity. Soil patchiness triggers 'root races' between competing plants, in which large quantities of assimilates are used to produce profuse roots (Passioura and Wetselaar, 1972), in an effort to secure exclusive benefit from resource-enriched spots (Hodge *et al.*, 1999), through a process known as resource pre-emption (Craine *et al.*, 2005; Bartelheimer *et al.*, 2006). It is generally agreed that such plastic responses represent foraging mechanisms, further enhanced via mycorrhizal symbioses (Hodge et al., 2001; Hodge, 2006), by which plants enhance nutrient uptake (Hodge, 2006).

At the individual level, interactions between root growth and the soil medium are of two types and include: (i) responses to the heterogeneity of local soil conditions, such as variations in soil strength (Masle and Passioura, 1987; Whalley et al., 1995), water and/or nutrient availability (Caldwell 1994; Dunbabin et al., 2002a, b) and soil structure (Ehlers et al., 1983; Logsdon and Allmaras, 1991); and (ii) variations in root soil contact (Tinker, 1976) due to changes in the geometry of either the soil or the root (such as diurnal variations in root diameter; Huck et al., 1970). Recent modelling of interspecific competition for bioavailable nutrients showed that this depends on soil and nutrient type as well as plant properties (Raynaud et al., 2008).

Influence of mechanical impedance and physical obstacles on root growth

The presence of zones of high mechanical resistance is one of the most common physical limitations to soil exploration by roots (Hoad et al., 1992). Mechanical impedance of the soil affects root elongation and branching and, in fine, root system architecture (Hamza and Anderson, 2005). Root growth proceeds via cell division and expansion in the apex and elongation zone, which results in forces being applied to soil particles, aggregates, gravels, etc. that are eventually displaced to a degree that depends mainly on soil bulk density and water content. Axial forces exerted by single growing roots vary as a function of time and soil depth, but generally are of the order of 0.6–1.0 N (Clark et al., 1996; Iijima et al., 2003). If the pores between solid constituents are sufficiently large, roots will readily penetrate the soil: it is generally considered that roots grow unimpeded in pores of equal diameter to that of the stele (Scholefield and Hall, 1985) but, in some plants, a thinning of both the cortex and

stele allows roots to grow in even smaller pores (Bengough et al., 1997).

When axial growth is hampered or prevented, localized root thickening ensues due to preferential, radial expansion of cortical cells (Souty, 1987; Croser et al., 2000; Bengough et al., 2006), a process through which roots can exert radial forces of the order of 1.0–2.0 N on surrounding grains (Kolb et al., 2009). Root thickening might reduce the axial stress experienced by the root cap (Abdalla et al., 1969; Graf and Cooke, 1980), and initiate/propagate cracking ahead of the cap (Dorgan et al., 2005). The latter hypothesis assumes that soil behaves like a homogeneous linear elastic medium, which does not always apply. In sandy soils or granular media, roots develop invaginations as they deform around particles that cannot be displaced (Goss, 1977). The potential for seedling roots to penetrate a mechanically resistant medium depends on plant species, with dicotyledons (having thicker radicles) having a higher penetrating potential than monocotyledons (Materechera et al., 1991).

Increased soil strength not only hampers root elongation, but also alters the average number of branching roots (Dexter, 1987a, b; Bennie, 1996; Bengough, 1997), with adverse consequences for resource acquisition (Lynch and Nielsen, 1996; Pagès et al., 2000; Pagès, 2002). Roots can, however, circumvent physical obstacles by self-inhibition via the release of allelopathic exudates (Falik et al., 2005), a process likely to increase plant performance by limiting resource allocation to less promising parts of the root system. In soils of high mechanical impedance, soil structure provides preferential pathways for root penetration (Tardieu and Katerji, 1991; Stirzaker et al., 1996). Therefore, one generally considers roots as flexible organs that follow tortuous pathways through the soil and apparently seek out the path of least resistance (Stewart et al., 1999; Moran et al., 2000; Hartmann et al., 2008a, b). However, this must be modulated in view of results showing that roots can penetrate compact clods even in the presence of soft soil (Konopka et al., 2008). While tap root geometry can improve plant anchorage, a long and thin tap root being preferable (Goodman et al., 2001), soil strength is reportedly the determining factor of lodging occurrence (Goodman and Ennos, 1999).

In very soft soil, limited root–soil contact may restrict plant uptake of water and nutrients (Herkelrath et al., 1977; Kooistra et al., 1992). Similarly, Moran et al. (2000) observed that fine wheat roots (<0.1 mm radius) growing in loose topsoil preferentially explored the denser parts of the soil volume (aggregates), where root–soil contact is better (Fig. 5.2). Lack of oxygen or hydromorphy are other physical factors affecting root development. It has been proposed that, as roots sense adverse soil conditions, they may send signals (Dodd, 2005) – including hormones to the shoots – so that the plant adjusts its physiology to compensate for a deteriorating or restrictive environment, especially if the plant's water supply is at risk (Passioura, 2002).

2.2 Effect of root growth and development on soil properties

The roles of plants in modifying soil structure have been extensively reviewed by Angers and Caron (1998) and include: the formation of macropores; soil fragmentation and aggregate formation; soil strengthening through exudation and drying; and maintenance of microflora and fauna that contribute to structural formation and stabilization through carbon input to the soil. Root carbon represents a major input of carbon to the soil (Rasse et al., 2005): large amounts of photosynthates allocated to roots can be lost to the soil as cap cells, mucilages, soluble exudates and lysates, as well as decaying tissues (Hutsch et al., 2002; Hawes et al., 2003; Nguyen, 2003).

Roots significantly and diversely alter soil physico-chemical properties through water and nutrient uptake and the release of chemical compounds – root exudates – and of dead organic matter. While most of these processes occur within the soil, some plant roots also influence soil properties from outside the soil. The Caucasian snow-bed plant *Corydalis conorhiza* provides an example of such unusual interactions: this plant forms extensive networks of specialized above-ground roots that grow against gravity and capture nitrogen that would otherwise partly run off down-slope over a frozen surface (Onipchenko et al., 2009).

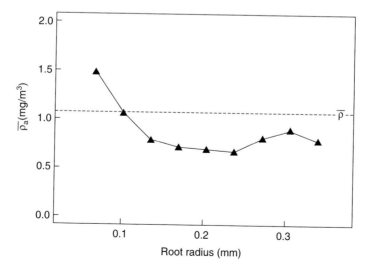

Fig. 5.2. Average soil density (ρ_a) in the immediate vicinity of wheat roots as a function of their radius, which indicates where the roots are growing in the soil density space. Confidence limits are too small to be represented. The broken line represents soil bulk density (adapted from Moran et al., 2000).

Impact of roots on soil porosity

As they grow, root apices reorganize the spatial arrangement of soil particles, an effect further strengthened via the secretion of organic compounds, typically extracellular polysaccharides, which bind soil particles together (Cheshire, 1979; Tisdall and Oades, 1982; Dorioz *et al.*, 1993). Although variable depending on soil type, these processes result in a general packing effect around growing roots (Bruand *et al.*, 1996; Clemente *et al.*, 2005), which is at the origin of the formation of some macropores (Jaillard and Callot, 1987; Young *et al.*, 1998). Several field studies showed that deep-rooted species could significantly improve soil macroporosity (Cresswell and Kirkegaard, 1995), even in the presence of a hardpan (Lesturgez *et al.*, 2004; Fig. 5.3).

The level of improvement depends on plant species and can affect both the micro- and macro-porosity at a wide range of soil depths, as for example with Lablab and mung bean, or be limited to a mere fractioning of compact clods near the surface, as

with sorghum (Pillai and McGarry, 1999). Growing root hairs are also capable of deforming moderately resistant clays: root hairs of pea seedlings were able to create perforations 0.5 mm long through re-moulded clay with an initial voids ratio (i.e. the ratio of the volume of voids to the volume of solids) of at least 1.1 (Champion and Barley, 1969). Root water uptake induces gradients in soil water content, which, depending on soil texture and mineralogy, can lead to cracking (Bruckler *et al.*, 1991; Lafolie *et al.*, 1991) and also contribute to soil aggregation (Tri and Monnier, 1973).

Carbon input to soil and its consequences

In association with the local rearrangement of soil particles that growing roots create in their immediate vicinity, there is also the development of a chemically and microbiologically differentiated environment, generally known as the rhizosphere (Darrah, 1993; Hinsinger, 1998). In some plant species, particularly graminates,

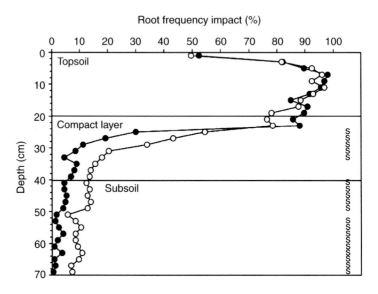

Fig. 5.3. Root frequency impacts of a maize crop in a *Stylosanthes hamata*/maize (*Zea mays* cv. SW 3601) rotation (–•–), i.e. *S. hamata* grown for 4 months, followed by maize for 4 months and fallow for another 4 months during the dry season; and in a continuous *S. hamata* cultivation (–○–), consisting of *S. hamata* being grown 24 months before being converted to maize production in the third year. S, significant difference at the depth increment studied (n = 5, p < 0.05) (adapted from Lesturgez *et al.*, 2004).

root-influenced soil can take the form of rhizosheaths, which are physically bound to parts of the root system (Watt et al., 1994; McCully, 1995). Root growth results in carbon input at different soil depths, including deep layers. Together with litter input, root production is believed to be the dominant input of organic carbon to soil and to correspond to a carbon stock more than twice that existing above ground (Jackson et al., 1996). Amongst root types, fine roots may be the prominent sink for carbon in soils: for example, Jackson et al. (1997) estimated from a comprehensive review of data on fine roots that these could represent 33% of global annual net primary production. It is also possible that fine root production would preferentially occur at soil depths where longer-term sequestration is more likely (Gill and Burke, 2002; Norby et al., 2004), which has important implications in the context of global change. Roots also alter the mineralization of soil organic matter (SOM) through the so-called 'priming' effect (Cheng and Coleman, 1990; Guenet et al., 2010). For example, carbon deposition by roots may enhance microbial growth resulting in an increase in SOM decomposition (Ingham and Molina, 1991). In contrast, plant uptake for nutrients might suppress soil microbial growth in nutrient-limiting soils, hence reducing SOM decomposition (Hu et al., 2001).

Root exudates and coarse particles play a major role in soil aggregation and, therefore, in physical fertility improvement (Materechera et al., 1992). Roots increase the proportion of aggregates as well as the quality of existing aggregates (Materechera et al., 1994). While root activity is often considered to play a minor role in soil aggregation and rehabilitation compared with soil macrofauna (Young et al., 1998), it was found that the opposite was true in a degraded vertisol resulting from intensive mono-cropping (Blanchart et al., 2004). Such a root-related improvement increased the soil OM content and had a dramatic effect on carbon storage (Chevallier et al, 2004). Aggregation and stabilization are promoted by exudates of high viscosity, but their impact depends on clay type, effects being greater for soils with 2:1 clays than 1:1 clays, with no effect for sandy soils (Barré and Hallett, 2009).

Effect of roots on soil water, soil aeration and associated processes

Plants can affect soil moisture and soil hydraulic properties both directly, as a consequence of water uptake, and indirectly, through the modification of soil structure that ensues (Gerke and Kuchenbuch, 2007). Root water uptake is associated with the formation and movement of a water extraction front and of high gradients of soil water content next to the roots (Fig. 5.4); such water uptake patterns are influenced by root

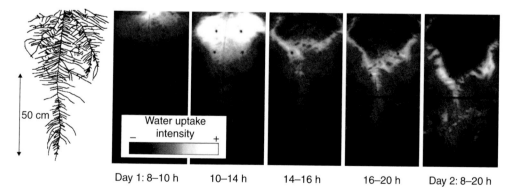

Fig. 5.4. Light transmission difference images of a water extraction front expanding as a 'moving sink' through a sandy growth medium (elapse time expressed as days after cessation of irrigation). Tap-rooted system of a c. 50-day-old narrow-leaf lupin (*Lupinus angustifolius*) (adapted from Garrigues et al., 2006).

system architecture (Garrigues et al., 2006). It has also been reported that root exudates may clog pores or become hydrophobic on soil particle surfaces, hence locally modifying soil hydrodynamic properties (Hallett et al., 2003). Under specific environmental conditions, plant water uptake can induce a rapid downward movement of the water table that results in air entry and plant-controlled sediment oxidation (Dacey and Howes, 1984). Due to their low tortuosity and high connectivity, only a few root biopores can induce dramatic increases in hydraulic conductivity. For example, with lucerne 1.7 and 1.8% increases in total and macro-porosity, respectively, resulted in a 57% increase in saturated conductivity (Rasse et al., 2000), which produced a four-fold increase in the infiltration rate through a heavily compacted soil over a 2-year period (Meek et al., 1992).

In response to waterlogging, many plants – including crops such as barley, rice and wheat (Justin and Armstrong, 1991; Drew et al., 2000; Evans, 2003) – can develop aerenchyma, i.e. specialized tissue that allows gas (notably oxygen) to be transported from the shoot to the root, either via simple diffusion or possibly under pressure flow (Jackson and Armstrong, 1999). This process has important implications for soil microbial activity, as some oxygen is released in the immediate vicinity of roots (Hodge et al., 2009).

Root growth in deep soil layers and rocks

A vast array of plant species have the inherent capability to develop deep or far-reaching roots and, even though physical barriers can restrict the growth of individual roots, the overall extension of root systems into weathered and/or fractured bedrock underlying shallow soils is commonplace, contributing significantly to plant moisture supplies (Fisher and Stone, 1968; Scholl, 1976; Calder et al., 1997; Jackson et al., 2000). Some plant species that have adapted to shallow-soil, drought-prone environments appear to have a specialized root system that allows them to explore a large rock surface area, thereby presumably increasing

their opportunity to encounter cracks in the underlying rock (Poot and Lambers, 2008). Deep rooting is not restricted to perennials, and this has been reported to play a significant role in N uptake by white cabbage (*Brassica oleracea* L.) at 2.5 m below the soil surface (Kristensen and Thorup-Kristensen, 2004). Depending on the underlying hydrogeological conditions, the subsoil can represent an important reservoir of water and nutrients for a broad range of plant species (Lehmann, 2003). In a comprehensive review, Stone and Kalisz (1991) listed 37 examples of tree root penetration between 10 and 60 m below the soil surface, with most or all of the distance being in rock, as well as 30 cases of roots in contact with water tables at considerable soil depth.

Root activity at depth can affect biogeochemical cycles and influence the soil's clay content and composition. Compared with oak, the pine tree develops a dense root system in both the litter and underlying soil horizons, which enables it to intercept potassium as soon as it is released from the litter. Consequently, under pine trees, less potassium is available for subsequent fixation by inter-layered clay minerals (Tice et al., 1996). The replacement of Hawaiian rainforest by pasture changed the proportion of non-crystalline to crystalline minerals through a concurrent alteration of the soil hydrological regime and a partitioning of silica into more stable biogenic forms (Kelly et al., 1998). More generally, the extractive activity of root systems could explain the dominance of illitic clay assemblage observed in temperate grassland soils or at the surface of soils developed on parent basalt materials that do not contain any mica (Barré et al., 2009); and the layers containing illitic clay could be seen and managed as a potassium reservoir that can be refuelled by fallow plants after intensive cropping (Barré et al., 2007).

There is good evidence that higher plants, together with soil microorganisms, play a major role in the weathering of rocks and formation of soils (Lambers et al., 2009) through the release of protons (Hinsinger et al., 1993) and siderophores. Furthermore, uptake processes such as that of K can become the driving force for the weathering of Mg-bearing silicates and concurrent formation of clay minerals in the absence of any pH change (Hinsinger and Jaillard, 1993). Proteoid (or cluster) roots, which are specially evolved roots that form clusters of closely spaced, short lateral rootlets adapted to phosphorus-deficient soils, are also known for their efficiency at dissolving phosphate rocks (Hinsinger and Gilkes, 1995). Under neutral conditions, plant root activity can rapidly (within 36 days) increase the weathering of basalt rocks (by 100–500-fold what would occur in the absence of roots (Hinsinger et al., 2001)).

Impact of root development on soil stabilization

Protection against soil losses results from the combined effects of canopy cover and roots (Gyssels et al., 2005). Grass roots are known to be very effective at reducing soil detachment rates (De Baets et al., 2006). A wide range of plant species play, via their root system, an important role in soil protection, particularly when above-ground biomass temporarily disappears as a result of dry season, grazing, fires, etc. (Cerda, 1998; De Baets et al., 2009).

Root systems also significantly stabilize sloping lands as the tensile strength of each root increases the global shear strength of the soil (Mattia et al., 2005; Tosi, 2007). In some cases, such as bamboo, the above-ground biomass does not properly reflect the actual landslide control capacity (Stokes et al., 2007). Root system architecture and spatial extension – in particular, lateral and vertical extensions – play a pivotal role in soil reinforcement. The depth of reinforcement is variable and generally increases with the size of the above-ground part of the plant (De Baets et al., 2008a; Docker and Hubble, 2009; Mickovski et al., 2009). Laterally, most soil reinforcement occurs close to the tree stem (< 1.0 m); this has important implications for soil failure control in the case of planting designs that include wide inter-rows or following thinning operations that create large empty gaps (Danjon et al., 2008; Genet et al., 2008; Mickovski and van Beek, 2009).

Roots of different sizes/diameters have different protective functions: while thick roots act like soil nails, thin roots provide the major contribution to soil cohesiveness (Burylo et al., 2009; Fatahi et al., 2009; Stokes et al., 2009). The binding effect resulting from root exudation is two orders of magnitude less than that resulting from the extra soil strength associated with soil colonization by roots (De Baets et al., 2008b).

Complex root–soil interactions often render landslide prediction difficult: for example, while the development of a deep tap root might mitigate the risk of soil failure, the development of macropores and preferential drainage below the tap root can, in some instances, greatly reduce or nullify the direct impact of roots on soil stabilization (Cammeraat et al., 2005).

3 Importance of Plant Roots at the Community Scale

3.1 Interactions between roots and root systems in multi-species systems

Interspecific interactions trigger plastic responses that can profoundly alter rooting patterns of plants compared with what they would be in a mono-specific stand. It was thus found that, while walnut and poplar trees grown in unmixed stands develop fine root profiles that decrease with increasing depth and distance from the tree trunk, when intercropped with grain crops these trees exhibit altered fine root profiles that are uniform or increase with soil depth (Mulia and Dupraz, 2006). Such a displacement of root density from the topsoil to layers below 1 m depth by intercrops has significant implications for water and nutrient cycling. Similarly, a comparison of wheat (*Triticum aestivum* L.)/maize (*Zea mays* L.) intercropping with wheat and maize monocrops showed that intercropping forced wheat roots to spread, at high root length density (RLD), under maize plants, while roots of maize intercropped with wheat were limited laterally but had a greater RLD than sole-cropped maize; the greater soil exploration and apparent root compatibility led to yield increases in spite of potential root competition for nutrients (Li *et al.*, 2006). Root segregation can provide competitive advantages for water and nutrient uptake as a result of an optimized utilization of the soil volume. Resource availability may also trigger and determine the relative importance of root segregation in plant communities, which appears to be most important in resource-limited environments (Schenk *et al.*, 1999). At a more local scale, mixing species also induces the aggregation of a given plant's roots towards that of neighbour plants, even at the cost of root development in unoccupied soil, indicating the importance of contesting of resources in competition (Bartelheimer *et al.*, 2006).

3.2 Agronomic implications of multi-specific systems

At the community level, diverse root systems corresponding to a multi-specific vegetation cover have the capacity to grow and colonize environments over an extensive range of lateral and vertical distances (Kutschera, 1960). From an agronomic perspective, combining plants with complementary rooting characteristics is central to the design of sustainable cropping systems: through the induced use of soil compartments that would not normally be colonized in a mono-crop system, particularly the subsoil (Lehmann, 2003), carefully designed multi-specific agrosystems can lead to higher productivity, improved control of pests and diseases, enhanced ecological services and greater economic profitability (Malezieux *et al.*, 2009). Plant species diversity is also one of the most effective means of enhancing soil stability in sloping terrain prone to landslides (Roering *et al.*, 2003; Pohl *et al.*, 2009).

The interplay between roots and soil corresponds to an iterative process: successive generations of roots tend to reuse paths of least mechanical resistance, such as preexisting cracks, biopores or casts excreted by soil macrofauna (Rasse and Smucker,

1998), the make-up of which is further modified with each new root generation. With time, the co-location of roots and macropores (McKenzie *et al.*, 1995; Volkmar, 1996; Stewart *et al.*, 1999) leads to the formation of a specific environment that significantly differs, both chemically and biologically, from the bulk soil (Pierret *et al.*, 1999; Pankhurst *et al.*, 2002). From an agronomic perspective, it has been demonstrated that, prior to introducing a crop on degraded land, so-called 'primer plants' selected based on their root traits could be used to improve soil properties (Yunusa and Newton, 2003).

3.3 Impact of root communities on climate

From the viewpoint of global atmospheric circulation, there is a growing body of literature showing that interactions between aquifers and vegetation cover, via deep roots, result in alterations of moisture conditions in the lower troposphere, eventually leading to changes in the seasonal pattern of air temperatures over large areas of the globe (Feddes *et al.*, 2001) and/or in precipitation regimes (Salati and Vose, 1984). It has been proposed that, through a process known as hydraulic redistribution (HR), by which root systems vertically and horizontally displace soil water from moister to drier zones in the soil, transpiration of the Amazonian forest can be sustained at high rates during the dry season, thus influencing air temperatures and directly linking plant root functioning with climate (Lee *et al.*, 2005).

Indirectly, deep rooting under some plant communities represents a vastly unknown potential for climate change mitigation via carbon sequestration. Whether deep rooting occurs concomitant with fine root proliferation remains to be established with any accuracy. However, as the relative importance of labile carbon pools decreases with soil depth (Trumbore, 1997) deep rooting might translate into increased potential for carbon sequestration, even if the actual quantities of deposited carbon are less than those prevailing in shallower soil horizons.

References

Abdalla, A.M., Hettiaratchi, D.R.P. and Reece, A.R. (1969) The mechanics of root growth in granular media. *Journal of Agricultural Engineering Research* 14, 236–248.

Angers, D.A. and Caron, J. (1998) Plant-induced changes in soil structure: processes and feedbacks. *Biogeochemistry* 42, 55–72.

Barré, P. and Hallett, P.D. (2009) Rheological stabilization of wet soils by model root and fungal exudates depends on clay mineralogy. *European Journal of Soil Science* 60, 525–538.

Barré, P., Velde, B. and Abbadie, L. (2007) Dynamic role of 'illite-like' clay minerals in temperate soils: facts and hypotheses. *Biogeochemistry* 82, 77–88.

Barré, P., Berger, G. and Velde, B. (2009) How element translocation by plants may stabilize illitic clays in the surface of temperate soils. *Geoderma* 151, 22–30.

Bartelheimer, M., Steinlein, T. and Beyschlag, W. (2006) Aggregative root placement: A feature during interspecific competition in inland sand-dune habitats. *Plant and Soil* 280, 101–114.

Bengough, A.G. (1997) Modelling rooting depth and soil strength in a drying soil profile. *Journal of Theoretical Biology* 186, 327–338.

Bengough, A.G., Croser, C. and Pritchard, J. (1997) A biophysical analysis of root growth under mechanical stress. *Plant and Soil* 189, 155–164.

Bengough, A.G., Bransby, M.F., Hans, J., McKenna, S.J., Roberts, T.J. and Valentine, T.A. (2006) Root responses to soil physical conditions; growth dynamics from field to cell. *Journal of Experimental Botany* 57, 437–447.

Bennie, A.T.P. (1996) Growth and mechanical impedance. In: Waisel, Y., Eshel, A. and Kafkafi, U. (eds) *Plant Roots: the Hidden Half.* Marcel Dekker, New York, pp. 453–470.

Bidel, L.P.R., Pagès, L., Rivière, L.-M., Pelloux, G. and Lorendeau, J.-Y. (2000) MassFlowDyn I: A carbon transport and partitioning model for root system architecture. *Annals of Botany* 85, 869–886.

Blanchart, E., Albrecht, A., Chevallier, T. and Hartmann, C. (2004) The respective roles of roots and earthworms in restoring physical properties of Vertisol under a *Digitaria decumbens* pasture (Martinique, WI). *Agriculture Ecosystems and Environment* 103, 343–355.

Bruand, A., Cousin, I., Nicoullaud, B., Duval, O. and Bégon, J.C. (1996) Backscattered electron scanning images of soil porosity for analyzing soil compaction around roots. *Soil Science Society of America Journal* 60, 895–901.

Bruckler, L., Lafolie, F. and Tardieu, F. (1991) Modeling root water potential and soil–root water transport: II. Field comparisons. *Soil Science Society of America Journal* 55, 1213–1220.

Burylo, M., Rey, F., Roumet, C., Buisson, E. and Dutoit, T. (2009) Linking plant morphological traits to uprooting resistance in eroded marly lands (Southern Alps, France). *Plant and Soil* 324, 31–42.

Calder, I.R., Rosier, P.T.W., Prasanna K.T. and Parameswarappa, S. (1997) Eucalyptus water use greater than rainfall input – a possible explanation from southern India. *Hydrology and Earth System Sciences* 1, 249–256.

Caldwell M.M. (1994) Exploiting nutrients in fertile soil microsites. In: Caldwell, M.M and Pearcy, R.W. (eds) *Exploitation of Environmental Heterogeneity by Plants. Ecophysiological Processes Above- and Belowground.* Academic Press, Inc., Beltsville, Maryland, pp. 325–347.

Cammeraat, E., van Beek, R. and Kooijman, A. (2005) Vegetation succession and its consequences for slope stability in SE Spain. *Plant and Soil* 278, 135–147.

Cannon, W.A. (1949) A tentative classification of root systems. *Ecology* 30, 452–458.

Cerda, A. (1998) Soil aggregate stability under different Mediterranean vegetation types. *Catena* 32, 73–86.

Champion, R.A. and Barley, K.P. (1969) Penetration of clay by root hairs. *Soil Science* 108, 402–407.

Cheng, W. and Coleman, D.C. (1990) Effect of living roots on soil organic matter decomposition. *Soil Biology and Biochemistry* 22, 781–787.

Cheshire, M.V. (1979) *Nature and Origin of Carbohydrates in Soils.* Academic Press, London.

Chevallier, T., Blanchart, E., Albrecht, A. and Feller, C. (2004) The physical protection of soil organic carbon in aggregates: a mechanism of carbon storage in a Vertisol under pasture and market gardening (Martinique, West Indies). *Agriculture Ecosystems and Environment* 103, 375–387.

Clark, L.J., Whalley, W.R., Dexter, A.R., Barraclough, P.B. and Leigh, R.A. (1996) Complete mechanical impedance increases the turgor of cells in the apex of pea roots. *Plant Cell and Environment* 19, 1099–1102.

Clausnitzer, V. and Hopmans, J.W. (1994) Simultaneous modeling of transient three-dimensional root growth and soil water flow. *Plant and Soil* 164, 299–314.

Clemente, E.P., Schaefer, C., Novais, R.F., Viana, J.H. and Barros, N.F. (2005) Soil compaction around *Eucalyptus grandis* roots: a micromorphological study. *Australian Journal of Soil Research* 43, 139–146.

Collet, C., Löf, M. and Pagès, L. (2005) Root system development of oak seedlings analysed using an architectural model. Effects of competition with grass. *Plant and Soil* 258, 1–17.

Coutts, M.P. (1987) Developmental processes in tree root systems. *Canadian Journal of Forestry Research* 17, 761–767.

Craine, J.M., Fargione, J. and Sugita, S. (2005) Supply pre-emption, not concentration reduction, is the mechanism of competition for nutrients. *New Phytologist* 166, 933–940.

Cresswell, H.P. and Kirkegaard, J.A. (1995) Subsoil amelioration by plant roots: the process and the evidence. *Australian Journal of Soil Research* 33, 221–239.

Croser, C., Bengough, A.G. and Pritchard, J. (2000) The effect of mechanical impedance on root growth in pea (*Pisum sativum*). II. Cell expansion and wall rheology during recovery. *Physiologia Plantarum* 109, 150–159.

Dacey, J.W.H. and Howes, B.L. (1984) Water uptake by roots controls water table movement and sediment oxidation in short Spartina marsh. *Science* 224, 487–489.

Danjon, F., Barker, D.H., Drexhage, M. and Stokes, A. (2008) Using three-dimensional plant root architecture in models of shallow-slope stability. *Annals of Botany* 101, 1281–1293.

Darrah, P.R. (1993) The rhizosphere and plant nutrition: a quantitative approach. *Plant and Soil* 156, 1–20.

De Baets, S., Poesen, J., Gyssels, G. and Knapen, A. (2006) Effects of grass roots on the erodibility of topsoils during concentrated flow. *Geomorphology* 76, 54–67.

De Baets, S., Poesen, J., Reubens, B., Wemans, K., De Baerdemaeker, J. and Muys, B. (2008a) Root tensile strength and root distribution of typical Mediterranean plant species and their contribution to soil shear strength. *Plant and Soil* 305, 207–226.

De Baets, S., Torri, D., Poesen, J., Salvador, M.P. and Meersmans, J. (2008b) Modelling increased soil cohesion due to roots with EUROSEM. *Earth Surface Processes and Landforms* 33, 1948–1963.

De Baets, S., Poesen, J., Reubens, B., Muys, B., De Baerdemaeker, J. and Meersmans, J. (2009) Methodological framework to select plant species for controlling rill and gully erosion: application to a Mediterranean ecosystem. *Earth Surface Processes and Landforms* 34, 1374–1392.

Dexter, A.R. (1987a) Mechanics of root-growth. *Plant and Soil* 98, 303–312.

Dexter, A.R. (1987b) Compression of soil around roots. *Plant and Soil* 97, 401–406.

Diggle, A.J. (1988) ROOTMAP – a model in three-dimensional coordinates of the growth and structure of fibrous root systems. *Plant and Soil* 105, 169–178.

Docker, B.B. and Hubble, T.C.T. (2009) Modelling the distribution of enhanced soil shear strength beneath riparian trees of south-eastern Australia. *Ecological Engineering* 35, 921–934.

Dodd, I.C. (2005) Root-to-shoot signalling: assessing the roles of 'up' in the up and down world of long-distance signalling *in planta*. *Plant and Soil* 724, 251–270.

Dorgan, K.M., Jumars, P.A., Johnson, B., Boudreau, B.P. and Landis, E. (2005) Burrow extension by crack propagation. *Nature* 433, 475.

Dorioz, J.M., Robert, M. and Chenu, C. (1993) The role of roots, fungi, and bacteria on clay particle organization: An experimental approach. *Geoderma* 56, 179–194.

Doussan, C., Pagès, L. and Vercambre, G. (1998) Modelling the hydraulic architecture of root systems: An integrated approach of water absorption. I. Model description. *Annals of Botany* 81, 213–223.

Drew, M.C. (1975) Comparison of the effects of a localized supply of phosphate, nitrate, ammonium, and potassium on the growth of the seminal root system, and the shoot, in barley. *New Phytologist* 75, 479–490.

Drew, M.C., He, C.-J. and Morgan, P.W. (2000) Programmed cell death and aerenchyma formation in roots. *Trends in Plant Science* 5, 123–127.

Dunbabin, V., Diggle, A.J. and Rengel, Z. (2002a) Simulation of field data by a basic three-dimensional model of interactive root growth. *Plant and Soil* 239, 39–54.

Dunbabin, V., Diggle, A.J., Rengel, Z. and van Hungten, R. (2002b) Modelling the interactions between water and nutrient uptake and root growth. *Plant and Soil* 239, 19–38.

Ehlers, W.V., Hesse, F.F. and Bohm, W. (1983) Penetration resistance and root growth in tilled and untilled loam soil. *Soil and Tillage Research* 3, 261–275.

Evans, D.E. (2003) Aerenchyma formation. *New Phytologist* 161, 35–49.

Falik, O., Reides, P., Gersani, M. and Novoplansky, A. (2005) Root navigation by self inhibition. *Plant, Cell and Environment* 28, 562–569.

Fatahi, B., Khabbaz, H. and Indraratna, B. (2009) Parametric studies on bioengineering effects of tree root-based suction on ground behaviour. *Ecological Engineering* 35, 1415–1426.

Feddes, R.A., Hoff, H., Bruen, M., Dawson, T.E., de Rosnay, P., Dirmeyer, P. *et al.* (2001) Modelling root water uptake in hydrological and climate models. *Bulletin of the American Meteorological Society* 82, 2797–2809.

Fisher, R.F. and Stone, E.L. (1968) Soil and plant moisture relations of red pine growing on a shallow soil. *Soil Science Society of America Proceedings* 32, 725–728.

Fitter, A.H. (1987) An architectural approach to the comparative ecology of plant root systems. *New Phytologist* 106, 61–77.

Garrigues, E., Doussan, C. and Pierret, A. (2006) Water uptake by plant roots: I – Formation and propagation of a water extraction front in mature root systems as evidenced by 2D light transmission imaging. *Plant and Soil* 283, 83–98.

Genet, M., Kokutse, N., Stokes, A., Fourcaud, T., Cai, X.H., Ji, J.N. *et al.* (2008) Root reinforcement in plantations of *Cryptomeria japonica* D. Don: effect of tree age and stand structure on slope stability. *Forest Ecology and Management* 256, 1517–1526.

Gerke, H.H. and Kuchenbuch, R.O. (2007) Root effects on soil water and hydraulic properties. *Biologia* 62, 557–561.

Gill, R.A. and Burke, I.C. (2002) Influence of soil depth on the decomposition of *Bouteloua gracilis* roots in the shortgrass steppe. *Plant and Soil* 241, 233–242.

Goodman, A.M. and Ennos, A.R. (1999) The effects of soil bulk density on the morphology and anchorage mechanics of the root systems of sunflower and maize. *Annals of Botany* 83, 293–302.

Goodman, A.M., Crook, M.J. and Ennos, A.R. (2001) Anchorage mechanics of the tap root system of winter-sown oilseed rape (*Brassica napus* L.). *Annals of Botany* 87, 397–404.

Goss, M.J. (1977) Effect of mechanical impedance on root growth in barley (*Hordeum vulgare* L.). Effect on elongation and branching of seminal roots. *Journal of Experimental Botany* 28, 96–111.

Graf, G.L. and Cooke, J.R. (1980) Soil stress around a root growing tip. *Proceedings of the American Society of Agricultural Engineers* 80, 10–24.

Guenet, B., Danger, M., Abbadie, L. and Lacroix, G. (2010) Priming effect: bridging the gap between terrestrial and aquatic ecology. *Ecology* 91, 2850–2861.

Gyssels, G., Poesen, J., Bochet, E. and Li, Y. (2005) Impact of plant roots on the resistance of soils to erosion by water: a review. *Progress in Physical Geography* 29, 189–217.

Hallett, P.D., Gordon, D.C. and Bengough, A.G. (2003) Plant influence on rhizosphere hydraulic properties: direct measurements using a miniaturized infiltrometer. *New Phytologist* 157, 597–603.

Hamza, M.A. and Anderson, W.K. (2005) Soil compaction in cropping systems – a review of the nature, causes and possible solutions. *Soil and Tillage Research* 82, 121–145.

Hartmann, C., Lesturgez, G., Sindhusen, P., Ratana-Anupap, S., Hallaire, V., Bruand, A. *et al.* (2008a) Consequences of slotting on the pore characteristics of a sandy soil in northeast Thailand. *Soil Use and Management* 24, 100–107.

Hartmann, C., Poss, R., Noble, A.D., Jongskul, A., Bourdon, E., Brunet, D. *et al.* (2008b) Subsoil improvement in a tropical coarse textured soil: Effect of deep-ripping and slotting. *Soil and Tillage Research* 99, 245–253.

Hawes, M.C., Bengough, A.G., Cassab, G. and Ponce, G. (2003) Root caps and rhizosphere. *Journal of Plant Growth Regulation* 2, 352–367.

Herkelrath, W.M., Miller, E.E. and Gardner, W.R. (1977) Water uptake by plants: II. The root contact model. *Soil Science Society of America Journal* 41, 1039–1043.

Hinsinger, P. (1998) How do plant roots acquire mineral nutrients? Chemical processes involved in the rhizosphere. *Advances in Agronomy* 64, 225–265.

Hinsinger, P. and Gilkes, R.J. (1995) Root-induced dissolution of phosphate rock in the rhizosphere of lupins grown in alkaline soil. *Australian Journal of Soil Research* 33, 477–489.

Hinsinger, P. and Jaillard, B. (1993) Root-induced release of interlayer potassium and vermiculitization of phlogopite as related to potassium depletion in the rhizosphere of ryegrass. *Journal of Soil Science* 44, 525–534.

Hinsinger, P., Elsass, F., Jaillard, B. and Robert, M. (1993) Root-induced irreversible transformation of a trioctahedral mica in the rhizosphere of rape. *Journal of Soil Science* 44, 535–545.

Hinsinger, P., Barros, O.N.F., Benedetti, M.F., Noack, Y. and Callot, G. (2001) Plant-induced weathering of a basaltic rock: Experimental evidence. *Geochimica et Cosmochimica Acta* 65, 137–152.

Hinsinger, P., Bengough, A.G., Vetterlein, D. and Young, I.M. (2009) Rhizosphere: biophysics, biogeochemistry and ecological relevance. *Plant and Soil* 321, 117–152.

Hoad, S.P., Russel, G., Lucas, M.E. and Bingham, I.J. (1992) The management of wheat, barley, and oat root systems. *Advances in Agronomy* 74, 193–246.

Hodge, A. (2006) Plastic plants and patchy soils. *Journal of Experimental Botany* 57, 401–411.

Hodge, A., Robinson, D., Griffiths B.S. and Fitter, A.H. (1999) Nitrogen capture by plants grown in N-rich organic patches of contrasting size and strength. *Journal of Experimental Botany* 50, 1243–1252.

Hodge, A., Campbell, C.D. and Fitter, A.H. (2001) An arbuscular mycorrhizal fungus accelerates decomposition and acquires nitrogen directly from organic material. *Nature* 413, 297–299.

Hodge, A., Berta, G., Doussan, C., Merchan, F. and Crespi, M. (2009) Plant root growth, architecture and function. *Plant and Soil* 321, 153–187.

Hu, S., Chapin, F.S., Firestone, M.K., Field, C.B. and Chiariello, N.R. (2001) Nitrogen limitation of microbial decomposition in a grassland under elevated CO_2. *Nature* 409, 188–191.

Huck, M.G., Klepper, B. and Taylor, H.M. (1970) Diurnal variations in root diameter. *Plant Physiology* 45, 529–530.

Hutsch, B.W., Augustin, J. and Merbach, W. (2002) Plant rhizodeposition: An important source of carbon turnover in soils. *Journal of Plant Nutrition and Soil Science* 165, 397–407.

Iijima, M., Higuchi, T., Barlow, P.W. and Bengough, A.G. (2003) Root cap removal increases root penetration resistance in maize (*Zea mays* L.). *Journal of Experimental Botany* 54, 2105–2109.

Ingham, E.R. and Molina, R. (1991) Interactions among mycorrhizal fungi, rhizosphere organisms, and plants. In: Barbosa, P., Krischik, V.A. and Jones, C.G. (eds) *Microbial Mediation of Plant–Herbivore Interactions*. John Wiley, Chichester, UK, pp. 169–197.

Jackson, M. and Armstrong, W. (1999) Formation of aerenchyma and the processes of plant ventilation in relation to soil flooding and submergence. *Plant Biology* 1, 274–287.

Jackson, R.B., Manwaring, J.H. and Caldwell, MM. (1990). Rapid physiological adjustment of roots to localized soil enrichment. *Nature* 344, 58–60.

Jackson, R.B., Canadell, J., Ehleringer, J.R., Mooney, H.A., Sala, O.E. and Schulze, E.D. (1996) A global analysis of root distributions for terrestrial biomes. *Oecologia* 108, 389–411.

Jackson, R.B., Mooney, H.A. and Schulze, E.D. (1997) A global budget for fine root biomass, surface area, and nutrient contents. *Proceedings of the National Academy of Sciences* 94, 7362–7366.

Jackson, R.B., Sperry, J.S. and Dawson, T.E. (2000) Root water uptake and transport: using physiological processes in global predictions. *Trends in Plant Science* 5, 482–488.

Jaillard, B. and Callot, G. (1987) Mineralogical segregation of soil mineral constituents under the action of roots. In: Fedoroff, N., Bresson, L.M. and Courty, M.A (eds) *Soil Micromorphology. Proceedings of the 7th International Working Meeting on Soil Micromorphology*, Paris, July 1985. AFES, Plaisir, Paris, pp. 371–375.

Justin, S.H.F.W. and Armstrong, W. (1991) Evidence for the involvement of ethylene in aerenchyma formation in adventitious roots of rice (*Oryza sativa*). *New Phytologist* 118, 49–62.

Kahn, F. (1978) Analyse structurale des systèmes racinaires des plantes ligneuses de la forêt tropicale dense humide. *Candollea* 32, 321–358.

Kelly, E.F., Chadwick, O.A. and Hilinski, T.E. (1998) The effect of plants on mineral weathering. *Biogeochemistry* 42, 21–53.

Kolb, E., Genet, P., Lecoq, L.E., Hartmann, C., Quartier, L. and Darnige, T. (2009) Root growth in mechanically stressed environment: in situ measurements of radial root forces measured by a photoelastic technique. In: Thibault, B. (ed.) *Proceedings of the 6th Plant Biomechanics Conference*, 16–21 November, Cayenne, French Guinea.

Konopka, B., Pagès, L. and Doussan, C. (2008) Impact of soil compaction heterogeneity and moisture on maize (*Zea mays* L.) root and shoot development. *Plant, Soil and Environment* 54, 509–519.

Kooistra, M.J., Schoonderbeek, D., Boone, F.R., Veen, B.W. and Van Noordwijk, M. (1992) Root-soil contact of maize, as measured by a thin-section technique. II. Effects of soil compaction. *Plant and Soil* 139, 119–129.

Krasilnikov, P.K. (1968) On the classification of the root system of trees and shrubs. In: Ghilanov, M.S. *et al.* (eds) *Methods of Productivity Studies in Root Systems and Rhizosphere Organisms*. USSR Academy of Sciences, Leningrad, Russia, pp. 106–114.

Kristensen, H.L. and Thorup-Kristensen, K. (2004) Uptake of N-15 labeled nitrate by root systems of sweet corn, carrot and white cabbage from 0.2–2.5 meters depth. *Plant and Soil* 265, 93–100.

Kutschera, L. (1960) *Wurzelatlas mitleleuropaïsher Ackerunkräuter und Kulturpflanzen*. Verlag, Frankfurt am Main, Germany.

Lafolie, F., Bruckler, L. and Tardieu, F. (1991) Modeling root water potential and soil–root water transport: I. Model presentation. *Soil Science Society of America Journal* 55, 1203–1212.

Lambers, H., Mougel, C., Jaillard, B. and Hinsinger, P. (2009) Plant-microbe-soil interactions in the rhizosphere: an evolutionary perspective. *Plant and Soil* 12, 763.

Laskowski, M., Grieneisen, V.A., Hofhuis, H., ten Hove, C.A., Hogeweg, P., Maré'e, A.F.M. *et al.* (2008) Root system architecture from coupling cell shape to auxin transport. *PLoS Biology* 6, e307.

Lee, J.E., Oliveira, R.S., Dawson, T.E. and Fung, I. (2005) Root functioning modifies seasonal climate. *Proceedings of the National Academy of Sciences* 49, 17576–17581.

Lehmann, J. (2003) Subsoil root activity in tree-based cropping systems. *Plant and Soil* 255, 319–331.

Lesturgez, G., Poss, R., Hartmann, C., Bourdon, E., Noble, A. and Ratana-Anupap, S. (2004) Roots of *Stylosanthes hamata* create macropores in the compact layer of a sandy soil. *Plant and Soil* 260, 101–109.

Li, L., Sun, J., Zhang, F., Guo, T., Bao, X., Smith, F.A. *et al.* (2006) Root distribution and interactions between intercropped species. *Oecologia* 147, 280–290.

Logsdon, S.D. and Allmaras, R.R. (1991) Maize and soybean root clustering as indicated by root mapping. *Plant and Soil* 131, 169–176.

Lynch, J. and Nielsen, K.L. (1996) Simulation of root system architecture. In: Waisel, Y., Eshel, A. and Kafkafi, U. (eds) *Plant Roots: the Hidden Half.* Marcel Dekker, New York, pp. 247–257.

Malezieux, E., Crozat, Y., Dupraz, C., Laurans, M., Makowski, D., Ozier-Lafontaine, H. *et al.* (2009) Mixing plant species in cropping systems: concepts, tools and models. A review. *Agronomy for Sustainable Development* 29, 43–62.

Masle, J. and Passioura, J.B. (1987) The effect of soil strength on the growth of young wheat plants. *Australian Journal of Plant Physiology* 14, 643–656.

Materechera, S.A., Dexter, A.R. and Alston, A.M. (1991) Penetration of very strong soils by seedling roots of different plant-species. *Plant and Soil* 135, 31–41.

Materechera, S.A., Dexter, A.R. and Alston, A.M. (1992) Formation of aggregates by plant-roots in homogenized soils. *Plant and Soil* 142, 69–79.

Materechera, S.A., Kirby, J.M., Alston, A.M. and Dexter, A.R. (1994) Modification of soil aggregation by watering regime and roots growing through beds of large aggregates. *Plant and Soil* 160, 57–66.

Mattia, C., Bischetti, G.B. and Gentile, F. (2005) Biotechnical characteristics of root systems of typical Mediterranean species. *Plant and Soil* 278, 23–32.

McCully, M.E. (1995) How do real roots work? Some new views of root structure. *Plant Physiology* 109, 1–6.

McKenzie, R.H., Dormaar, J.F., Schaalje, G.B. and Stewart, J.W.B. (1995) Chemical and biological changes in the rhizosphere of wheat and canola. *Canadian Journal of Soil Science* 75, 439–447.

Meek, B.D., Rechel, E.R., Carter, L.M., Detar, W.R. and Urie, A.L. (1992) Infiltration rate of a sandy loam soil – effects of traffic, tillage, and plant roots. *Soil Science Society of America Journal* 56, 908–913.

Mickovski, S.B. and van Beek, L.P.H. (2009) Root morphology and effects on soil reinforcement and slope stability of young vetiver (*Vetiveria zizanioides*) plants grown in semi-arid climate. *Plant and Soil* 324, 43–56.

Mickovski, S.B., Hallett, P.D., Bransby, M.F., Davies, M.C.R., Sonnenberg, R. and Bengough, A.G. (2009) Mechanical reinforcement of soil by willow roots: impacts of root properties and root failure mechanism. *Soil Science Society of America Journal* 73, 1276–1285.

Monshausen, G.B. and Gilroy, S. (2009) The exploring root – root growth responses to local environmental conditions. *Current Opinion in Plant Biology* 12, 766–772.

Moran, C.J., Pierret, A. and Stevenson, A.W. (2000) X-ray absorption and phase contrast imaging to study the interplay between plant roots and soil structure. *Plant and Soil* 223, 99–115.

Mulia, R. and Dupraz, C. (2006) Unusual fine root distributions of two deciduous tree species in southern France: What consequences for modelling of tree root dynamics? *Plant and Soil* 281, 71–85.

Nguyen, C. (2003) Rhizodeposition of organic C by plants: Mechanisms and controls. *Agronomie* 23, 375–396.

Norby, R.J., Ledford, J., Reilly, C.D., Miller, N.E. and O'Neill, E.G. (2004) Fine-root production dominates response of a deciduous forest to atmospheric CO_2 enrichment. *Proceedings of the National Academy of Sciences* 101, 9689–9693.

Onipchenko, V.G., Makarov, M.I., van Logtestijn, R.S.P., Ivanov, V.B., Akhmetzhanova, A.A., Tekeev, D.K. *et al.* (2009) New nitrogen uptake strategy: specialized snow roots. *Ecology Letters* 12, 758–764.

Ottenschläger, I., Wolff, P., Wolverton, C., Bhalerao, R.P., Sandberg, G., Ishikawa, H. *et al.* (2003) Gravity-regulated differential auxin transport from columella to lateral root cap cells. *Proceedings of the National Academy of Sciences* 100, 2987–2991.

Pagès, L. (1995) Growth patterns of the lateral roots in young oak (*Quercus robur* L.) trees. Relationship with apical diameter. *New Phytologist* 130, 503–509.

Pagès, L. (2002). Modelling root system architecture. In: Waisel, Y., Eshel, A. and Kafkafi, U. (eds) *Plant Roots: the Hidden Half*. Marcel Dekker, New York, pp. 359–382.

Pagès, L. and Aries, F. (1988) SARAH: modèle de simulation de la croissance, du développement, et de l'architecture des systèmes racinaires. *Agronomie* 8, 889–896.

Pagès, L., Asseng, S., Pellerin S. and Diggle. A. (2000) Modelling root system growth and architecture. In: Smit, A.L., Bengough, A.G., Engels, C., van Noordwijk, M., Pellerin, S. and van de Geijn, S.C. (eds) *Root Methods: a Handbook*. Springer, Paris, pp. 113–146.

Pagès, L., Vercambre, G., Drouet, J.-L., Lecompte, F., Collet, C. and Le Bot, J. (2004) RootTyp: a generic model to depict and analyse the root system architecture. *Plant and Soil* 258, 103–119.

Pagès, L., Serra, V., Draye, X., Doussan, C. and Pierret, A. (2009) Estimating root elongation rates from morphological measurements of the root tip. *Plant and Soil* 319, 185–207.

Pankhurst, C.E., Pierret, A., Hawke, B.G. and Kirby, J.M. (2002) Microbiological and chemical properties of soil associated with macropores at different depths in a red-duplex soil in NSW Australia. *Plant and Soil* 238, 11–20.

Passioura, J.B. (2002) Soil conditions and plant growth. *Plant, Cell and the Environment* 25, 311–318.

Passioura, J.B. and Wetselaar, R. (1972) Consequences of banding nitrogen fertilizers in soil. II. Effects on the growth of wheat roots. *Plant and Soil* 36, 461–473.

Pierret, A., Moran, C.J. and Pankhurst, C.E. (1999) Differentiation of soil properties related to the spatial association of wheat roots and soil macropores. *Plant and Soil* 211, 51–58.

Pillai, U.P. and McGarry, D. (1999) Structure repair of a compacted vertisol with wet-dry cycles and crops. *Soil Science Society of America Journal* 63, 201–210.

Pohl, M., Alig, D., Korner, C. and Rixen, C. (2009) Higher plant diversity enhances soil stability in disturbed alpine ecosystems. *Plant and Soil* 324, 91–102.

Poot, P. and Lambers, H. (2008) Shallow-soil endemics: adaptive advantages and constraints of a specialized root-system morphology. *New Phytologist* 178, 371–381.

Pregitzer, K.S., King, J.A., Burton, A.J. and Brown, S.E. (2000) Responses of tree fine roots to temperature. *New Phytololgist* 147, 105–115.

Rasse, D.P. and Smucker, A.J.M. (1998) Root recolonization of previous root channels in corn and alfalfa rotations. *Plant and Soil* 204, 203–212.

Rasse, D.P., Smucker, A.J.M. and Santos, D. (2000) Alfalfa root and shoot mulching effects on soil hydraulic properties and aggregation. *Soil Science Society of America Journal* 64, 725–731.

Rasse, D.P., Rumpel, C. and Dignac, M.-F. (2005) Is soil carbon mostly root carbon? Mechanisms for a specific stabilisation. *Plant and Soil* 269, 341–356.

Raynaud, X., Jaillard, B. and Leadley, P.W. (2008) Plants may alter competition by modifying nutrient bioavailability in rhizosphere: A modeling approach. *American Naturalist* 171, 44–58.

Richter, G.L., Monshausen, G.B., Krol, A. and Gilroy, S. (2009) Mechanical stimuli modulate lateral root organogenesis. *Plant Physiology* 151, 1855–1866.

Roering, J.J., Schmidt, K.M., Stock, J.D., Dietrich, W.E. and Montgomery, D.R. (2003) Shallow landsliding, root reinforcement, and the spatial distribution of trees in the Oregon Coast Range. *Canadian Geotechnical Journal* 40, 237–253.

Salati, E. and Vose, P.B. (1984) Amazon basin: a system in equilibrium. *Science* 225, 129–138.

Schenk, H.J., Callaway, R.M. and Mahall, B.E. (1999) Spatial root segregation: are plants territorial? *Advances in Ecological Research* 28, 145–180.

Scholefield, D. and Hall, D.M. (1985) Constricted growth of grass roots through rigid pores. *Plant and Soil* 85, 153–162.

Scholl, D.G. (1976) Soil moisture flux and evapotranspiration determined from soil hydraulic properties in a chaparral stand. *Soil Science Society of America Journal* 40, 14–18.

Souty, N. (1987) Mechanical-behavior of growing roots. 1. Measurement of penetration force. *Agronomie* 7, 623–630.

Stewart, J.B., Moran, C.J. and Wood, J.T. (1999) Macropore sheath: quantification of plant root and soil macropore association. *Plant and Soil* 211, 59–67.

Stirzaker, R.J., Passioura, J.B. and Wilms, Y. (1996) Soil structure and plant growth: Impact of bulk density and biopores. *Plant and Soil* 185, 151–162.

Stokes, A., Lucas, A. and Jouneau, L. (2007) Plant biomechanical strategies in response to frequent disturbance: Uprooting of *Phyllostachys nidularia* (Poaceae) growing on landslide-prone slopes in Sichuan, China. *American Journal of Botany* 94, 1129–1136.

Stokes, A., Atger, C., Bengough, A.G., Fourcaud, T. and Sidle, R.C. (2009) Desirable plant root traits for protecting natural and engineered slopes against landslides. *Plant and Soil* 324, 1–30.

Stone, E.L. and Kalisz, P.J. (1991) On the maximum extent of tree roots. *Forest Ecology and Management* 46, 59–102.

Tardieu, F. and Katerji, N. (1991) Plant-response to the soil-water reserve: Consequences of the root-system environment. *Irrigation Science* 12, 145–152.

Tice, K.R., Graham, R.C. and Wood, H.B. (1996) Transformations of 2:1 phyllosilicates in 41-year-old soils under oak and pine. *Geoderma* 70, 49–62.

Tinker, P.B. (1976) Transport of water to plant roots in soil. *Philosophical Transactions of the Royal Society B* 273, 445–461.

Tisdall, J.M. and Oades, J.M. (1982) Organic matter and water stable aggregates in soils. *Journal of Soil Science* 33, 141–163.

Tosi, M. (2007) Root tensile strength relationships and their slope stability implications of three shrub species in the Northern Apennines (Italy). *Geomorphology* 87, 268–283.

Tri, B.H. and Monnier, G. (1973) Etude quantitative de la granulation des sols sous prairies de graminées [in French]. *Annales Agronomiques* 24, 401–424.

Trumbore, S.E. (1997) Potential responses of soil organic carbon to global environmental change. *Proceedings of the National Academy of Sciences* 94, 8284–8291.

Tsutsumi, D., Kosugi, K. and Mizuyama, T. (2004) Three-dimensional modeling of hydrotropism effects on plant root architecture along a hillslope. *Vadose Zone Journal* 3, 1017–1030.

Vercambre, G., Pagès, L., Doussan, C. and Habib, R. (2003) Architectural analysis and synthesis of the plum tree root system in an orchard using a quantitative modelling approach. *Plant and Soil* 51, 1–11.

Volkmar, K.M. (1996) Effects of biopores on the growth and N-uptake of wheat at three levels of soil moisture. *Canadian Journal of Soil Science* 76, 453–458.

Waisel, Y. and Eshel, A. (2002) Functional diversity of various constituents of a single root system. In: Waisel, Y., Eshel, A. and Kafkafi, U. (eds) *Plant Roots: the Hidden Half.* Marcel Dekker, New York, pp. 157–174.

Watt, M., McCully, M.E. and Canny, M.J. (1994) Formation and stabilization of rhizosheaths in *Zea mays* L.: Effect of soil water content. *Plant Physiology* 106, 79–86.

Weaver, J.E. and Clements, F.E. (1938) *Plant Ecology.* McGraw-Hill, New York and London.

Whalley, W.R., Dumitru, E. and Dexter, A.R. (1995) Biological effects of soil compaction. *Soil and Tillage Research* 35, 53–68.

Wolverton, C., Mullen, J.L., Ishikawa, H. and Evans, M.L. (2000) Two distinct regions of response drive differential growth in *Vigna* root electrotropism. *Plant Cell and the Environment* 23, 1275–1280.

Young, I.M., Blanchart, E., Chenu, C., Dangerfield, M., Fragoso, C., Grimaldi, M. *et al.* (1998) The interaction of soil biota and soil structure under global change. *Global Change Biology* 4, 703–712.

Yunusa, I.A.M. and Newton, P.J. (2003) Plants for amelioration of subsoil constraints and hydrological control: the primer-plant concept. *Plant and Soil* 257, 261–281.

6   Biota–Mineral Interactions

David A.C. Manning[*]

1 Introduction

Within soils, minerals (in the geological sense) are dynamic constituents. Some (such as quartz) predominantly provide a physical framework and hence their role is to influence soil texture. Others are more reactive chemically, dissolving to yield essential nutrients or precipitating as sinks for biological products. Mineral stabilities and surface properties control to varying extents the availability and behaviour of all nutrients, including N and C, and provide a physical framework for the living space of soil organisms.

The purpose of this chapter is to highlight the diversity of the dynamic role that minerals play in soil systems, using examples of specific reactions that affect mainly the feldspars and their decomposition products, and the carbonates. It is impossible to provide a comprehensive review within the space available, and so reference is made to a range of sources that provide reviews, sometimes with other purposes. Perhaps the most wide-ranging description of the characterization and interpretation of mineral occurrences within soils is provided by Dixon and Weed (1989), and for more detailed information about rock-forming

minerals reference should also be made to Deer *et al.* (1992).

Within a soil, minerals have three sources. First, they may be residual, derived from the underlying rock, or, secondly, they may enter the soil following erosion and transport from another location. In these two cases, the role of a reactive mineral is to act as a source of elements to the soil solution as it undergoes corrosion and dissolution. Thirdly, clay and carbonate minerals especially may form *in situ* (bearing in mind that clays washed into a specific soil may be derived from a weathering profile elsewhere), acting as sinks for some elements, or through ion exchange as buffers for others.

2 The Geological Parent Material

The continental crust, from which all soils derive their parent material, is dominated by a limited number of major chemical components (Table 6.1). These occur in rocks mainly as silicate minerals (reflecting the importance of Si and O as the dominant chemical constituents of the Earth's crust) and as carbonate minerals. Oxide, sulfate, sulfide and phosphate minerals occur widely but in minor amounts. In addition to

*david.manning@newcastle.ac.uk

summarizing chemical data for the continental crust (taking the Canadian Shield as an example; Shaw *et al.*, 1967, 1976), Table 6.1 also gives an estimate of its mineralogical composition, recalculated from the chemical analysis on an anhydrous basis (CIPW norm) or allowing for hydrated components, such as micas and chlorite (mesonorm; Shaw *et al.*, 1967). It also contains estimates made by Nesbitt and Young (1984) based on the predicted weathering behaviour of igneous rocks. Using any of these methods of calculation, the feldspars are the dominant mineral species in crustal rocks (40–50% by weight, allowing for hydrated minerals), and it is the weathering of these that gives rise to the overwhelming majority of aluminous clays (kaolinite, smectite and illite).

The importance of feldspars as nutrient sources has been recognized for almost 100 years. De Turk (1919) and Haley (1923) address the availability of K for plant nutrition from feldspars, and Lewis and Eisenmenger (1948) specifically develop the relationship between K availability and plant health. But feldspars are not simply sources of K. They contain significant amounts of N, as ammonium in substitution for K. Although N is not normally thought of as a 'geological' material, buddingtonite (the ammonium feldspar; $NH_4AlSi_3O_8$) occurs naturally (e.g. Ramseyer *et al.*, 1993), and Holloway and Dahlgren (1999) show that 30–50% of soil N may have a geological source. Ammonium is known to be present in amounts up to 450 mg/kg within feldspars in unaltered granites and 700 mg/kg in micas from pegmatites or hydrothermally altered rocks (e.g. Hall, 1988). Within sedimentary sequences, ammonium occurs within interlayer sites ('fixed ammonium') in clay minerals associated with hydrocarbons or coal, and reaches 2000 mg/kg (Williams *et al.*, 1992). Feldspars also act as

Table 6.1. Constitution of the continental crust, by percentage (taking the Canadian Shield as an example). Total Fe reported as FeO_t.

Constituent	Rudnick and Gao (2003)	Shaw *et al.* (1967, 1976)	Shaw *et al.* (1967, 1976)	Nesbitt and Young (1984)
SiO_2	66.6	64.93	–	–
TiO_2	0.64	0.52	–	–
Al_2O_3	15.4	14.63	–	–
FeO_t	5.04	3.97	–	–
MnO	0.1	0.07	–	–
MgO	2.48	2.24	–	–
CaO	3.59	4.12	–	–
Na_2O	3.27	3.46	–	–
K_2O	2.8	3.1	–	–
P_2O_5	0.15	0.15	–	–
Total	100.07	97.19	–	–
	CIPW norm	CIPW norm	Mesonorm	Calculated
Quartz	22.47	20.33	24.4	23.2
Feldspar	63.66	63.26	47.9	42.9
plagioclase	44.50	44.07	39.3	30.9
orthoclase	19.16	19.19	8.6	12.9
Hydrous Na–K	–	–	7.6	5.0
Anhydrous Fe–Mg	9.99	10.41	–	1.4
Hydrous Fe–Mg	–	–	14.5	10.9
Oxides	3.81	2.72	1.4	1.6
Apatite	0.35	0.35	–	–
Others	–	–	4.7	3.0

a reservoir for P. In addition to containing inclusions of the calcium phosphate mineral apatite, it is possible for feldspars to contain P in substitution for Si within their crystal lattice, according to the berlinite substitution: $2Si = Al + P$. This substitution occurs in granitic feldspars, which can contain up to 1% P_2O_5 in rare examples, and more commonly 0.2–0.3% P_2O_5 (Kontak et al., 1996). This is not taken into account in the calculated mineralogical compositions given in Table 6.1, where it is assumed that apatite is the only P-bearing mineral (to simplify the calculation). However, because of the preponderance of feldspars within the crust it is quite possible that the majority (50–90%) of the crustal P reservoir is in fact accommodated by feldspar. This arises as a fundamental characteristic of the crystal structure of the feldspars.

From the point of view of soil biological systems, feldspars are an attractive storehouse of nutrients, certainly offering K, providing a readily accessible (but dispersed and dilute) source of P and delivering some N as a bonus. In this chapter, attention will focus on interaction between feldspars and biological systems within soil, illustrating the textural consequences of their alteration within soils, some of which can be ascribed directly to biological processes. For a very much more comprehensive review of silicate mineral weathering in soil environments, Wilson (2004) describes weathering of ferromagnesian minerals in addition to the feldspars and sheet silicates.

3 Dissolution of the Framework Silicates

Quartz and the feldspars are the dominant framework silicates (minerals with a complex three-dimensional (3D) aluminosilicate structure) found within soil environments. Quartz is the simplest of the framework silicates, having very limited scope for chemical variation within its overall crystal structure. To all intents and purposes, quartz is chemically unreactive under near-Earth surface conditions; its solubility in water at near-surface conditions is approximately 6 mg/l (as Si), and it has the lowest dissolution rate of the framework silicates (White, 1995). It contributes to the physical properties of soils, providing open space and compressive 'strength'. But quartz has additional functions. It provides a chemically stable surface on which biofilm may accumulate, and like other minerals it passes through the guts of soil (and sediment) fauna.

Like quartz, feldspars contribute to the structure of the soil, as residual grains, but they are also actively corroded. Figure 6.1 shows a typical weathered feldspar surface from a grain from a peaty soil developed on granite, south-west England. At low magnification (Fig. 6.1a) the sample shows an irregular surface with clearly visible filaments. The nature of the sample surface is qualitatively fractal; as magnification is increased, the complexity of the surface also increases (Fig. 6.1b) and fungal filaments can still be seen. The architecture of the grain surface is remarkably complex, and it evidently represents a spatial environment amenable to be frequented by soil microorganisms.

The texture shown in Fig. 6.1 arises from the dissolution of the feldspar, which affects some parts of the crystal more than others. In bulk, feldspars are particularly sensitive to corrosion involving plant root exudates, especially the low-molecular weight organic acids (LMWOA) oxalate and citrate (e.g. Bevan and Savage, 1989; Manning et al., 1992; van Hees et al., 2002). To compare the dissolution behaviour of feldspars in soil environments, two approaches to determine mineral dissolution rates have been adopted: (i) laboratory dissolution experiments, in which crushed unweathered feldspars are used (e.g. Chou and Wollast, 1985; Holdren and Speyer, 1985); and (ii) modelling of observed soil chemistry in terms of calculated mineral dissolution rates (e.g. White et al., 1996).

In general terms, the measured or calculated dissolution rate for the feldspars under soil conditions (25°C) varies between $10^{-11.0}$ and $10^{-13.0}$ mol/m^2/s for laboratory experiments using unweathered feldspars, and between $10^{-12.0}$ and $10^{-16.8}$ mol/m^2/s for estimates derived from chronosequence, catchment and other field-based studies

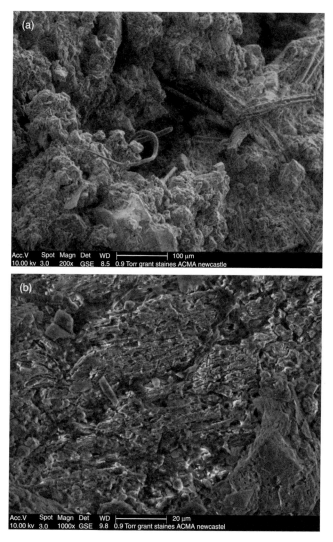

Fig. 6.1.   Surface of a corroded alkali feldspar from peat on granite from Cornwall, UK. (a) Hyphae clearly visible at low magnification (image width approximately 0.6 mm); (b) fine-scale structure of the feldspar surface at high magnification, with penetrating fungal (?) hyphae (image width approximately 0.1 mm). Images taken with environmental scanning electron microscope, uncoated samples.

(White *et al.*, 2001). The reason why laboratory and field-derived estimates differ so much is believed to be due to differences in our ability to measure the surface area, as well as kinetic factors (White and Brantley, 1995). Measurement of surface area in feldspars is fraught with difficulty, because of the extremely complex geometry of the feldspar surface (Fig. 6.1). Gas adsorption

methods (BET) involve the use of nitrogen or argon as adsorptive gases, but these are limited by their ability to penetrate the smallest pores (<0.4 nm are inaccessible). Geometrical measurements of surface area take into account surface roughness (Blum, 1994).

Lee and Parsons (1995) describe the 3D geometrical heterogeneity of feldspars from peaty soils developed on a granite in the

English Lake District. Detailed scanning and transmission electron microscopy showed the development of nanometer-sized etch pits according to a basic crystallographic control. In natural feldspars formed within igneous rocks, the composition of the alkali feldspar that first forms includes a proportion of sodium (for example, $K_{0.7}Na_{0.3}AlSi_3O_8$; Parsons et al., 1998). As it cools, the crystal structure becomes unstable and separate exsolution lamelli of the plagioclase albite ($NaAlSi_3O_8$) form within a matrix of alkali feldspar that still contains some sodium but not as much as before. These exsolution lamelli form pervasively throughout the feldspar grain, giving it the perthitic texture, and are typically up to 0.4 μm thick and 1–2 μm apart (Parsons et al., 1998).

Imaging of exsolution phenomena within alkali feldspars shows a range of textures. First, etch pits form at the interface between the exsolution lamelli and the host alkali feldspar. These are locations within the crystal that are strained, and so more amenable to dissolution. Lee and Parsons (1995) impregnated feldspars with resin to make a cast of the internal voids, which was revealed after curing by dissolving away the feldspar using hydrofluoric acid. This showed very clearly that the etch pits observed on surfaces using scanning electron microscopy are the surface expression of a 3D network of tubes that permeates the feldspar grain. Individual tubes developed at the albite–alkali feldspar interface are typically 0.4–0.6 μm in diameter (Parsons et al., 1998), with cross-linking tubes approximately 0.2 μm across.

In parallel with the observations made by Lee and Parsons, Jongmans et al. (1997) reported tunnel-like features in feldspars from podzol E horizons. These tubes differed from those reported by Lee and Parsons (1995) by being larger (3–10 μm in diameter), by being irregular in shape and by having rounded ends (Hoffland et al., 2002). These geometrical considerations led Jongmans et al. (1997) to suggest that the tubes arise as a consequence of fungal mycelia penetrating the feldspar. The relationship between fungal tunnels and the

crystallographically controlled etch pits is discussed by Hoffland et al. (2003), who suggest that fungal hyphae exploit the network of etch pits to penetrate the feldspar crystal structure. In estimates of tunnel volume relative to the bulk of the feldspar grain, Smits et al. (2005) show that this is less than 1%, increases with age and that the proportion of tunnel is less for alkali feldspars than for the plagioclases.

The detailed investigations of the feldspars summarized above reflect the considerable amount of attention that these minerals have received in view of their common occurrence in soils and potential as nutrient sources. Numerous other studies describe the weathering of other minerals (reviewed in detail by Wilson, 2004); work on the feldspars, however, especially well illustrates the direct spatial relationships between soil organisms and the mineralogical framework in which they live.

4 Newly Formed Minerals within Soils

In addition to the formation of space as a consequence of corrosion, mineral reactions within soils also involve precipitation of newly formed minerals that may (i) reduce pore space and (ii) be produced via biological processes. Mineral transformations typically involve close spatial relationships between a parent material and newly formed minerals, while mineral precipitates form directly from solution, filling pre-existing pore space.

As far as silicate minerals are concerned, the key minerals that may form in soils are the clay minerals. For example, kaolinite and illite are produced by the weathering of feldspars, according to reactions of the following type:

$$3KAlSi_3O_8 + 2H^+ = KAl_3Si_3O_{10}(OH)_2 + 6SiO_2 + 2K^+$$
K-feldspar   in solution   illite/muscovite   in solution

$$2KAl_3Si_3O_{10}(OH)_2 + 3H_2O + 2H^+ = 3Al_2Si_2O_5(OH)_4 + 2K^+$$
illite/muscovite   in solution   kaolinite   in solution

The dissolution of feldspars is directly associated with the formation of clays, given the very low solubilities of aluminium and silica in solution (Fig. 6.2). Aluminium solubility is lowest around near-neutral pH, and increases as pH rises or falls. Silica solubility remains low as pH falls, explaining (amongst other factors) why quartz remains as a residual mineral in acid soils such as podzols.

Texturally, the development of clay minerals may be associated spatially with the weathering or alteration of pre-existing minerals. It is well known, for example, that illite and kaolinite form in voids produced by the corrosion of feldspars. Figure 6.3 shows the surface of feldspars from regolith in south-west England, with few kaolinite plates. Within this framework, typical vermiform textures of kaolinite are formed, giving the classical textures shown in Fig. 6.4 (Psyrillos et al., 1999). When examined as polished sections using high-resolution scanning electron microscopy with backscattered electron imaging (Fig. 6.4b), the kaolinite verms can be seen to be composed

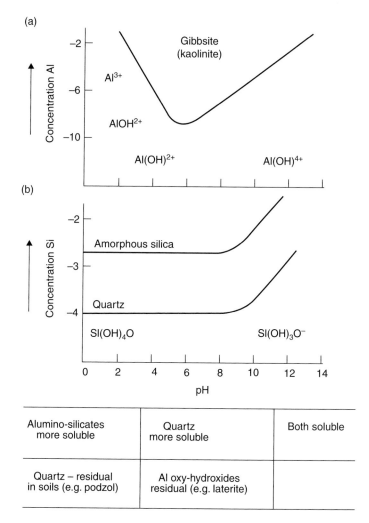

Fig. 6.2. Contrasting solubility of aluminium (Al) and silica (Si) in solution as a function of soil pH, with implications for soil mineral stability.

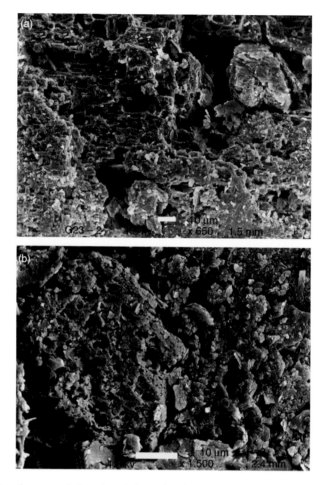

Fig. 6.3. Examples of textures of clay minerals formed within regolith alkali feldspar, Cornwall, UK. (a) 'Clean' corroded feldspar surface, with little clay mineral growth (10 μm scale bar; image width approx. 0.2 mm); (b) corroded surface with fine-grained kaolinite (10 μm scale bar; image width approx. 0.07 mm). Conventional scanning electron microscope secondary electron image; gold-coated sample (photomicrographs taken by A. Psyrillos).

of alternating sheets of kaolinite and illite, indicating that the verms were formed by alteration of pre-existing illite or muscovite (Fig. 6.5). The large vermiform kaolinite grains (10–50 μm) inherit the morphology of the parent mica or clay, whereas kaolinite believed to have formed by direct precipitation within voids in the regolith has a very different morphology, occurring as individual grains of the order of 1 μm across (Psyrillos *et al.*, 1999; Fig. 6.6).

An important process that leads to the transformation of clays and other minerals within soils is that of ingestion by soil invertebrates. At a very basic level, ingestion by beetle larvae and earthworms reduces the particle size of both quartz and alkali feldspars (e.g. Suzuki *et al.*, 2003). In more detail, Needham *et al.* (2004) have shown that passage through the gut of the common earthworm (*Lumbricus terrestris*) affects clay minerals ingested by the worm. On a molecular scale, the crystal structure becomes less well ordered, reflecting an increase in the layer spacing, and (as in the marine annelid *Arenicola marina*) newly

Fig. 6.4. Examples of textures typically shown by kaolinite within granitic regolith, Cornwall, UK.
(a) Typical vermiform kaolinite (scale bar 10 μm; image width approx. 0.04 mm; conventional scanning electron microscope secondary image; gold-coated sample); (b) back-scattered electron image of polished section of kaolinite verm from granite regolith, Cornwall; relic illite/muscovite, white; kaolinite, grey (scale bar 10 μm; image width approx. 0.2 mm; photomicrographs taken by A. Psyrillos).

formed ferrous clays develop at the expense of detrital ferric clays.

In the case of non-silicates, carbonates and oxalates are the most important mineral precipitates within soil systems. Precipitation of calcite within a soil profile takes place with increasing depth as rainfall increases (Jenny, 1980). In a transect through soil developed on calcareous loess in Kansas, USA, calcite accumulates at depths varying from 50 cm (250–500 mm annual rainfall) to 300 cm (750–1000 mm maximum annual rainfall). Similar relationships are observed for calcareous soils from California and Nevada (Jenny, 1980). Pedogenic carbonates vary in extent within soil systems from calcretes (which occlude all porosity to produce laterally extensive layers of calcite) to calcite precipitates associated with specific small-scale environments within a soil profile (such as calcic pendants, cutans, on the underside of pebbles; e.g. Wang and Anderson, 1998; Fig. 6.7). Although many calcretes are of chemical origin, relating to the evaporation of soil water and consequent precipitation of calcite ($CaCO_3$), much

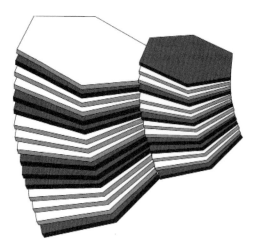

Fig. 6.5. Schematic diagram to illustrate expansion of illite/muscovite as part of the kaolinization process (reproduced with the kind permission of the Mineralogical Society from a paper by Psyrillos *et al.* (1999) published in Clay Minerals).

(a)

Fig. 6.6. Typical appearance of kaolinite formed by direct precipitation within regolith (scale bar 1 μm; image width approx. 0.025 mm; conventional scanning electron microscope secondary electron image; gold-coated sample; photomicrograph taken by A. Psyrillos).

Fig. 6.7. (a) Soil profile with calcic pendants (black chernozem; Dunvargen soil group, Calgary, Alberta, Canada; MacMillan, 1987); (b) close-up of calcic pendants beneath pebble (scale bar 5 cm).

soil calcite is precipitated as a consequence of biological activity, and its carbon records stable isotope signatures that demonstrate this (Cerling, 1984).

The precipitation of calcite on the undersides of pebbles to form carbonate pendants is well known from arid or semi-arid environments. Generally believed to form in a manner analogous to speleothems (stalactites), calcic pendants are predominantly composed of calcite but can contain detrital grains derived from the parent clast (which may fragment) and pedogenic clay minerals (e.g. Brock and Buck, 2005). The carbon-stable isotope characteristics of pendant carbon have been widely used as an indicator of vegetation change with time, assuming that pendants can be dated (Cerling and Quade, 1993; Wang and Anderson, 1998). In cryoarid soils, calcic pendants provide evidence of periglacial vegetation histories (Pustovoytov, 1998).

In addition to carbonates, calcium oxalate minerals are reported from forest litter (Graustein et al., 1977) and occur widely in soil systems in which fungi are active (Burford et al., 2003). Weddellite (Ca(COO)$_2$. xH$_2$O) is reported mostly for recent sediments and soil systems, and whewellite (Ca(COO)$_2$.H$_2$O) occurs in a surprisingly diverse range of geological environments as well as soils (Manning, 2000). Burford et al. (2003) review the role of fungi in rock weathering, and emphasize the importance of fungal exudates, especially oxalate, in corroding primary silicate minerals and clays. Drever and Stillings (1997) similarly emphasize the potential of oxalate as a complexing agent capable of enhancing dissolution of aluminosilicates.

Although the occurrence of oxalate minerals is widespread in diverse geological environments (Manning, 2000), they are generally not common. It is to be expected

that they become unstable during sediment burial and, if preserved, their carbon will be in the form of carbonate, in calcite or other carbonate minerals. Certainly much remains to be learnt about the oxalate pathway, and the fate of oxalate in detail, in soil systems within which fungal activity takes place.

5 The Soil Solution and Ion Exchange Processes

So far, this chapter has focused on the minerals that form or react within soil systems. It has ignored the soil solution, from which minerals precipitate (thus buffering the soil solution composition) and with which soils interact through ion exchange processes.

Figure 6.8 shows a sketch phase diagram that illustrates the relationships between the products of feldspar weathering and the composition of the soil solution. Similar diagrams can be drawn for other minerals present within soils (Garrels and Christ, 1965).

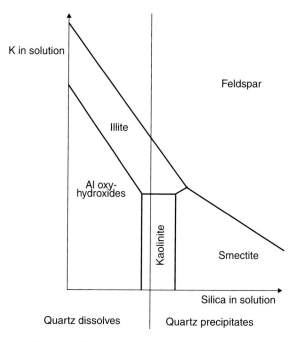

Fig. 6.8. Schematic phase diagram for the system SiO$_2$–K$_2$O–Al$_2$O$_3$–H$_2$O showing relationships between feldspar and clay minerals as a function of solution composition expressed in terms of potassium/hydrogen ion ratio and silica content.

This simplified diagram shows the proportions of potassium and silica in solution. For feldspar to be stable, both potassium and silica have to be high in concentration. In situations where K remains high and Si is removed, feldspar alters to illite and then to aluminium oxy-hydroxides. Conversely, if Si remains high, it alters to smectite clays. In most cases, an intermediate path is followed, and the sequence feldspar → illite → kaolinite is seen. The diagram shows that, in the presence of quartz, aqueous Si contents cannot be below values corresponding to quartz saturation (approximately 6 mg/l as SiO_2 in solution; Rimstidt and Barnes, 1980), and that the formation of lateritic soils that lack quartz requires sufficient rainfall (assuming rainwater to be chemically least rich in solutes, compared with surface and groundwaters) to remove quartz by solution. Referring to Fig. 6.2, this will be enhanced by near-neutral pH values for the soil solution, corresponding to the minimum solubilities for aluminium oxides.

The dominant cation exchange minerals are the clays, bearing in mind that zeolites are important in soils developed on some volcanic rocks. Cation exchange involves two processes: (i) exchange with loosely bound cations on clay surfaces, which is dependent on the clay sheets having a net negative charge; and (ii) exchange with inter-layer cations. Exchange with inter-layer cations can be modelled assuming equilibrium thermodynamics for specific compositions which are end members of a cation exchange series (i.e. ideal mixing of wholly exchanged components; Garrels and Christ, 1965; Garrels, 1984). Using this approach, it can be shown that cation contents of the soil solution are controlled by the fundamental thermodynamic properties of the soil clay mineral assemblage. Cation ratios in solution are fixed for any given clay composition, although their abundances may vary in accordance with changes in solution pH (Garrels and Christ, 1965). This applies not only to the alkalis and alkaline earths but also to ammonium, thus controlling the ammonium content of the soil solution. As an illustration of this, Manning and Hutcheon (2004) interpret the observed

ammonium contents for groundwaters of different origins and show that ammonium is controlled by cation exchange reactions involving clays, confirming relationships observed by Owen and Manning (1997).

In general, processes of anion exchange analogous to cation exchange involving clearly defined crystallographic sites within minerals are not regarded as significant in soil systems. The dominant exchangeable anion in clays and micas is hydroxyl, and this exchanges with fluoride – but not under pedogenic conditions. Apatite undergoes anion exchange involving hydroxyl, fluoride and carbonate, but again within hydrothermal or magmatic systems prior to incorporation into a soil. The phosphate ion is sorbed on to clay and aluminium oxy-hydroxide surfaces, where its tetrahedral geometry allows it to map on to tetrahedrally coordinated Al or Si.

6 Conclusions: Minerals as Dynamic Constituents of the Soil System

This chapter has focused on the ways in which feldspars in particular are associated with the formation of clays within soils, and has justified this approach on the basis of the preponderance of the feldspars as constituents of crustal rocks. It has noted that feldspars are host to potentially significant quantities of N and P within the crust, and so may play a vital role in plant nutrient supply. Additionally, the plagioclase feldspars are calcium-bearing, and it is the weathering of these (by virtue of their preponderance) that provides a dominant source of the calcium preserved in carbonates and oxalates precipitated as a consequence of or in association with biological activity. Thus the feldspar story and the pedogenic carbonate/oxalate story coalesce. Figure 6.9 summarizes this conceptually. Weathering of feldspars (and igneous/metamorphic micas) leads to the formation of clays and weathered micas capable of undergoing cation exchange with the soil solution. This influences the availability of K, and also N as ammonium. The dissolution of feldspars releases lattice-bound phosphate, which

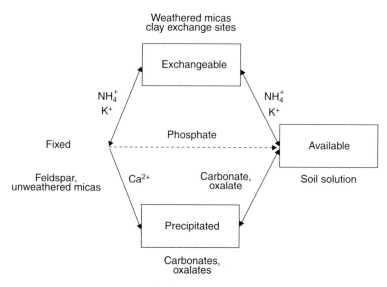

Fig. 6.9.   Diagrammatic summary of the role of feldspars, clays and carbonate minerals in the soil system.

(if soil solution Ca contents are sufficiently low) does not precipitate as discrete minerals but participates in weak sorption reactions, so that it is readily available for plant uptake. Cations including Ca and Na are also released by the weathering of feldspars. In most cases, Na plays little part in soil solution–mineral reactions, but Ca is released and participates in precipitation reactions to produce calcite within soil profiles, perhaps as deep as 2–3 m depending on rainfall, and calcium oxalate minerals within leaf litter as well as within plant tissue.

It can be seen from Fig. 6.9 that mineral reactions within soils relate to one another, with minerals that are inherently unstable at Earth surface conditions decomposing to give several new products. Interaction with biological processes is an integral part of the process. Fungal processes in particular accelerate the weathering of minerals such as feldspars through the exudation of organic acid anions, and by exploiting structural weaknesses that have developed in the crystal as a consequence of strain induced by cooling from magmatic to Earth surface temperatures. Conversely, the precipitation of carbonate minerals in particular within

soils provides a record of the influence of plant growth on the soil solution, providing an opportunity to capture carbon introduced into the soil as root exudates. Between these two extreme processes, soil invertebrates influence the texture and mineralogical composition of the soil through reactions that take place within their gut.

The purpose of this chapter has been to demonstrate the vitality of minerals within the soil system. They are dynamic participants in biological processes, and their contribution needs to be understood. This chapter presents a selective view of what is a field sufficiently large to fill a textbook. However, if soil function is also to be managed, the behaviour of soil minerals needs to be known, to ensure that natural processes are exploited to achieve desired outcomes in soil management and sustainable agriculture.

Acknowledgements

I thank Imerys Minerals Ltd for financial support of research that generated the images shown in Figs 6.3, 6.4 and 6.6.

References

Bevan, J. and Savage, D. (1989) The effect of organic acids on the dissolution of K-feldspar under conditions relevant to burial diagenesis. *Mineralogical Magazine* 53, 415–425.

Blum, A.E. (1994) Feldspars in weathering. In: Parsons, I. (ed.) *Feldspars and their Reactions*. NATO ASI Series C421, Kluwer, Dordrecht, The Netherlands, pp. 595–630.

Brock, A.L. and Buck, B.J. (2005) A new formation process for calcic pendants from Pahranagat Valley, Nevada, USA, and implications for dating Quaternary landforms. *Quaternary Research* 63, 359–367.

Burford, E.P., Fomina, M. and Gadd, G.M. (2003) Fungal involvement in bioweathering and biotransformation of rocks and minerals. *Mineralogical Magazine* 67, 1127–1155.

Cerling, T.E. (1984) The stable isotope composition of modern soil carbonate and its relationship to climate. *Earth and Planetary Science Letters*, 71, 229–240.

Cerling, T.E. and Quade, J. (1993) Stable carbon and oxygen isotopes in soil carbonates. In: Swart, P.K., Lohmann, K.C., McKenzie, J. and Savin S. (eds) *Climate Change in Continental Isotopic Records*. Geographical Monograph 78, American Geophysical Union, Washington, DC, pp. 217–231.

Chou, L. and Wollast, R. (1985) Steady-state kinetics and dissolution mechanisms of albite. *American Journal of Science* 285, 963–993.

Deer, W.A., Howie, R.A. and Zussman, J. (1992) *An Introduction to Rock-forming Minerals*. Longman Group Ltd., London, 528 pp.

De Turk, E. (1919) Potassium-bearing minerals as a source of potassium for plant growth. *Soil Science* 8, 269–301.

Dixon, J.B. and Weed, S.B. (eds) (1989) *Minerals in Soil Environments*, 2nd edn. Soil Science Society of America Book Series, Soil Science Society of America, Madison, Wisconsin.

Drever, J.I. and Stillings, L.L. (1997) The role of organic acids in mineral weathering. *Colloids and Surfaces* 120, 167–181.

Garrels, R.M. (1984) Montmorillonite/illite stability diagrams. *Clays and Clay Minerals* 32, 161–166.

Garrels, R.M. and Christ, C.L. (1965) *Solutions, Minerals, and Equilibria*. Harper & Row, New York, 450 pp.

Graustein, W.C., Cromack, K. and Sollins, P. (1977) Calcium oxalate: occurrence in soils and effect on nutrient and geochemical cycles. *Science* 198, 1252–1254.

Haley, D. (1923) Availability of potassium in orthoclase for plant nutrition. *Soil Science* 15, 167–180.

Hall, A. (1988) The distribution of ammonium in granites from South-West England. *Journal of the Geological Society of London* 145, 37–41.

Hoffland, E., Giesler, R., Jongmans, A.G. and van Breeman, N. (2002) Increasing feldspar tunneling by fungi across a North Sweden podzol chronosequence. *Ecosystems* 5, 11–22.

Hoffland, E., Giesler, R., Jongmans, A.G. and van Breeman, N. (2003) Feldspar tunneling by fungi along natural productivity gradients. *Ecosystems* 6, 739–746.

Holdren, G.R. and Speyer, P.M. (1985) pH dependent changes in the rates and stoichiometry of dissolution of an alkali feldspar at room temperature. *American Journal of Science* 285, 994–1019.

Holloway, J.M. and Dahlgren, R.A. (1999) Geologic nitrogen in terrestrial biogeochemical cycling. *Geology* 27, 567–570.

Jenny, H. (1980) *The Soil Resource*. Springer-Verlag, New York, 377 pp.

Jongmans, A.G., van Breeman, N., Lundström, U.S., van Hees, P.A.W., Finlay, R.D., Srinivasan, M. *et al.* (1997) Rock-eating fungi. *Nature* 389, 682–683.

Juster, T.C., Brown, P.E. and Bailey, S.W. (1987) NH_4-bearing illite in very low grade metamorphic rocks associated with coal, northeastern Pennsylvania. *American Mineralogist* 72, 555–565.

Kontak, D.J., Martin, R.F. and Richard, L. (1996) Patterns of phosphorus enrichment in alkali feldspar, South Mountain Batholith, Nova Scotia, Canada. *European Journal of Mineralogy* 8, 805–824.

Lee, M.R. and Parsons, I. (1995) Microtextural controls of weathering of perthitic feldspars. *Geochimica et Cosmochimica Acta* 59, 4465–4488.

Lewis, C.J. and Eisenmenger, W. (1948) Relationship of plant development to the capacity to utilize potassium in orthoclase feldspar. *Soil Science* 65, 495–500.

MacMillan, R.A. (1987) *Soil Survey of the Calgary Urban Perimeter*. Alberta Soil Survey Report No. 45, Alberta Research Council, Edmonton, Canada, 244 pp.

Manning, D.A.C. (2000) Carbonates and oxalates in sediments and landfill: monitors of death and decay in natural and artificial systems. *Journal of the Geological Society of London* 157, 229–238.

Manning, D.A.C. and Hutcheon, I.E. (2004) Distribution and mineralogical controls on ammonium in deep groundwaters. *Applied Geochemistry* 19, 1495–1503.

Manning, D.A.C., Gestsdottir, K. and Rae, E.I.C. (1992) Feldspar dissolution in the presence of organic acid anions under diagenetic conditions: an experimental study. *Organic Geochemistry* 19, 483–492.

Needham, S.J., Worden, R.H. and McIlroy, D. (2004) Animal-sediment interactions: the effect of ingestion and excretion by worms on mineralogy. *Biogeosciences* 1, 113–121.

Nesbitt, H.W. and Young, G.M. (1984) Prediction of some weathering trends of plutonic and volcanic rocks based on thermodynamic and kinetic considerations. *Geochimica et Cosmochimica Acta* 48, 1523–1534.

Owen, J.A. and Manning, D.A.C. (1997) Silica in landfill leachates: implications for clay mineral stabilities. *Applied Geochemistry* 12, 267–280.

Parsons, I., Lee, M.R. and Smith, J.V. (1998) Biochemical evolution II: Origin of life in tubular microstructures on weathered feldspar surfaces. *Proceedings of the National Academy of Sciences* 95, 15173–15176.

Psyrillos, A., Howe, J.H., Manning, D.A.C. and Burley, S.D. (1999) Geological controls on kaolin particle shape and consequences for mineral processing. *Clay Minerals* 34, 193–208.

Pustovoytov, K. (1998) Pedogenic carbonate cutans as a record of the Holocene history of relic tundra-steppes of the Upper Kolyma Valley (North-Eastern Asia). *Catena* 34, 185–195.

Ramseyer, K., Diamond, L.W. and Boles, J.R. (1993) Authigenic K-NH_4-feldspar in sandstones; a fingerprint of the diagenesis of organic matter. *Journal of Sedimentary Research* 63, 1092–1099.

Rimstidt, J.D. and Barnes, H.L. (1980) The kinetics of silica-water reactions. *Geochimica et Cosmochimica Acta* 44, 1683–1699.

Rudnick, R.L. and Gao, S. (2003) The composition of the Continental Crust. In: Rudnick, R.L., Holland, H.D. and Turekian, K.K. (eds) *The Crust, vol. 3: Treatise on Geochemistry*. Elsevier-Pergamon, Oxford, UK, pp. 1–64.

Shaw, D.M., Reilly, G.A., Muysson, J.R., Pattenden, G.E. and Campbell, F.E. (1967) An estimate of the chemical composition of the Canadian Precambrian Shield. *Canadian Journal of Earth Sciences* 4, 829–853.

Shaw, D.M., Dostal, J. and Keays, R.R. (1976) Additional estimates of continental surface Precambrian shield composition in Canada. *Geochimica et Cosmochimica Acta* 40, 73–83.

Smits, M.M., Hoffland, E., Jongmans, A.G. and van Breeman, N. (2005) Contribution of mineral tunnelling to total feldspar weathering. *Geoderma* 125, 59–69.

Suzuki, Y., Matsubara, T. and Hoshino, M. (2003) Breakdown of mineral grains by earthworms and beetle larvae. *Geoderma* 112, 131–142.

van Hees, P.A.W., Lundström, U.S. and Mörth, C.-M. (2002) Dissolution of microcline and labradorite in a forest O horizon extract: the effect of naturally occurring organic acids. *Chemical Geology* 189, 199–211.

Wang, D. and Anderson, D.W. (1998) Stable carbon isotopes of carbonate pendants from Chernozemic soils of Saskatchewan, Canada. *Geoderma* 84, 309–322.

White, A.F. (1995) Chemical weathering rates of silicate minerals in soils. In: White, A.F. and Brantley, S.L. (eds) *Chemical Weathering Rates of Silicate Minerals, Reviews in Mineralogy*, vol. 31, pp. 407–461.

White, A.F. and Brantley, S.L. (1995) Chemical weathering rates of silicate minerals. In: *Reviews in Mineralogy* 31, p. 584, Mineralogical Society of America, Madison, Wisconsin.

White, A.F., Blum, A.E., Schulz, M.S., Bullen, T.D., Harden, J.W. and Peterson, M.L. (1996) Chemical weathering rates of a soil chronosequence on granitic alluvium: I. Quantification of mineralogical and surface area changes and calculation of primary silicate reaction rates. *Geochimica et Cosmochimica Acta* 60, 2533–2550.

White, A.F., Bullen, T.D., Schulz, M.S., Blum, A.E., Huntington, T.G. and Peters, N.E. (2001) Differential rates of feldspar weathering in granitic regoliths. *Geochmica et Cosmochimica Acta* 65, 847–869.

Williams, L.B., Wilcoxon, B.R., Ferrell, R.E. and Sassen, R. (1992) Diagenesis of ammonium during hydrocarbon maturation and migration, Wilcox Group, Louisiana, U.S.A. *Applied Geochemistry* 7, 123–134.

Wilson, M.J. (2004) Weathering of the primary rock-forming minerals: processes, products and rates. *Clay Minerals* 39, 233–266.

7 How do the Microhabitats Framed by Soil Structure Impact Soil Bacteria and the Processes that they Regulate?

Patricia A. Holden*

1 Introduction to Soil Bacterial Microhabitats

Soil microbial ecology concerns understanding microbes in soil, the habitats they occupy, the functions they perform and the interactions between these. Whilst modern molecular soil microbial ecology has progressively, and significantly, advanced the larger understanding of 'who's there and where are they?' (Ritz et al., 1994; Jesus et al., 2009), phylogenetic characterizations of soil community diversity and composition mostly produce spatially averaged, not location-specific, assessments. This is partly because grams of surface soil are typically used to extract sufficient DNA for PCR amplification of genes encoding 16S rRNA (e.g. Dunbar et al., 1999; LaMontagne et al., 2003; Nicol et al., 2003). Yet soil bacteria reside in numerous microhabitats that are heterogeneously distributed over small spatial scales (Killham et al., 1993; Grundmann et al., 2001; Dechesne et al., 2003, 2005; Vogel et al., 2003; Pallud et al., 2004; Gonod et al., 2006). As developed conceptually by others, bacteria do not continuously coat particle surfaces and thus bacteria on particles and in intervening pores must exist in spatially segregated patches. This has been the conclusion of other studies

(Fair et al., 1994; Grundmann and Debouzie, 2000; Chenu et al., 2001; Ranjard and Richaume, 2001; Zhou et al., 2002; Gonod et al., 2003; Nunan et al., 2003; Tokunaga et al., 2003; Pallud et al., 2004). But what, physically, defines a microbial patch? Is a patch simply microbes attached to a surface? What is the nature of the bacterial habitat at the bacterial scale? This chapter concerns bacterial growth habits and their importance to nutrient processing.

Environmental scanning electron microscopy (ESEM) images of surface soil reveal its complex micro-scale texture (Fig. 7.1), to which abiotic (organomineral) and biotic constituents contribute. However, unlike plant ecologists who, in the context of landscape ecology (Pickett and Cadenasso, 1995), can visually identify plant species and describe their spatial associations with each other and with other landscape elements, soil microbial ecologists have hitherto lacked the technologies to map soil microbial taxa with other features in the microenvironment. This situation could change, however, as new approaches, e.g. coupling SEM, energy dispersive spectroscopy (EDS) and tagged whole-cell oligonucleotide probes using metal or fluorescent epitopes (Gerard et al.,

*holden@bren.ucsb.edu

Fig. 7.1. A representative environmental scanning electron microscope (ESEM) image of surface soil from a California annual grassland showing the surface heterogeneity at the micro scale. The site, soil and sampling descriptions are as previously described (Fierer *et al.*, 2003). Soil was acquired from 10 cm depth, sieved (2 mm) and mixed by rolling prior to imaging. The image was acquired in wet mode using an FEI Co. XL30 ESEM with field emission gun (FEG) and a gaseous secondary electron detector (GSED). No sample preparation (e.g. drying or coating with a conducting film) was performed prior to imaging; multiple images were acquired from several specimens, each consisting of soil deposited on to a carbon-coated sticky tab (Ted Pella, Inc.; Redding, California).

2005; Kenzaka *et al.*, 2005), are refined and applied to soil-based studies.

The organization of soil at the micro scale, including cellular structures, can also be inferred from light microscopy (Chenu *et al.*, 2001) and transmission electron microscopy (TEM) images of embedded thin sections of soil (Foster, 1988), and SEM images (Foster *et al.*, 1983; Foster, 1988), with quantitative image interpretations using geo-statistical approaches (Nunan *et al.*, 2002; Young and Crawford, 2004). Also, the characteristics of microhabitats occupied by soil bacteria can be inferred from studies involving sequential cellular extraction (Lindahl and Bakken, 1995; Bockelmann *et al.*, 2003), and *ex situ* cultivation in porous media (Roberson and Firestone, 1992; Chenu, 1993; Holden, 2001; Holden and Pierce, 2001), on membranes (Soroker, 1990; Holden *et al.*, 1997; Auerbach *et al.*, 2000; Steinberger and Holden, 2004; Priester *et al.*, 2006) and in two-dimensional physical models of actual

pore structures, called 'micromodels' (Wan and Wilson, 1994; Wan *et al.*, 1994). From such studies, several conceivable bacterial growth habits can be inferred (Fig. 7.2): (i) free-living at the air–water interface or in pore water; (ii) attached to particle surfaces; (iii) attached to particle surfaces and embedded in extracellular polymeric substances (EPS); or (iv) attached to soil organic matter (SOM) deposits on particle surfaces and embedded in EPS. In all cases, water films play a key role in governing the nature of these habitats. Water films may vary in thickness with varying soil matric water potential at the micro scale (Jury *et al.*, 1991), and are always present at some scale in unsaturated soil (Papendick and Campbell, 1981). In this chapter, the characteristics and evidence for each of the habitats are discussed. Each has consequences for local bacterial physiology and mass transfer of substrates, and thus influences local nutrient processing. The possible consequences for soil nutrient

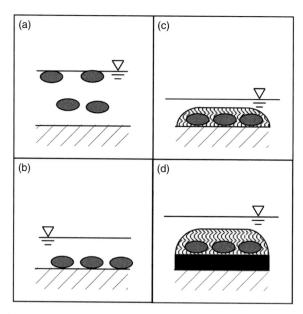

Fig. 7.2. Conceptual representation of soil bacterial growth habitats: (a) free-living at the air–water interface or in water films; (b) attached to solid surfaces within water films; (c) encapsulated in hydrated EPS and attached to solid soil surfaces; (d) colonizing SOM on solid surfaces where bacteria are encapsulated in hydrated EPS with surrounding water films.

processing observed at the larger scale are discussed, and an attempt is made to demonstrate the relevance of micro-scale bacterial habitats to larger-scale observations.

2 Cells on Surfaces in Films of Water, and at the Air–Water Interface

According to the conceptual model (Fig. 7.2), one habitat for microbes in soil is in water films, either at the air–water interface or 'floating' within films whose thickness varies with soil water content. This is essentially a planktonic habit, which is common in aquatic biomes. Microbes readily attach to solid–liquid (Characklis and Marshall, 1990; Costerton *et al.*, 1995; Marshall, 1999), liquid–liquid (Holden *et al.*, 2002) and air–liquid (Wan and Wilson, 1994; Schafer *et al.*, 1998) interfaces. Attachment to solid–liquid interfaces is preceded by the formation of an organic macromolecular conditioning film (Costerton *et al.*, 1995). However, many organic macromolecules are ampiphiles or surface active, i.e. having both hydrophilic

and hydrophobic sites (Tanford, 1980), and readily accumulate at air–water and other interfaces (Israelachvili, 1992).

The air–water interface is thus enriched in nutrients and supports microbes in a habitat called, in aquatic environments, the 'neuston' (Plusquellec *et al.*, 1991; Grammatika and Zimmerman, 2001; Zuev *et al.*, 2001). Empirical evidence for the neuston as a soil habitat is by observing bacteria in micromodels operated in unsaturated conditions (Wan and Wilson, 1994). As discussed by Walsby and Dunton (Walsby and Dunton, 2006), many soil bacteria produce gas vesicles that could enable them to float to the air–water interface within a water film; microbial surface hydrophobicity would then contribute to air–liquid interfacial attachment. However, are bacteria likely to reside at the air–water interface and in water films for prolonged periods? In the short term, companion column and micromodel studies support the view that bacteria are pinned to solid surfaces in thin water films and at air–water interfaces when porous media are drained (Auset *et al.*, 2005). Under similar conditions in unsaturated soils, bacteria are non-motile (Gammack *et al.*, 1992) and become motile only when water films are of appropriate thickness (i.e. corresponding to c. −0.001 MPa and higher) (Dechesne *et al.*, 2008a), making chemotaxis in relatively dry soils an inefficient way for bacteria to access nutrients (Arora and Gupta, 1993). Thus, in the absence of an ongoing supply of nutrients to the neuston, it is unlikely that soil bacteria will persist at the air–water interface. Further, as per Or *et al.* (2007a), only under the wettest conditions in soil (i.e. at matric potentials less negative than −0.02 to −0.04 MPa) are water films at particle junctures actually thick enough to cover a bacterial cell. Thus, the planktonic lifestyle in soil is unrealistic except during transient wetting events when soil pores are mostly water-filled. Furthermore, heterotrophic bacteria provided with an ample carbon source within reasonably thick water films form expanding colonies (Dechesne *et al.*, 2008a), i.e. clusters of cells with their EPS, and surfaces added to an air–water interface in soil solutions are rapidly colonized by small patches of EPS-rich cells surrounded by clay particles (Lunsdorf *et al.*, 2000).

3 Cells in Hydrated EPS on Surfaces

If soil bacteria are only transiently planktonic and associated with the air–water interfaces, then they must be associated with solid surfaces, either as bare cells or in EPS (Fig. 7.2c). What is the rationale for these conceptual models, and are soil bacteria likely to be embedded in EPS at surfaces? Across all environments, bacteria are thought mainly to exist within biofilms that are broadly defined as surface-associated cells embedded in hydrated EPS of cellular origin (Characklis and Marshall, 1990). The term 'biofilm' (Williamson and McCarty, 1976a, b) is mostly associated with surface coatings in saturated environments (e.g. sediments, sewage treatment plants, pipes, teeth, etc.) (Costerton and Irvin, 1981; Costerton *et al.*, 1994; Costerton, 1995), but its definition does not specify either a spatial dimension or degree of surface coverage (Characklis and Marshall, 1990). The phrase 'colony biofilm'

(Oldak and Trafny, 2005; Venugopalan *et al.*, 2005) describes colonial bacterial growth on a moist, nutrient-rich surface, including solid media in the laboratory. The narrower phrase 'unsaturated biofilm' is specific to cells with EPS in transiently wet environments such as soil (Holden *et al.*, 1997; Auerbach *et al.*, 2000; Holden, 2001; Steinberger *et al.*, 2002; Steinberger and Holden, 2004, 2005; van de Mortel and Halverson, 2004; Or *et al.*, 2007a) or perhaps on plant leaf surfaces (Morris *et al.*, 2002). In such environments, dryness can profoundly influence bacterial EPS production (Roberson and Firestone, 1992). If one accepts that 'unsaturated biofilm' is a useful conceptual construct of a soil bacterial growth habit, then direct visual evidence would support that unsaturated biofilms in soil are predominantly colonial or patchy (Nunan *et al.*, 2003) and not continuous films that are observed in the laboratory when nutrients are non-limiting (Roberson and Firestone, 1992; Holden *et al.*, 1997). Regardless of the chosen terminology, the constituents of the growth habit are expected to be cells on surfaces, in hydrated EPS (Fig. 7.2c).

Winogradsky reported in 1925 that soil bacteria, cultivated on buried slides, occurred as colonies and in a gel-like matrix (Winogradsky, 1925). But are bacteria in soil always in EPS and on surfaces? According to the conceptual model (Fig. 7.2b), one bacterial growth mode could be surface-associated without EPS. Yet direct evidence for the latter in soil is lacking, as are robust explanations for the benefits of a wholly EPS-deficient lifestyle. To explore further, there are two types of evidence to consider: (i) direct, e.g. high-resolution micrographs of soil thin sections coupled with chemical analysis; and (ii) indirect, i.e. through experimentally simulating soil physico-chemical characteristics and determining their relationship to EPS production.

3.1 Characteristics and evidence for the model of bacteria in EPS on surfaces

Although soil bacteria are not expected to be in continuous biofilms per se, they are commonly found to occur in association with visible EPS (Foster, 1981; Chenu and Jaunet, 1992; Chenu and Tessier, 1995). The visual evidence in electron micrographs is compelling (Dudman, 1977), e.g. TEM micrographs of electron-sparse, presumably EPS-rich, regions that surround cell clusters and separate them from mineral particles (Foster, 1988; Lunsdorf et al., 2000); TEM images of clay platelets encrusting EPS-enveloped cells where the EPS is visible by specific electron-dense stains (Foster, 1981); and SEM images of EPS 'drifts' that cloak laboratory-cultivated colonies in drying sand (Roberson and Firestone, 1992). As discussed by Chenu (Chenu, 1995) and further summarized by Or (Or et al., 2007a), the amount of EPS in soil can be as much as 500% of the estimated microbial biomass. Comparing this upper estimate of soil microbial EPS with the apparent volumes that EPS occupies around cells in TEM images (e.g. Foster, 1981), it would appear that most soil bacteria are encapsulated by EPS, a conclusion consistent with the broad (i.e. Gram-positive and Gram-negative) distribution of bacterial groups that produce exopolysaccharides (Dudman, 1977), a major component of bacterial EPS. However, the direct chemical characterization of EPS around soil bacteria has not been performed, and thus the attribution of soil-extractable EPS to bacteria versus plants is still tentative.

For some time, it has been accepted that soil bacteria manufacture exopolysaccharides (Whistler and Kirby, 1956) that contribute to soil aggregation (Martin, 1946; Whistler and Kirby, 1956). Yet chemical evidence for microbially produced exopolymers in soils is equivocal because of similarly composed plant exudates and decomposition products (Chenu, 1995) that are co-resolved, for example, when soil extracts are analysed for carbohydrates (Martens and Frankenberger, 1991a). The ratio (arabinose + xylose):(galactose + mannose) has been proposed as one way to distinguish bacterial from plant polysaccharides by assuming that the ratio is 2:1 when plant sugars dominate and 1:2 when microbial sugars dominate (Oades, 1984). Using this

characterization when examining various particle size fractions, microbial sugars mainly appear to be associated with clays in surface soils (Glaser et al., 2000). Clays are often observed in close association with EPS, both in the presence (Foster, 1981; Lunsdorf et al., 2000) and absence (Chenu, 1993) of bacterial cells, supporting the general concept that microbial exopolysaccharides are the interface between cells and clay minerals (Chenu, 1995). The co-localization of exopolysaccharides and soil bacteria is also implied by improved extraction efficiencies of bacteria when soils are pretreated enzymatically to break extracellular polysaccharide linkages that encapsulate and bind bacteria to solid surfaces (Bockelmann et al., 2003). Thus, the model in Fig. 7.2c is supportable, i.e. that soil bacteria are within EPS matrices and surface-associated. Still, little is known about the composition and structure of soil microbial polysaccharides, the full composition of soil bacterial EPS and compositional aspects that would distinguish such material clearly from plant mucilage or decomposition products.

Although soil microbial EPS chemistry, as well as chemically derived physical properties, is relatively unknown, cultivating bacteria in the laboratory under soil-like conditions and analysing the exopolymers is an approach to infer possible EPS composition and the conditions that promote EPS production. As such, EPS macromolecular composition measurements of laboratory-cultivated unsaturated biofilms support the view that they mainly comprise polysaccharide, but also protein and DNA, where proportions vary with nutritional conditions (Steinberger and Holden, 2004, 2005), the presence of toxic metals (Priester et al., 2006) and bacterial taxa (Steinberger and Holden, 2005). In some cases, for example Bacillus subtilis in pure culture biofilm, the component protein and polysaccharides are each essential for biofilm formation (Branda et al., 2006). In addition to varying macromolecular proportions, EPS varies in its carbohydrate (Sutherland, 2001a, b) and protein (Martens and Frankenberger, 1991a) composition with varying organism and growth conditions. Among many influencing factors, EPS production is promoted by an increased C:N ratio (Priester *et al.*, 2007), by bioavailable C (Steinberger and Holden, 2004) and by desiccation (Roberson and Firestone, 1992), a major environmental variable in unsaturated soil.

3.2 Impacts of EPS on survival

In general, EPS constitute an extracellular barrier that protects cells from toxins, predation, starvation and dehydration, while facilitating adhesion, nutrient and water retention, and community interactions (Wolfaardt *et al.*, 1999). Most related knowledge is derived from studying biofilms cultured under fluid-flowing systems but, to the degree that bacteria in soil are within EPS (Fig. 7.2c), there may be useful analogies to soil bacteria. In the following sections, the various roles of EPS for soil bacterial survival are considered in more detail.

Toxicity

Exopolysaccharides are especially recognized for protecting bacteria (Dudman, 1977), i.e. by inhibiting the penetration of antibiotics (Hentzer *et al.*, 2001) and binding toxic metals (Bitton and Freihofer, 1978; Kachlany *et al.*, 2001; Lamelas *et al.*, 2006), which benefit entire biofilms (Teitzel and Parsek, 2003). Extracellular DNA can also potentially offer protection to bacteria in unsaturated biofilms, e.g. through binding of a toxic metal such as chromium (Priester *et al.*, 2006). The general explanations for protection are sorption and the resistance to diffusion that EPS affords (Stewart, 2003). Such resistance may be quite high in unchannelized, compact unsaturated biofilms where the diffusion coefficient is quite low (Holden *et al.*, 1997) compared with saturated, highly channelized biofilms where diffusion coefficients are similar to those measured in water (Stewart, 2003).

The benefits of EPS to toxicity resistance suggest that bacterial EPS production would be enhanced in the presence of toxicants. Enhanced EPS production is related to environmental chemistry (Dudman, 1977), with toxins generally enhancing production (Poynter and Mead, 1964). However, oxygen also stimulates EPS production (Bayer *et al.*, 1990), which may be relevant to the survival of micro-aerophiles in unsaturated soils (Holden *et al.*, 2001). EPS and the biofilm growth habit may facilitate very local cellular diversification from the combination of mutations that arises with oxidative stress, along with diverse physically structured micro-environments (Boles and Singh, 2008). However, EPS can quench reactive oxygen species (ROS) (Kiraly *et al.*, 1997; Bylund *et al.*, 2006), which may serve to limit the more severe effects of oxidative stress.

Grazing

To a degree, soil bacteria are protected from grazers by restrictive pore spaces (Wang *et al.*, 2005) and thin water films, although the relationships are likely to be more complex than simple exclusion (Hassink *et al.*, 1993). EPS could have a dual role in bacterial susceptibility to grazing, i.e. as either a deterrent or an attractant. As previously reviewed (Joubert *et al.*, 2006), EPS in aquatic systems are an important part of the food web. By extension, while a role for EPS may be physical protection to enveloped bacteria in biofilms, another role may be as a food source for protozoan predators, which, in turn, provides a mechanism of returning EPS nutrients to biofilm cells (Joubert *et al.*, 2006). This is a very different idea from that of EPS serving as a static, protective barrier. Planktonic prey can be preferred over biofilm cells: in a laboratory flow-cell study using biofilm-forming yeast and ciliates from the same soil, the protozoa selectively fed on planktonic yeast and biofilm EPS, with the endpoint being enhanced growth of biofilm organisms (Joubert *et al.*, 2006). Whether such preferential feeding occurs for soil bacteria is mostly unknown, and organisms that appear to exclude predators by EPS-trapping under laboratory conditions do not appear to resist predation in soil cultures (Djigal *et al.*, 2004). This would suggest that relationships between predation and soil bacterial EPS are more complicated than can be explained by simple preferential feeding.

Desiccation

Whether cell aggregates on plant leaves (Monier and Lindow, 2003) or micro-colonies on sand grains (Roberson and Firestone, 1992), the biofilm growth habit affords bacterial resistance to desiccation. Microbial exopolysaccharides hold, and slowly absorb and release, water (Chenu, 1993), thereby acting as hydrologic 'uncouplers' (Or et al., 2007a) that buffer bacteria against rapid hydration to the point of cells bursting ('upshock') upon rapid soil wetting (Kieft et al., 1987), and desiccation from severe soil drying (Harris, 1981). The role of exo-polysaccharides in bacterial desiccation resistance has long been discussed (Dudman, 1977). However, the seminal work of Roberson and Firestone (1992) first showed that bacteria actually respond to desiccation by producing exopolysaccharide. Ophir and Gutnick (1994) then showed that soil bacteria upregulate capsule-associated genes in response to desiccation. Rhodococcus, an actinomycete that is prevalent in desert soils, accumulates EPS when desiccated (Alvarez et al., 2004). Factors such as C availability and quality, in addition to desiccation, also affect EPS production in pathogenic Escherichia coli that colonize food surfaces (Ryu and Beuchat, 2004). Bradyrhizobium japonica also upregulates many genes in response to desiccation stresses, including those involved in extra-cellular polymer production and export (Cytryn et al., 2007). Pseudomonas putida, a common soil isolate, upregulates a number of genes, including expolysaccharide synthesis, in response to low water availability or matric stress in soil (van de Mortel and Halverson, 2004). The essentiality of EPS to bacterial function during both desiccation and freeze–thaw stress is shown for cyano-bacteria that continue to function in the presence, but not in the absence, of their EPS (Tamaru et al., 2005). When subjected to low matric water potentials, soil- and plant-asso-ciated pseudomonads are less water-stressed within their EPS where they remain rela-tively hydrated (Chang et al., 2007). Further-more, photosynthetic algal biofilm cells retain more of their photosynthetic activity when desiccated if they are in an EPS matrix, as biofilm (Kemmling et al., 2004).

Starvation

Many bacteria have been shown under labo-ratory conditions to alter their EPS level and composition in response to nutrient changes, e.g. Shewanella with varying elec-tron acceptors (Neal et al., 2007), and Pseu-domonas with different carbon sources (Steinberger and Holden, 2004), suggesting that EPS may be useful in mediating starva-tion. As reviewed previously (Wolfaardt et al., 1999), EPS has a high affinity for various exogenous carbon sources and inorganic nutrients. The 'egg box' configu-ration (De Kerchove and Elimelech, 2007) of EPS polymers reasonably invokes the expectation of its trapping abilities. But do bacteria use their EPS polymers directly as energy, carbon, or other nutrient sources? It has long been thought that bacteria do not metabolize their own exopolysaccharides (Sutherland, 1977, 1999a). However, Pseu-domonas aeruginosa produces a lysase spe-cific to its main exopolysaccharide involved in biofilm attachment (Boyd and Chakra-barty, 1994), and studies of oral biofilm bac-teria show similar findings (Kaplan et al., 2004). Polysaccharide-degrading enzymes (glycosyl hydrolase, in this case 'dispersin') can break down the EPS matrix, facilitating the loss of biofilm structure and the disper-sion of bacteria (Ramasubbu et al., 2005).

None the less, while polysaccharase production and accumulation in biofilms may be widespread among bacteria, direct evidence for bacteria using their own expolysaccharides as nutrients is lacking (Sutherland, 1999b). However, as biofilms are typically diverse (Cole et al., 2004), it remains possible that the breakdown of one species' EPS can occur by another species' enzyme (Kaplan et al., 2004). Also, as noted by others, EPS generally (Wingender et al., 1999a), as well as specifically in saturated (Whitchurch et al., 2002) and unsaturated (Steinberger and Holden, 2005) biofilms, contains ample DNA that has been shown to be readily used as a sole macronutrient source for Shewanella from terrestrial and other environments (Pinchuk *et al.*, 2008). There are also ample extracellular proteases among biofilm bacteria, the best studied being those produced by *P. aeruginosa* (Caballero *et al.*, 2001; Oldak and Trafny, 2005). Still unknown are the conditions in soil that would promote bacteria using EPS polymers as nutrient sources, and direct evidence for the phenomenon, but the C-, N- and P-rich composition of EPS suggests it could be an important nutrient reserve for soil bacteria.

3.3 Diffusional availability of nutrients

Most microbial processes in unsaturated soil are limited by aqueous phase nutrient diffusion, a situation that also facilitates high soil microbial diversity (Focht, 1992). For example, denitrification in aerated soils is limited by carbon diffusion to anaerobic microsites (Myrold and Tiedje, 1985). However, both extra-colony distance to substrate and local substrate solubility and bioavailability can cause diffusional re-supply constraints to bacteria growing on surfaces (Steinberger *et al.*, 2002). For example, to compensate for local diffusional constraints, bacteria that receive their nutrients from the surfaces they colonize may change their morphology to increase nutrient acquisition (Steinberger *et al.*, 2002).

Local-range diffusion

An unsaturated biofilm is comprised of cells attached to a surface, encapsulated in hydrated EPS and for which the surrounding fluid phase is transiently water and often air (Fig. 7.3). Unsaturated biofilms appear to be dense, unchannelled and compact (Holden *et al.*, 1997), with diffusion-controlled substrate delivery and metabolite removal. There are three potential modes of substrate supply (i.e. flux, or F_s) to unsaturated biofilms (Fig. 7.3): (i) the colonized surface is the source of a limiting substrate ($F_{s,s}$); (ii) the substrate is concentrated at some distance away and diffuses along a concentration gradient through water films ($F_{s,w}$); and (iii) the substrate is in the gas phase (e.g. methane or other VOC), which advects or diffuses to the air–biofilm interface then dissolves and diffuses into the biofilm ($F_{s,g}$) (Hilger *et al.*, 2000). Metabolites are also removed from the biofilm by diffusion (F_m). In the case of bacteria colonizing solid chitin- or elastin-rich polymers in nature, soluble nutrients from around and beneath the initial colonizers would presumably be necessary to prime colony growth, and then the surface itself could become a substrate when extracellular enzymes (i.e. chitinase and elastase) were produced.

For unsaturated biofilms, a common mode of substrate supply would be from

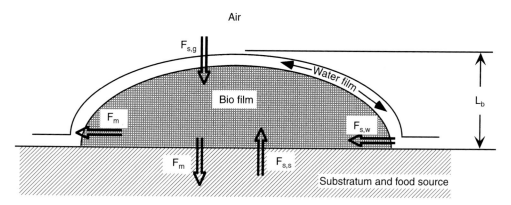

Fig. 7.3. Conceptual unsaturated bacterial biofilm of thickness L_b (not to scale) colonizing a solid substratum that provides nutrients. Open arrows represent flux (F, into or out of the biofilm) of metabolites (m), and dissolved (s,w), solid (s,s) or gas phase (s,g) substrates. Air surrounds the biofilm; a thin film of water coats the biofilm and uncolonized substrata.

below, whereby biofilms are colonizing a surface that they also degrade. Each flux (i.e. F_m, $F_{s,g}$, $F_{s,w}$ and $F_{s,s}$ in Fig. 7.3) is diffusional and can be modelled by Fick's Law (Crank, 1999):

$$F_y = D_{eff} \frac{dC}{dy} \qquad (7.1)$$

where (in generic dimensions of mass or m, length or L, and time or t):

F_y = diffusional flux of solute through the biofilm in direction y; $mL^{-2}t^{-1}$

D_{eff} = effective diffusivity of the diffusing solute through the biofilm; L^2t^{-1}

C = concentration of solute; mL^{-3}

y = distance in the direction of diffusion; L.

Diffusion through saturated biofilm micro-colonies (Stewart, 2003) is also Fickian and is thus also controlled by the effective diffusivity (D_{eff}) and the concentration gradient which depends on C and L_b (Fig. 7.3). D_{eff} is temperature-dependent, but also depends on the tortuosity, or channelled nature, of the matrix through which solutes diffuse. D_{eff} depends too on cellular packing because diffusion through biofilms, particularly of polar solutes, occurs around and between cells. For saturated biofilms, the magnitude of D_{eff} for a given solute is typically 20–80% of the aqueous molecular diffusivity (Stewart, 2003). Given that so much water comprises the open, channelled structures of saturated biofilms, it is no wonder that saturated biofilms resemble water in their resistance to diffusion (Stewart, 2003). The thickness of the biofilm, L_b, will vary with substrate type and availability, time and bacterial species, as well as with the amount of EPS that is formed, but is relatable to the 'dy' term in the denominator of Equation 7.1. Thus, as L_b increases, the flux of any substance through the biofilm decreases. Similarly, a high concentration of diffusing solute, whether a substrate or a metabolite, leads to a higher rate of diffusion (flux) into or out of the biofilm.

Ultimately, within-biofilm diffusion is important because the rate at which unsaturated biofilm bacteria grow and produce metabolites is intertwined with the diffusional processes depicted in Fig. 7.3. If bacteria in a biofilm grow rapidly at the expense of a labile substrate, then their growth rates will be determined by how fast nutrients diffuse to cells and how fast metabolites diffuse away. On the other hand, if bacterial cells replicate slowly at the expense of a particular (perhaps toxic, or weakly soluble) substrate, then the intrinsic rate of cellular growth and not diffusion will determine how fast biofilms grow. This balance between diffusion and 'reaction' can be expressed mathematically (Holden et al., 1997) as:

$$\frac{D_{eff}}{L_b}\left[\frac{dC}{dy}\right]_{y=L_b} = \left[\frac{q_b C}{K_s + C}\right]\eta_i \rho_b \qquad (7.2)$$

where:

q_b = maximum intrinsic rate of substrate metabolism per cell; m (# cells)$^{-1}t^{-1}$

K_s = Monod (growth) or Michaelis–Menten (non-growth)-associated half-saturation kinetic constant; mL^{-3}

η_i = internal 'effectiveness factor' (Bailey and Ollis, 1986); dimensionless

ρ_b = volumetric cellular density in the biofilm; (#cells) L^{-3}.

The internal effectiveness factor, η_i, is essentially a proportionality constant between the two processes of diffusion and reaction; if the diffusion is slow and thus controls the overall rate of biofilm metabolism, then η_i is very small (Bailey and Ollis, 1986). Equation 7.2 is thus a mathematical framework for considering how biofilm growth and biofilm diffusion are coupled. This type of model has been used in the past to evaluate saturated biofilm function (Characklis and Marshall, 1990), and it has also been used to evaluate experimental data showing that local, or within unsaturated biofilm, diffusion can control the rate at which a carbon source is metabolized (Holden et al., 1997).

However, in unsaturated soils, thin or discontinuous soil water films are the main factors limiting longer-range, or extra-colony aqueous phase, diffusion, and thus soil water content is a variable used to calculate diffusion coefficients for nutrients in soil (Paul and Clark, 1989; Jury *et al.*, 1991). Theoretically, low water content and associated thin films facilitate the patchy distribution of bacterial taxa in soil through limiting the substrate diffusion that reduces competition (Long and Or, 2005). The coexistence of otherwise competitive bacterial strains in a diffusion-limited system has also been shown experimentally (Dechesne *et al.*, 2008b). Also experimentally, diffusional limitations on soil microbial processes explain that soil nitrification is stimulated by supplying N as gaseous NH_3, which overcomes thin water film diffusional limitations to the re-supply of dissolved NH_4^+-N (Stark and Firestone, 1995).

Long-range constraints

The role of EPS in overcoming long-range diffusional constraints, involving EPS-mediated water retention, hydraulic conductivity, local diffusion and aggregate structuring, is discussed by Or *et al.* (2007b). As explained above, bacteria produce EPS in response to desiccation (e.g. Roberson and Firestone, 1992), which means that EPS will be relatively abundant in dry soils where it retains water and facilitates diffusion of dissolved solutes (Chenu and Roberson, 1996). Regarding the fact that it is very hygroscopic, EPS absorbs water rapidly and releases it slowly (Chenu, 1993), and facilitates prolonged cellular hydration and nutrient re-supply in drying soils (Or *et al.*, 2007b). In this way, EPS is a 'bridge' between patches of microbes and distant points that enables survival when water films are too thin substantially, rapidly, to conduct dissolved nutrients to microbial sinks. This can be seen visually in Fig. 7.4, where bacteria cultivated in moist sand form an EPS-rich colony bridge between two sand grains. However, EPS also occupies air-filled pore spaces, and thus hyper-production of EPS could also limit gas-phase substrate (e.g. O_2 or a volatile organic C source) diffusional re-supply; then again, under very dry conditions where the EPS matrix collapses, pore spaces may reopen (Holden *et al.*, 1995, 1997). Thus, the role of EPS in diffusional re-supply of nutrients is a dynamic one,

owing in large part to the interplay of EPS physical elasticity to wetting and drying, and the dynamic physiological responses microbes have to drying and re-wetting.

The conceptual model of cells on surfaces, embedded in EPS but spatially distant from SOM as depicted in Fig. 7.2c, implies that cells are passive receptors for their exogenous nutrients. What is the evidence for this growth model? When could it occur? Why would cells persist where SOM re-supply is limited by spatial separation from cells? To answer, evidence for spatial segregation of soil microorganisms and their substrates includes the presence of starvation markers in PLFA profiles (Fierer *et al.*, 2003), the stimulation of microbes by addition of exogenous C substrates (Chenu *et al.*, 2001), the relative proportions of biomass-C to SOM-C and the fact that much soil C, by carbon dating, can be hundreds to thousands of years old (Schöning and Kögel-Knabner, 2006). Yet a C limitation, imposed by spatial segregation of biofilms from SOM, would limit C-rich exopolysaccharide production (Sutherland, 1977; Hilger *et al.*, 2000). Oligotrophy, simulated under laboratory conditions by cultivating bacteria with ambient soil DOC, stimulates the growth of otherwise unculturable bacteria that apparently do not produce visible EPS (Ferrari *et al.*, 2005). This suggests that the predominance of growth habits of cells on surfaces, with or without EPS (Fig. 7.2b, c), could be time-dependent, i.e. with a preponderance of EPS surrounding cells (Fig. 7.2c) when SOM C is sufficiently available, either locally or by longer-range diffusion, and a temporal progression to less EPS (Fig. 7.2b) when the rate of C use for EPS production exceeds the rate of C re-supply.

3.4 Physiological state and diversity

Soils harbour phylogenetically diverse microbial communities distributed across small spatial scales (Becker *et al.*, 2006), that vary depending on whether within or outside aggregates (Mummey *et al.*, 2006). As explained by Frey (2007), sufficient soil water has an effect of lowering diversity by

(a)

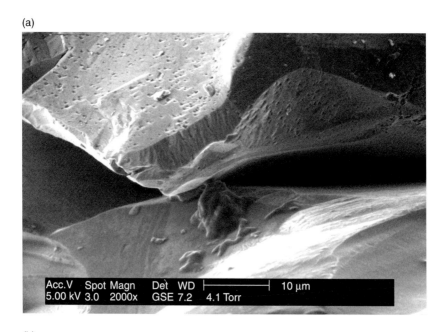

Acc.V   Spot Magn   Det WD |———————| 10 µm
5.00 kV 3.0   2000x   GSE 7.2   4.1 Torr

(b)

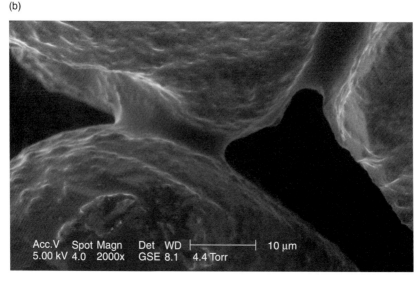

Acc.V   Spot Magn   Det WD |———————| 10 µm
5.00 kV 4.0   2000x   GSE 8.1   4.4 Torr

Fig. 7.4. ESEM images of *Pseudomonas aeruginosa* PG201 cultivated in moist sand with aqueous mineral media and glucose as the carbon source, where (a) two sand grains are bridged by a micro-colony embedded in EPS and (b) copious EPS is observed to span mineral grains. Specimen preparation, instrument and imaging conditions are as per those in Fig. 7.1, except that part (b) was stained with ruthenium red as per Priester *et al.* (2007). Note the connection point between EPS-embedded cells and the upper grain, where a radius of curvature is observed in the biological matrix.

overcoming spatial isolation between populations located in patches. Improved nutrient re-supply in wet versus dry soils (Stark and Firestone, 1995) and enhanced bacterial motility (Dechesne *et al.*, 2008b) are two observed mechanisms by which niche equalization, and thus diversity reduction, might occur. Low water-mediated diffusional constraints on nutrient re-supply encourage heterogeneous patch development in soils (Long and Or, 2005) and reduce the competition between otherwise competing populations (Dechesne *et al.*, 2008b). This suggests that the extremely high microbial diversity in soil depends largely on a hydrologically mediated spatial isolation of heterogeneously distributed populations. Curiously, conventional protocols for soil bacterial enrichment have called for slurrying the soil, adding nutrients and incubating while shaking under selective conditions (Wollum, 1982). However, if soil bacteria are mostly attached to solid surfaces, slurrying would probably select for planktonic bacteria. Other approaches for cultivating soil bacteria include provision of a solid substratum as well as limiting nutrients and, as a result, exotic lineages become newly cultivatable (Ferrari *et al.*, 2005). Theoretically, aerobic biofilm cells have a competitive advantage from producing EPS in that cell descendants are pushed towards their electron acceptor (oxygen), while oxygen re-supply to neighbours is diffusion-limited and neighbours 'suffocate' (Xavier and Foster, 2007).

3.5 Cell–cell communication

Unsaturated biofilms, having EPS and tightly packed cells as barriers to diffusional efflux, provide a mechanism for cells in soil to retain quorum sensing (QS) compounds within a patch, close to cells that can benefit from the gene auto-induction enabled by QS compounds. Acyl homoserine lactone (AHL) is a well-studied type of QS chemical that is manufactured by the common soil microbe, *P. aeruginosa*. The ready metabolism of AHL (Leadbetter and Greenberg, 2000) by an organism commonly found in the subsurface, *Variovorax paradoxus* (LaMontagne *et al.*, 2003), suggests that this type of QS compound is present and influential on soil bacterial communities even at depth. Theoretically, colonial growth, spatial isolation of colonies, nutrient availability and within-biofilm diffusional restriction are prerequisites to auto-inducer accumulation (Hense *et al.*, 2007). As these four

characteristics are descriptive of bacterial life in soil, the presence of active QS via AHL or other auto-inducing compounds in soils is conceivable. Experimental support for AHL-mediated, QS-controlled bacterial processes in soil is reported by DeAngelis *et al.* (2008), where AHLs were found to be abundant in rhizosphere soils. Many (25%) rhizosphere soil-isolated bacteria that produced exoenzymes also produced AHL. For several bacterial isolates, QS was quenched by genetically transforming the strains to produce lactonase; for some of the transformants, extracellular chitinase activity was also greatly diminished (DeAngelis *et al.*, 2008). This suggests that exoenzyme activity for some isolates was QS-controlled. That these particular transformants also showed EPS-rich biofilm morphologies in the wild type (DeAngelis *et al.*, 2008) may indicate either that EPS production was promoted by cell–cell signalling in the rhizosphere, that the presence of EPS around rhizosphere bacteria enhanced QS and associated QS-mediated exoenzyme activity, or both. In either case, the physico-chemical characteristics of EPS and patchy colonial growth could enhance the local retention of QS chemicals around cells, which would promote QS-mediated processes.

What are the specific relationships of EPS and QS in soil to diffusion, both at the pore scale and within biofilms? Thin water films in dry soils limit diffusional re-supply of dissolved nutrients (Stark and Firestone, 1995), and thus it is possible that soil bacteria hyper-produce EPS because of nutrient imbalances that lead to increased exopolysaccharide synthesis as in liquid culture (Sutherland, 1977). An alternative explanation is that auto-inducers involved in EPS synthesis could accumulate when biofilms dry because of a thin film-mediated reduction in AHL diffusional efflux. The restriction on AHL diffusional efflux has already been postulated to explain greater QS-mediated transcriptional activity in the compact, basement layers of saturated biofilms (De Kievit *et al.*, 2001). Quorum sensing requires build-up and retention, i.e. low diffusional loss, of cell–cell signalling compounds, but this is inherently tied to diffusional constraints (Redfield, 2002). In static biofilms, QS compound retention leads to higher virulence gene expression relative to that in high-flow biofilms (De Kievit *et al.*, 2001); hydrodynamics similarly affects expression of genes encoding chitinase and elastase (Purevdorj *et al.*, 2002). The concept of 'efficiency sensing' casts the idea of density and quorum sensing into a spatial realm where distances between populations are important. This concept shows mathematically that it is advantageous for cells to be in clonal micro-colonies where they can benefit from their own signals (Hense *et al.*, 2007). In unsaturated biofilms, restricted diffusional efflux could result in local AHL retention and transcription of genes encoding for EPS polysaccharides. Thus, one might expect enhanced quorum sensing in unsaturated soils, owing in large part to the low water content-mediated restrictions on diffusion and the role of EPS in retaining QS-type compounds. AHL QS compounds are indeed present and influential in soil (Burmolle *et al.*, 2003), which may indirectly lend further support for the growth habit of cells attached to surfaces and embedded in EPS (Fig. 7.2c).

3.6 Extracellular enzymes

The importance of extracellular enzymes in soil processes is long established (Pollock, 1962; Skujins, 1976), and their theoretical importance to terrestrial ecosystem function is of contemporary interest (Schimel and Weintraub, 2003). Methodologically, exoenzyme activity is assessed after homogenizing and slurrying soils (Sinsabaugh *et al.*, 1991). Yet exoenzymes appear to be spatially structured in soils (Marx *et al.*, 2005), probably owing to the high degree of spatial patchiness in soil bacterial communities (O'Donnell *et al.*, 2007). Where are extracellular enzymes relative to bacterial growth habits? How does exoenzyme location affect activity and what is the relevance of EPS? In 1962, Pollock suggested that exoenzymes are either cell-bound, i.e. either in cells or tightly associated with the cell surface, or extracellular, i.e. freely dispersed in

the medium surrounding cells (Pollock, 1962). Burns (1982) added more definition and nominated the soil enzyme spatial locations as cytoplasmic, periplasmic, extracellular but cell-bound (i.e. in the mucilage that originates from either plants or microbes), and extracellular in the surrounding solution, where exoenzymes are either in or attached to dead cells and may be very transient, associated with minerals and humics, or temporarily associated with an enzyme–substrate complex (Burns *et al.*, 1978). The association of exoenzymes and their substrates with extracellular polysaccharides was suggested as a mechanism to ensure enzyme activity near cells (Burns *et al.*, 1978). Assessing the susceptibility of soil catalytic activity to γ-irradiation was used to devolve the relationships to exoenzyme spatial location (Burns *et al.*, 1978). Burns also stated that the method of soil enzyme extraction can alter the perceptions of exoenzyme location, and of observed activity (Burns *et al.*, 1978). For example, if enzymes are in the soil solution, drying the soil as a pretreatment can cause sorption to soil solids such as minerals or humics, and sorption is likely to change the activity from what was originally present in undisturbed soil (Burns *et al.*, 1978).

Following Allison (2006), a number of the soil exoenzyme locations nominated by Burns (1982) may not be important to soil processes if they promote attachment to soil minerals. When attached to minerals, although enzymes are stabilized against degradation, they may not be active *in situ* because they are sterically unavailable for catalysing reactions (Allison, 2006). The activity of extracellular enzymes in soil is lowered when they are bound to minerals or 'stabilized', as evidenced by the significant loss (between 30 and 90%) of activity for urease and phosphatase depending on the clay mineral binding the enzyme (Gianfreda *et al.*, 2002). If stabilization is extensive, then extracellular enzyme activity may be a less important mode of catalysis relative to intracellular activity (Allison, 2006). On the other hand, the EPS matrix around cells may be a zone conducive to exoenzyme retention and activity (Burns,

1982): EPS shields cells from clay minerals and this may serve to maintain extracellular enzyme activity near cells. Evidence for minerals binding to outer layers of EPS is indicated in biofilm thin sections from stone surfaces (Kemmling *et al.*, 2004). Furthermore, within stone surface biofilms, Kemmling *et al.* (2004) found that amylase activity was not greatly affected by the temperature of drying (80°C) but instead was mainly enhanced by desiccation during seven drying/re-wetting cycles (350 min) when various exopolysaccharides were present, including bacterial and algal alginate. This activity increase could be due to the spatial proximity of substrates to enzymes that might be afforded with exopolymers, but the mechanism is not clear.

One drying and re-wetting cycle resulted in nearly 80% of the enzyme function being retained relative to the initial state, which was significantly more than the activity retained by just heating alone (Kemmling *et al.*, 2004). Repeated drying and re-wetting of the enzyme in either bacterial or algal alginate resulted in more activity as compared with the enzyme in buffer alone (Kemmling *et al.*, 2004). The increase was, on average, 100–200% with either of these alginates present. Other studies support the concept of biofilm EPS protecting exoenzyme activity. For example, in *P. aeruginosa*, lipase and alginate are co-excreted (as summarized by Wingender *et al.*, 1999a) and co-localized in biofilms. Alginate stabilizes the lipase; furthermore, extracellular lipase activity is higher in mucoid biofilms that produce alginate versus non-mucoid cells with or without alginate, again as summarized by Wingender *et al.* (1999b). Also, in wastewater treatment biofilms, hydrolytic decomposition of polymeric substrates occurs in matrices associated with cells, but not in the external aqueous solution (Confer and Logan, 1998). As summarized by Sutherland (2001), biofilm matrices harbour active enzymes of many types. If we accept the soil bacterial growth habit model of Fig. 7.2c, then EPS in soil is catalytically very important. Whilst the concept is intriguing, its importance has not been expressed quantitatively, i.e. through demonstration of how

exoenzymes in soil micro-colony EPS contribute to overall bacterial catalysis.

The specific involvement of EPS in extracellular enzyme activity in soil, as discussed above, is currently supported. Might the growth habit of cells in EPS on surfaces (Fig. 7.2c) also promote exoenzyme production? As described in the previous section on 'cell–cell communication', the accumulation of AHL in biofilm EPS and the related promotion of QS-mediated processes in bacterial biofilms are also supported. What remains to be shown definitively is that EPS, with its potentially positive role in AHL retention in unsaturated biofilms or soil colonies, promotes or is required for the production of extracellular enzymes in soil. However, for biofilm *P. aeruginosa*, AHL-mediated auto-induction increases the production of extracellular enzymes involved in depolymerization of N-rich compounds (Passador *et al.*, 1993), and thus it is conceivable that the soil bacterial growth habit of cells in EPS on surfaces (Fig. 7.2c) would be a zone of active QS and QS-mediated processes, including exoenzyme production.

4 Cells in Hydrated EPS on SOM-rich Surfaces

One might conclude, based on direct evidence and indirect explanations for when EPS is produced, that soil bacteria predominantly live in EPS as either colonies or discontinuous biofilms, with minerals encrusting the EPS exterior (Chenu and Stotzky, 2002). Each colony is a patch (Grundmann, 2004) or hot spot, and only under sufficiently wet conditions do bacteria become motile (Dechesne *et al.*, 2008b) or move by advection with infiltrating water. Where, then, are these patches located within the soil matrix? When bacteria become motile or are displaced from a patch, where do they form new colonies? As with the model of bacteria associated with EPS on surfaces, evidence for co-localization of patches with SOM comes directly from soil characterization by visual, chemical and biological means, and from the study of bacterial responses to C addition.

4.1 Characteristics and evidence for the model of cells in hydrated EPS on SOM-rich surfaces

By directly counting bacteria in soils, Gray et al. (1968) quantified that up to 64% of bacteria are associated with organic matter. Also, in EM images of soil thin sections by Foster (1988), bacterial colonies appear mostly located near SOM or faecal pellets. These observed associations are logical: in order to survive in soil, bacteria must have access to SOM in order to derive energy from such material (Chenu and Stotzky, 2002). The associations of bacteria and SOM would probably vary with varying bacterial inhabitants and SOM composition. For example, products of plant decomposition that are generated and packaged by soil fauna are colonized by bacteria to different degrees, perhaps reflecting varying ages and qualities of this particulate and (source of) dissolved SOM (Davidson and Grieve, 2006). Nevertheless, the microbial patch would consist generally of bacteria in their exopolymers, possibly encrusted in minerals and certainly near SOM. This idea is supported by recent X-ray synchrotron methods whereby, at sub-micron spatial resolution, functional groups associated with microbial biopolymers are mapped at specific locations that are near, but do not co-occur with, other forms of carbon (e.g. the rest of the bulk soil SOM as well as carbon black) (Lehmann et al., 2008).

4.2 Proximity to SOM and the role of diffusion

But how close must bacteria be to SOM in order to derive energy or carbon from it? When bacteria colonize a nutrient-rich surface, they are not initially limited by substrate and thus may resemble planktonic cells as in Fig. 7.2b (Steinberger et al., 2002). When a diffusional limitation to nutrient resupply is imposed, cells compensate morphologically to increase their ability to collect nutrients (Steinberger et al., 2002). When C is in ample supply, bacteria manufacture exopolysaccharides (Sutherland,

1977; Hilger et al., 2000). Clear and consistent visual evidence for EPS around soil bacteria (e.g. Foster, 1988) suggests that soil bacteria have sufficient C in their vicinity to manufacture exopolysaccharides. Bacterial exopolymers are produced at relatively balanced C:N ratios, but production increases significantly when, for example, the C:N ratio increases to 200 on a molar basis (Priester et al., 2007). When carbon is added through the solution phase (e.g. glucose) to soil, bacteria respond by increasing their population size while colonizing surfaces are exposed to the substrate (Chenu et al., 2001). In this regard, microbes are responding to locally available C. The spatial nature of this response is evidenced by bacteria growing less in diffusionally constrained clayey soils while responding more by substrate addition in convectively driven sandy soils (Chenu et al., 2001). Thus, where C is available, colonies grow; conversely, where colonies exist, C must have been, or must be, available. However, the C that bacteria require does not have to be in the immediate vicinity of cells. Rather, its relevance to supporting bacterial biomass depends on factors that promote its bioavailability.

Several reports support the view that soil microbial patches can be sustained by dissolved C that is supplied by diffusing to cells from distant sources. This would result in concentrated patches of microbes that are surrounded by less dense, more diffuse patches (Nunan et al., 2003). Gradients in microbial abundance and in community composition are indeed observed away from zones of actively degrading detritus in soil (Nicolardot et al., 2007). For bacteria within aggregates, diffusion of C over a long range (i.e. milli- to centimetre scales) is a means by which exogenous C may be available (Smucker et al., 2007). In wet soils, C diffuses in water films; in dry soils, the EPS portion of micro-colonies facilitates the hydraulic connectivity (Or et al., 2007a), thus maintaining paths for C diffusion from distant SOM to micro-colonies. Inherently, the relationships between diffusion and water films, and of EPS, desiccation and diffusion, suggest a strong dependence of the microbial growth habit on soil water dynamics.

Table 7.1. Depth variations of soil characteristics in a California grassland. Source: Fierer *et al.* (2003)

Characteristic	Surface (0–5 cm)	Subsurface (200 cm)
Substrate–induced respiration (SIR) (mg C-CO$_2$/g soil/h)	42	0.24
C (%)	2.8	0.11
Moisture (%)	0.6–29.7	16.7–19.3
Temperature (°C)	2–60.1	14.5–23.9
Profile C (%)	34 (0–15 cm)	66 (0.15–2.0 m)

4.3 Wetting/drying and microbial habitat dynamics

As suggested by Grundmann (2004), the distribution of microbes at the micro scale is temporally, as well as spatially, variable. Local SOM depletion adjacent to a patch may be a factor that influences bacterial dispersal from that patch (Grundmann, 2004). Therefore, a temporal progression in microbial growth habits may be considered. For example, it is conceivable that, when soil microbes are newly deposited on to surfaces via advection during water infiltration into soil pores, they would proliferate preferentially on or very near surfaces intimately associated with SOM. However, bacteria would then grow until the more labile resources within such SOM were sufficiently depleted. During growth from nearby SOM, EPS production is favoured by a C-rich environment and thus a biofilm or EPS-rich micro-colony would form. Once SOM-C is locally depleted, the microbes would enter into a starvation mode. At this point, bacterial population sizes would reduce and EPS could be a starvation reserve for cellular persistence. Polysaccharases may loosen the bonds between EPS strands that tether colonies to surfaces. Due to local depletion of some limiting nutrients, wetting and drying would be necessary to prime the cells into respiring and growing again, and increase in micro-colony patch size would occur only with an influx of a reasonable amount. This could occur with either SOM translocation during soil wetting, soil bacterial movement to SOM during soil wetting or long-range diffusion of DOC, where each would be mediated by thick water films. Thus, there is a temporal dimension to the theme motivating this section in that bacteria would be in intimate contact with SOM, but such contact could be transient as local nutrient depletion occurs and soil water fluctuations redistribute soil microbes relative to their substrates.

4.4 Does the model differ for surface versus subsurface soils?

In surface soils bacteria colonize where flow preferentially occurs (Bundt *et al.*, 2001), i.e. where assimilable C is supplied and resupplied. More than likely, particulate SOM is deposited concomitantly with flow, and then microbes colonize proximally to such depositions (Nunan *et al.*, 2003). Microbial communities in subsurface soils can phylogenetically represent a subset of surface communities (LaMontagne *et al.*, 2003), and thus bacteria in the subsurface may respond to environmental conditions similarly to surface soil bacteria. However, subsurface soil differs from surface soil in important ways that might influence the bacterial growth habit (Table 7.1).

As summarized previously (Holden and Fierer, 2004), subsurface soils harbour far fewer microbes but do contain large overall reserves of organic C. Much of this C is hundreds to thousands of years old (Trumbore *et al.*, 1995), suggesting that it has historically been unavailable to microbes through either spatial segregation from microbial patches or by its biochemical recalcitrance. In surface soils, bacteria are mostly in aggregates (Nunan *et al.*, 2003). The relatively high concentration of microbes and abundance of organic C in surface soils leads to a rather connected set of micro-compartments, including micro-aggregates, that can accumulate DOC, particulate OM and bacteria, feeding off each. This ties to the model of Smucker *et al.* (2007) that suggests a rather fluid feedback between initially C-bathed surfaces, uptake of C by bacteria and EPS production, plus diffusion of C into small pores. Since surface soil has a relatively steady stream of C input from plants, this allows for maintenance of EPS around cells on an ongoing basis (Smucker *et al.*, 2007). In the absence of plant inputs, there is not an ongoing supply of C and thus micro-sites in the subsurface can quickly become C-depleted. The carbon required for EPS production is therefore not continuously available to bacteria in the subsurface. This makes it plausible that cells do not exist particularly associated with EPS at depth. Further, the temperature and moisture extremes that promote EPS production in surface soils do not occur in subsurface soils (Table 7.1). While the concept has not been directly tested, in the absence of desiccation, subsurface soil environments could therefore deselect for exopolysaccharide production, unless other selective pressures promote it.

What has been shown is that bacteria in the subsurface are mostly on the faces of pores, and not in aggregates (Nunan *et al.*, 2003). On pore faces bacteria would be exposed and susceptible to grazing, but the ratio of grazer abundance to bacterial abundance decreases extensively with depth through the soil profile (Fierer *et al.*, 2003), making predation less likely. Thus, it would seem that most of the factors influencing EPS accumulation around bacteria in surface soils are less pronounced in the subsurface. However, in order to persist, cells inoculated to the subsurface from the surface during infiltration events must have an ongoing source of C. This would suggest, just as in surface soils (Chenu and Stotzky, 2002), that subsurface bacteria will locate proximal to assimilable SOM. Near SOM, C is available to fuel ongoing metabolism for at least cellular maintenance, but the local availability of C would also promote EPS production. Thus, the model of Fig. 7.2d

would most certainly apply to bacteria in the subsurface, albeit where microbial patches are more sparsely distributed. However, either direct evidence from either high-resolution TEMs of soil thin sections or indirect chemical evidence from laboratory studies of subsurface soil microbes (i.e. demonstrating their EPS production under subsurface-like conditions) would be required to support this assertion.

4.5 Methodological opportunities for understanding bacterial growth habits

New methods such as X-ray tomography appear quite useful for resolving the close spatial associations between soil bacteria and other colloids (Thieme *et al.*, 2003). Also, synchrotron radiation methods to resolve the nm-scale spatial distribution of various forms of C in soil (Lehmann *et al.*, 2008) appear quite exciting for providing spatially resolved chemical evidence for the model of Fig. 7.2d. The combination of micropedological specimen (e.g. polished thin sections) preparation approaches and fluorescence *in situ* hybridization (FISH) reveals the intimate associations of individual bacterial cells of specific taxa with the organics and minerals in soils (Eickhorst and Tippkötter, 2008). Penetration of molecular probes into samples is a concern in FISH (Eickhorst and Tippkötter, 2008), and EPS could influence the process. However, EPS is transparent to such methods and thus we do not learn about the cellular growth habit while simultaneously learning 'who' is associated with 'which' fractions of the soil fabric. Metals used as stains in EM methods will reveal EPS around cells (Priester *et al.*, 2007), and metal-tagged oligonuceotides can be visualized with EM (Gerard *et al.*, 2005) and X-ray microscopy (Ménez *et al.*, 2007) to probe specific bacterial taxa and simultaneously chemically characterize the mineral surfaces they colonize. However, the simultaneous visual resolution of surface chemistry, EPS and cells by taxa is undeveloped as yet for soil science. To fully explain 'who' is 'where' and perhaps 'why', an ongoing goal would be to visualize (i) cells on a phylogenetic basis (DNA-based or RNA-based probes); (ii) EPS (stained by metals); (iii) organic matter (also perhaps stained by metals); and (iv) mineral structures (electron dense), which are all achievable, most likely, by EM or other high-resolution instrumentation of samples interrogated with the appropriate probe(s). There is apparently yet another scale of organization that is of interest, i.e. the scale of cellular organization in EPS and of EPS macromolecules in a single colony. The latter is addressed, at least for a saturated biofilm, by Lawrence et al. (2003), who showed that the macromolecules in biofilms are resolvable spatially in the context of cell locations. This suggests there may be an opportunity to resolve an internal pattern of cellular organization in individual patches, if it exists.

5 Do these Micro-scale Architectures and Processes Create the Heterogeneity we Observe as Nutrient Cycling at Larger Scales, and How?

As per Schimel and Gulledge (1998), the structure and function of microbial communities are expected measurably to influence the biogeochemical reactions that they catalyse, which, in turn, affect ecosystem processes. Thus, explicitly accounting for microbiology in ecosystem process models is warranted (Schimel, 2001). Balser et al. (2006) provide several researched case studies that demonstrate how microbiology and ecosystem processes are linked; they also suggest how ecological science conducted at macro- and micro-scales could be coordinated in the future to further reveal linkages. Inherently, to conceptually and quantitatively relate microbial to ecosystem processes requires consideration of multiple spatial scales (Balser et al., 2006), i.e. from nm- and μm-sized scales around microbes and biofilms, to mm- and cm-sized scales over which diffusion occurs in pores, to m-sized scales over which profile variations in C and microbial populations

are significant (Holden and Fierer, 2005), and to m- and km-sized scales over which variations in bulk soil characteristics influence communities (Fierer and Jackson, 2006). Importantly, Fierer and Jackson (2006) showed that, even when soils across broad geographical regions are studied, bacterial community composition and diversity are most strongly correlated to a local factor, i.e. pH. Thus, not all spatial scales are necessarily important when relating soil microbiology to ecosystem function. But community composition, which also varies spatially and locally in soils (Nicolardot et al., 2007), can also affect soil microbial function, e.g. even denitrification, which is notoriously a polyphyletic function (Rich et al., 2003). Acknowledging this, Wallenstein et al. (2006) advance the idea that proximal, i.e. functional community composition, and distal, i.e. nutrient availability and environmental, factors work together to exert control over denitrification. As such, the spatial relegation of denitrification to soil microsites (Parkin, 1987) is probably due to both distal and proximal controls. Could the microbial growth habit, through its effects on diffusional re-supply of nutrients and on microbial physiology including survival, be the mediator of distal and proximal effects on soil microbial processes?

As summarized herein, bacteria are spatially and patchily distributed in soils. Variations in spatial distribution also apparently vary by microbial population function (e.g. oxidizing ammonia or degrading a pesticide (Dechesne et al., 2003)). Conceivably, variations in microbial growth habits, i.e. related to relative EPS production variations (as before), add another dimension to micro-scale variability: that of the habitat and its function. How then can processes of larger-scale significance, such as denitrification, N fixation and C mineralization, be controlled by how microbes structure their immediate environment? Two sample scenarios are discussed below: metabolism of gas phase hydrocarbons in soil and mineralization of organic C in soil, as controlled by wetting and drying.

5.1 Gas phase C mineralization and the soil microhabitat

Soil bacteria typically acquire carbon either as SOM through the aqueous phase or, in the case of autotrophs, as CO_2 through the gas phase. In very specific cases, organic C may be supplied in the gas/vapour phase as well. An example is methane (CH_4), which is used by methanotrophic bacteria in soil as the sole C and energy source (Mancinelli, 1995). While obligate methanotrophs are constrained by their specific requirement of a rather narrow C-source that is derived from methanogenesis, the re-supply of available methane-C, relative to dissolved sugars, is rapid by diffusion of gases through soil pores. This suggests that, in the presence of ample methane and relatively open soil pores, aerobic methanotrophic bacteria in soil experience C-rich conditions that favour exopolysaccharide production. In two studies, Hilger *et al.* modelled (1999) and experimentally tested (2000) their hypothesis that methanotrophs in landfill cover soil would respond to high concentrations of methane by hyper-producing EPS. In their model, the accumulation of EPS would eventually hinder the Fickian diffusion (i.e. Equation 7.1) of methane through EPS to biofilm methanotrophs. The balance of diffusion to reaction (Equation 7.2) would eventually tip such that diffusional re-supply of methane would become the limiting factor in methane oxidation. Profiles of methane, oxygen and other factors in landfill soil were reasonably well predicted assuming that methanotrophs were in EPS, attached to solid surfaces (Fig. 7.2c; Hilger *et al.*, 1999). Over time, landfill cover soil methanotrophs accumulated EPS and consumed increasingly less methane (Hilger *et al.*, 2000), a result predictable from the model of methane-derived EPS creating a mass transfer barrier to methane re-supply (Hilger *et al.*, 1999). In this way, the configuration of the local microbe-scale bacterial growth habit directly influenced the rate at which an important ecosystem-scale process, i.e. biological methane consumption, occurred.

While the demonstration of these results remains to be tested in the field, the immediate implication is that the microbial growth habit and its physico-chemical characteristics could act to constrain the rate of methane oxidation in surface soils. An even more specialized example of this type could be soil bacterial metabolism in subsurface soils of a gas phase hydrocarbon pollutant such as toluene, which is often present in deep soils surrounding leaking underground fuel storage tanks. A direct measurement of diffusion through unsaturated biofilms suggests that EPS could, similarly to the example of methane re-supply above, create a mass transfer barrier to toluene re-supply (Holden *et al.*, 1997), which could then exert control over toluene attenuation rates observed over larger scales.

5.2 Mineralization of organic C through soil drying and re-wetting

Within minutes of re-wetting dry soil, copious amounts of CO_2 evolve due to either to microbes newly mineralizing organic C or extracellular enzyme activity. Following this rapid respiration response, available substrate declines and CO_2 emissions typically decline again to near-basal levels. This is commonly referred to as the 'Birch effect' in recognition of the phenomenon's discoverer (Birch, 1958). In between soil re-wetting events, respiration continues, as the association of microbes with particulate organic matter (POM) in surface soils leads to a stable supply of DOC in the soil solution (Smucker *et al.*, 2007). The wetting rate of dry soils may dictate whether newly mobilized C is from either dead microbial biomass, intracellular solutes or intra-aggregate occlusions that are newly exposed (Park *et al.*, 2007). If drying and re-wetting cycles liberate carbon from aggregates, and if that liberated C sorbs to minerals, then drying and re-wetting could lead to C sequestration which could account for lower CO_2 emissions (Park *et al.*, 2007).

How does the microbial growth habit influence the patterns of respiration observed with variations in soil moisture? The answers to this question are mainly considered in relation to a laboratory study of re-wetting effects on soil respiration in surface and subsurface soils (Xiang *et al.*, 2008). During a period of prolonged drying in both surface and subsurface soils, water-extractable DOC increased (Fig. 7.5). Accumulation of DOC over long dry periods could occur from ongoing extracellular enzyme-mediated decomposition of POM (Miller *et al.*, 2005) and slow microbial metabolism due to desiccation stress (Harris, 1981). In order for extracellular enzymes to do this, they must be protected, and EPS surrounding bacteria is the place where that can happen (Burns, 1982). In dry soils desiccated bacteria would produce more EPS (Roberson and Firestone, 1992), which would maintain a bridge (Or *et al.*, 2007b) between SOM and EPS-encapsulated exoenzymes (Fig. 7.6). Thus, for DOC to accumulate during dry periods (Fig. 7.5), the co-localization of cells, EPS and SOM is strongly implied, both for surface and subsurface soils. Conversely, the microbial growth habit of Fig. 7.2d facilitates what is observed at larger scales as DOC accumulation.

Upon re-wetting, both in surface and in C-depleted subsurface soils, respiration rapidly increases (Fierer and Schimel, 2002; Xiang *et al.*, 2008), but moisture alone is not enough to sustain this enhanced respiration, particularly in subsurface soils (Xiang *et al.*, 2008). Rather, in more C-depleted subsurface soils, cycling through drying and re-wetting liberates CO_2 much more than it maintains soil moisture. In the end, in the subsurface, extractable organic C is the same in moist and dried–re-wetted soils. Thus, extractable OC increases over time, but more mineralization occurs with drying and re-wetting cycles. This would suggest that C depolymerization is kinetically decoupled from dissolved C mineralization (Schimel and Weintraub, 2003). Particularly in subsurface soils, the physical segregation between microbes and most soil C means that drying and re-wetting are needed to mobilize dissolved C to microbes, move microbes towards SOM or both (Xiang *et al.*, 2008). But the immediate flush of CO_2 that is observed upon soil re-wetting means that C is immediately available for mineralization,

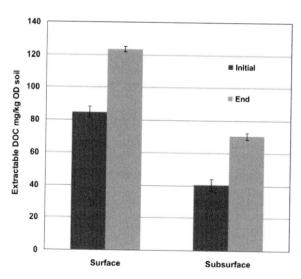

Fig. 7.5. Extractable dissolved organic carbon (DOC) accumulation in surface and subsurface soils during a prolonged drying period in the laboratory study by Xiang *et al.* (2008). Increases in DOC over time occurred despite very low CO_2 emissions, suggesting that exoenzymes were depolymerizing soil organic matter (SOM) in dry soil. Accumulations of DOC were large in both the surface and subsurface soils, suggesting that, at depth, exoenzymes were similarly spatially proximate to SOM, active and thus potentially protected from mineral stabilization by their location in microbial extracellular polymeric substances (EPS). Error bars represent the standard error of the mean of three independent experimental replicates ($n = 3$).

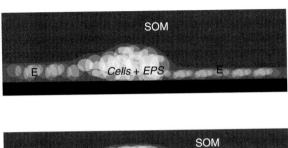

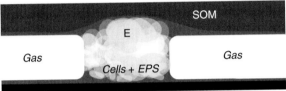

Fig. 7.6. Conceptual representation of a spatial relationship between soil organic matter (SOM) and free exoenzymes (E) where, under initially wet conditions (top), contact is enabled through water films (top left and right). When soils dry (bottom), water films thin, pores become gas filled and bacteria produce more extracellular polymeric substances (EPS) as a desiccation response. EPS provide continued hydration and contact between exoenzymes and soil organic matter, allowing for continued SOM decomposition in dry soil even when bacteria are water-stressed.

a situation that would occur if extracellular enzyme activity – as would be facilitated in microbial EPS near SOM – is ongoing during intervening dry periods (Miller *et al.*, 2005).

5.3 Microbial growth habits: response and feedbacks to climate change

Although somewhat speculative, evidence would support the view that the bacterial growth habit alterations with feedbacks to bacterial processes will partly comprise ecosystem responses to changing above-ground, climate-driven processes. With increasing atmospheric CO_2 (Keeling *et al.*, 1989), plant communities are expected to respond to changing nutrient inputs by altering structural allocations and process rates (Curtis, 1996). One manifestation is that elevated CO_2 will increase C in plant roots (Norby *et al.*, 1986; Curtis *et al.*, 1990). Higher root C is likely to be propagated into soils through root deposition (Ineson *et al.*, 1996; Hungate *et al.*, 1997). Under conditions of CO_2-C enrichment, the additional below-ground C results in increased microbial biomass and the imposition of N limitation (Hu *et al.*,

2001), both of which could lead to an increase in EPS production by soil bacteria. Overall increases in soil EPS would certainly change soil structural properties that could feed back to further influence rates of biogeochemical cycling. For example, in surface soils, bacterial EPS contributes to soil aggregation (Cheshire, 1979; Oades, 1984; Amellal *et al.*, 1998; Alami *et al.*, 2000), which benefits soil water-holding properties and improves the stability of surface soil to erosional forces (Hillel, 1980). When considering the overall, undoubtedly complex, implications and feedbacks of climate change to microbial ecology (Bardgett *et al.*, 2008), one could assert that microbial growth habits, as the transducers of soil process, will partly modulate the microbial responses to environmental factors that vary with climate change and thus the observable ecosystem-scale microbial feedbacks in a changing climate.

6 Conclusions

The growth habitat of bacteria in soils is of considerable interest. Increasingly, there is recognition that the spatial patchiness of bacteria in soil, while extensive and heterogeneous, should be devolved in detail if factors affecting soil bacterial processes are to be better described and process predictions, primarily related to C cycling, improved. However, it is not only where bacteria are concentrated that should be known, i.e. relative to soil depth or pore architecture, but also the physical construct of the bacterial local microenvironment. Do cells thrive planktonically, i.e. floating in water films in soil? It would appear not. Do they tend to be in cell-produced exopolymers (EPS)? Evidence from a wide range of studies support that soil bacteria are likely to be in EPS, even in deep soils where the average concentration of SOM in bulk samples is low relative to surface soils. Furthermore, evidence also supports co-localization, with some temporal variation, of bacterial cells in EPS-rich patches with SOM.

Beyond the physical construct of the soil bacterial microenvironment and the logic supporting it, bacterial actions take place in that construct and thus understanding of its composition, size and physicochemical characteristics that affect local diffusion and reaction is extremely important. Experimental evidence exists to support the view that soil bacteria respond to their local substrate supply and increase EPS around cells when conditions favour that. This, in turn, appears to alter substrate supply, for example by excluding C-rich vapours from EPS-clogged pore spaces or by physically maintaining links between exoenzyme-rich EPS and SOM spatially segregated from cells. Thus, EPS and proximity to SOM would appear to define how fast, and to what extent, bacteria carry out their processes. Feedbacks are paramount: substrates and environmental factors influence bacteria through EPS, but bacteria respond by altering EPS, which, in turn, feed back to alter external substrate availability and environmental influences. There are challenges remaining towards quantifying these feedbacks so that they can be accounted for in C- and other nutrient-cycling models, including (i) better discriminating between plant- and bacterial-derived EPS; (ii) visualizing EPS simultaneously with phylogenetically and activity-resolved cells; (iii) measuring exoenzyme activity *in situ* in soil bacterial EPS; (iv) directly observing bacterial responses to EPS-derived protective influences in soils; and (v) observing in as near to real time as possible the progression of bacterial establishment on SOM-rich surfaces through EPS accumulation and local SOM depletion with variations in water availability. Technological advances in microscopy, spectrometry, molecular probing and model experimental systems for simultaneous use of high-resolution tools applied towards these issues will enable fundamental advances in the understanding of bacterial micro-scale growth habits and their effects on soil bacterial processes. Mathematical modelling as a framework for designing experiments and interpreting the balance of diffusion and reaction in observed process rates at the micro scale is a mainstay of this science, and a factor in scaling-up from the micro scale to, perhaps, the scale of ecosystems.

Acknowledgements

Preparation of this work was supported by US EPA STAR Awards # R831712 and R833323; the University of California Toxic Substances Research and Training Program: Lead Campus program in Nanotoxicology; the Office of Science (BER); US Department of Energy, Grant No. DE-FG02-05ER63949; the Kearney Foundation; the National Science Foundation (NSF: #997772, 9977874 and 0444712); and the National Science Foundation and the Environmental Protection Agency under Cooperative Agreement Number EF 0830117. Any opinions, findings and conclusions or recommendations expressed in this material are those of the author and do not necessarily reflect the views of the National Science Foundation or the Environmental Protection Agency. This work has not been subjected to EPA review and no official endorsement should be inferred. Mary Firestone and Josh Schimel are acknowledged for helpful discussions. John Priester assisted with reviewing this work and in preparing the specimen in Fig. 7.4b. Shurong Xiang produced data for Fig. 7.5 (reproduced from Xiang *et al.*, 2008). Jose Saleta performed the ESEM imaging (UCSB, http://www.bren.ucsb.edu/facilities/MEIAF/).

References

Alami, Y., Achouak, W., Marol, C. and Heulin, T. (2000) Rhizosphere soil aggregation and plant growth promotion of sunflowers by an exopolysaccharide-producing *Rhizobium* sp. strain isolated from sunflower roots. *Applied and Environmental Microbiology* 66, 3393–3398.

Allison, S.D. (2006) Soil minerals and humic acids alter enzyme stability: implications for ecosystem processes. *Biogeochemistry* 81, 361–373.

Alvarez, H.M., Silva, R.A., Cesari, A.C., Zamit, A.L., Peressutti, S.R., Reichelt, R. *et al.* (2004) Physiological and morphological responses of the soil bacterium *Rhodococcus opacus* strain PD630 to water stress. *FEMS Microbiology Ecology* 50, 75–86.

Amellal, N., Burtin, G., Bartoli, F. and Heulin, T. (1998) Colonization of wheat roots by an exopolysaccharide-producing *Pantoea agglomerans* strain and its effect on rhizosphere soil aggregation. *Applied and Environmental Microbiology* 64, 3740–3747.

Arora, D.K. and Gupta, S. (1993) Effect of different environmental conditions on bacterial chemotaxis toward fungal spores. *Canadian Journal of Microbiology* 39, 922–931.

Auerbach, I.D., Sorensen, C., Hansma, H.G. and Holden, P.A. (2000) Physical morphology and surface properties of unsaturated *Pseudomonas putida* biofilms. *Journal of Bacteriology* 182, 3809–3815.

Auset, M., Keller, A.A., Brissaud, F. and Lazarova, V. (2005) Intermittent filtration of bacteria and colloids in porous media. *Water Resources Research* 41.

Bailey, J.E. and Ollis, D.F. (1986) *Biochemical Engineering Fundamentals.* McGraw-Hill, New York.

Balser, T.C., McMahon, K.D., Bart, D., Bronson, D., Coyle, D.R., Craig, N. *et al.* (2006) Bridging the gap between micro- and macro-scale perspectives on the role of microbial communities in global change ecology. *Plant and Soil* 289, 59–70.

Bardgett, R.D., Freeman, C. and Ostle, N.J. (2008) Microbial contributions to climate change through carbon cycle feedbacks. *ISME Journal* 2, 805–814.

Bayer, A.S., Eftekhar, F., Tu, J. and Nast, C.C. (1990) Oxygen-dependent up-regulation of mucoid exopolysaccharide (alginate) production in *Pseudomonas aeruginosa*. *Infection and Immunity* 58, 1344–1349.

Becker, J.M., Parkin, T., Nakatsu, C.H., Wilbur, J.D. and Konopka, A. (2006) Bacterial activity, community structure, and centimeter-scale spatial heterogeneity in contaminated soil. *Microbial Ecology* 51, 220–231.

Birch, H.F. (1958) The effect of soil drying on humus decomposition and nitrogen availability. *Plant and Soil* 10, 9–31.

Bitton, G. and Freihofer, V. (1978) Influence of extracellular polysaccharides on toxicity of copper and cadmium toward *Klebsiella aerogenes*. *Microbial Ecology* 4, 119–125.

Bockelmann, U., Szewzyk, U. and Grohmann, E. (2003) A new enzymatic method for the detachment of particle associated soil bacteria. *Journal of Microbiological Methods* 55, 201–211.

Boles, B.R. and Singh, P.K. (2008) Endogenous oxidative stress produces diversity and adaptability in biofilm communities. *Proceedings of the National Academy of Sciences of the United States of America* 105, 12503–12508.

Boyd, A. and Chakrabarty, A.M. (1994) Role of alginate lyase in cell detachment of *Pseudomonas aeruginosa*. *Applied and Environmental Microbiology* 60, 2355–2359.

Branda, S.S., Chu, F., Kearns, D.B., Losick, R. and Kolter, R. (2006) A major protein component of the *Bacillus subtilis* biofilm matrix. *Molecular Microbiology* 59, 1229–1238.

Bundt, M., Widmer, F., Pesaro, M., Zeyer, J. and Blaser, P. (2001) Preferential flow paths: Biological 'hot spots' in soils. *Soil Biology and Biochemistry* 33, 729–738.

Burmolle, M., Hansen, L.H., Oregaard, G. and Sorensen, S.J. (2003) Presence of N-acyl homoserine lactones in soil detected by a whole-cell biosensor and flow cytometry. *Microbial Ecology* 45, 226–236.

Burns, R.G. (1982) Enzyme activity in soil: location and a possible role in microbial ecology. *Soil Biology and Biochemistry* 14, 423–427.

Burns, R.G., Gregory, L.J., Lethbridge, G. and Pettit, N.M. (1978) Effect of gamma-irradiation on soil enzyme stability. *Experientia* 34, 301–302.

Bylund, J., Burgess, L.A., Cescutti, P., Ernst, R.K. and Speert, D.P. (2006) Exopolysaccharides from *Burkholderia cenocepacia* inhibit neutrophil chemotaxis and scavenge reactive oxygen species. *Journal of Biological Chemistry* 281, 2526–2532.

Caballero, A.R., Moreau, J.M., Engel, L.S., Marquart, M.E., Hill, J.M. and O'Callaghan, R.J. (2001) *Pseudomonas aeruginosa* protease IV enzyme assays and comparison to other *Pseudomonas* proteases. *Analytical Biochemistry* 290, 330–337.

Chang, W.S., van de Mortel, M., Nielsen, L., de Guzman, G.N., Li, X.H. and Halverson, L.J. (2007) Alginate production by *Pseudomonas putida* creates a hydrated microenvironment and contributes to biofilm architecture and stress tolerance under water-limiting conditions. *Journal of Bacteriology* 189, 8290–8299.

Characklis, W.G. and Marshall, K.C. (eds) (1990) *Biofilms. Ecological and Applied Microbiology*. John Wiley and Sons, Inc., New York.

Chenu, C. (1993) Clay- or sand-polysaccharide associations as models for the interface between microorganisms and soil: water related properties and microstructure. *Geoderma* 56, 143–156.

Chenu, C. (1995) Extracellular polysaccharides: an interface between microorganisms and soil constituents. In: Huang, P.M, Berthelin, J., Bollag, J.-M., McGill, W.B. and Page, A.L. (eds) *Environmental Impact of Soil Component Interactions*. CRC Press, Inc., Boca Raton, Florida, pp. 217–233.

Chenu, C. and Jaunet, A.M. (1992) Cryoscanning electron microscopy of microbial extracellular polysaccharides and their association with minerals. *Scanning* 14, 360–364.

Chenu, C. and Roberson, E.B. (1996) Diffusion of glucose in microbial extracellular polysaccharide as affected by water potential. *Soil Biology and Biochemistry* 28, 877–884.

Chenu, C. and Stotzky, G. (2002) Interactions between microorganisms and soil particles: an overview. In: Huang, P.M., Bollag, J.M. and Senesi, N. (eds) *Interactions Between Soil Particles and Soil Microorganisms*. John Wiley and Sons, Chichester, UK, pp. 4–40.

Chenu, C. and Tessier, D. (1995) Low temperature scanning electron microscopy of clay and organic constituents and their relevance to soil microstructures. *Scanning Microscopy* 9, 989–1010.

Chenu, C., Hassink, J. and Bloem, J. (2001) Short-term changes in the spatial distribution of microorganisms in soil aggregates as affected by glucose addition. *Biology and Fertility of Soils* 34, 349–356.

Cheshire, M.V. (1979) *Nature and Origin of Carbohydrates in Soils*. Academic Press, London, New York.

Cole, A.C., Semmens, M.J. and LaPara, T.M. (2004) Stratification of activity and bacterial community structure in biofilms grown on membranes transferring oxygen. *Applied and Environmental Microbiology* 70, 1982–1989.

Confer, D.R. and Logan, B.E. (1998) Location of protein and polysaccharide hydrolytic activity in suspended and biofilm wastewater cultures. *Water Research* 32, 31–38.

Costerton, J.W. (1995) Overview of microbial biofilms. *Journal of Industrial Microbiolology* 15, 137–140.

Costerton, J.W. and Irvin, R.T. (1981) The bacterial glycocalyx in nature and disease. *Annual Review of Microbiology* 35, 299–324.

Costerton, J.W., Lewandowski, Z., De Beer, D., Caldwell, D., Korber, D. and James, G. (1994) Biofilms, the customized microniche. *Journal of Bacteriology* 176, 2137–2142.

Costerton, J.W., Lewandowski, Z., Caldwell, D.E., Korber, D.R. and Lappinscott, H.M. (1995) Microbial biofilms. *Annual Review of Microbiology* 49, 711–745.

Crank, J. (1999) *The Mathematics of Diffusion*. Clarendon Press, Oxford, UK.

Curtis, P.S. (1996) A meta-analysis of leaf gas exchange and nitrogen in trees grown under elevated carbon dioxide. *Plant Cell and Environment* 19, 127–137.

Curtis, P.S., Baldmann, L.M., Drake, B.G. and Whigman, D.F. (1990) Elevated atmospheric CO_2 effects on belowground processes in C_3 and C_4 estuarine communities. *Ecology* 71, 2001–2006.

Cytryn, E.J., Sangurdekar, D.P., Streeter, J.G., Franck, W.L., Chang, W.S., Stacey, G. *et al.* (2007) Transcriptional and physiological responses of *Bradyrhizobium japonicum* to desiccation-induced stress. *Journal of Bacteriology* 189, 6751–6762.

Davidson, D.A. and Grieve, I.C. (2006) Relationships between biodiversity and soil structure and function: Evidence from laboratory and field experiments. *Applied Soil Ecology* 33, 176–185.

DeAngelis, K.M., Lindow, S.E. and Firestone, M.K. (2008) Bacterial quorum sensing and nitrogen cycling in rhizosphere soil. *FEMS Microbiology Ecology* 66, 197–207.

Dechesne, A., Pallud, C., Debouzie, D., Flandrois, J.P., Vogel, T.M., Gaudet, J.P. *et al.* (2003) A novel method for characterizing the micro scale 3D spatial distribution of bacteria in soil. *Soil Biology and Biochemistry* 35, 1537–1546.

Dechesne, A., Pallud, C., Bertolla, F. and Grundmann, G.L. (2005) Impact of the micro scale distribution of a *Pseudomonas* strain introduced into soil on potential contacts with indigenous bacteria. *Applied and Environmental Microbiology* 71, 8123–8131.

Dechesne, A., Or, D., Gulez, G. and Smets, B.F. (2008a) The porous surface model, a novel experimental system for online quantitative observation of microbial processes under unsaturated conditions. *Applied and Environmental Microbiology* 74, 5195–5200.

Dechesne, A., Or, D. and Smets, B.F. (2008b) Limited diffusive fluxes of substrate facilitate coexistence of two competing bacterial strains. *FEMS Microbiology Ecology* 64, 1–8.

De Kerchove, A.J. and Elimelech, M. (2007) Formation of polysaccharide gel layers in the presence of Ca^{2+} and K^+ ions: measurements and mechanisms. *Biomacromolecules* 8, 113–121.

De Kievit, T.R., Gillis, R., Marx, S., Brown, C. and Iglewski, B.H. (2001) Quorum-sensing genes in *Pseudomonas aeruginosa* biofilms: their role and expression patterns. *Applied and Environmental Microbiology* 67, 1865–1873.

Djigal, D., Sy, M., Brauman, A., Diop, T.A., Mountport, D., Chotte, J.L. *et al.* (2004) Interactions between *Zeldia punctata* (Cephalobidae) and bacteria in the presence or absence of maize plants. *Plant and Soil* 262, 33–44.

Dudman, W.F. (1977) The role of surface polysaccharides in natural environments. In: Sutherland, I.W. (ed.) *Surface Carbohydrates of the Prokaryotic Cell*. Academic Press, London, pp. 357–414.

Dunbar, J., Takala, S., Barns, S.M., Davis, J.A. and Kuske, C.R. (1999) Levels of bacterial community diversity in four arid soils compared by cultivation and 16S rRNA gene cloning. *Applied and Environmental Microbiology* 65, 1662–1669.

Eickhorst, T. and Tippkötter, R. (2008) Detection of microorganisms in undisturbed soil by combining fluorescence in situ hybridization (FISH) and micropedological methods. *Soil Biology and Biochemistry* 40, 1284–1293.

Fair, R.J., Jamieson, H.M. and Hopkins, D.W. (1994) Spatial-distribution of nitrifying (ammonium-oxidizing) bacteria in soil. *Letters in Applied Microbiology* 18, 162–164.

Ferrari, B.C., Binnerup, S.J. and Gillings, M. (2005) Micro-colony cultivation on a soil substrate membrane system selects for previously uncultured soil bacteria. *Applied and Environmental Microbiology* 71, 8714–8720.

Fierer, N. and Jackson, R.B. (2006) The diversity and biogeography of soil bacterial communities. *Proceedings of the National Academy of Sciences of the United States of America* 103, 626–631.

Fierer, N. and Schimel, J.P. (2002) Effects of drying–re-wetting frequency on soil carbon and nitrogen transformations. *Soil Biology and Biochemistry* 34, 777–787.

Fierer, N., Schimel, J.P. and Holden, P.A. (2003) Variations in microbial community composition through two soil depth profiles. *Soil Biology and Biochemistry* 35, 167–176.

Focht, D.D. (1992) Diffusional constraints on microbial processes in soil. *Soil Science* 154, 300–307.

Foster, R.C. (1981) Polysaccharides in soil fabrics. *Science* 214, 665–667.

Foster, R.C. (1988) Microenvironments of soil microorganisms. *Biology and Fertility of Soils* 6, 189–203.

Foster, R.C., Rovira, A.D. and Cock, T.W. (1983) *Ultrastructure of the Root–Soil Interface*. American Phytopathological Society, St Paul, Minnesota.

Frey, S.D. (2007) Spatial distribution of soil organisms. In: Paul, E.A. (ed.) *Soil Microbiology, Ecology, and Biochemistry*. Academic Press, Oxford, UK, pp. 283–300.

Gammack, S.M., Paterson, E., Kemp, J.S., Cresser, M.S. and Killham, K. (1992) Factors affecting the movement of microorganisms in soils. In: Stotzky, G. and Bollag, J.-M. (eds) *Soil Biochemistry*. Marcel Dekker, Inc., New York, pp. 263–305.

Gerard, E., Guyot, F., Philippot, P. and Lopez-Garcia, P. (2005) Fluorescence in situ hybridisation coupled to ultra small immunogold detection to identify prokaryotic cells using transmission and scanning electron microscopy. *Journal of Microbiological Methods* 63, 20–28.

Gianfreda, L., Rao, M.A., Sannino, F., Saccomandi, F. and Violante, A. (2002) Enzymes in soil: properties, behaviour and potential applications. In: Violante, A., Huang, P.M., Bollag, J.M. and Gianfreda, L. (eds) *Soil Mineral–Organic Matter–Microorganism Interactions and Ecosystem Health*. Elsevier, Amsterdam, pp. 301–327.

Glaser, B., Balashov, E., Haumaier, L., Guggenberger, G. and Zech, W. (2000) Black carbon in density fractions of anthropogenic soils of the Brazilian Amazon region. *Organic Geochemistry* 31, 669–678.

Gonod, L.V., Chenu, C. and Soulas, G. (2003) Spatial variability of 2,4-dichlorophenoxyacetic acid (2,4-D) mineralisation potential at a millimetre scale in soil. *Soil Biology and Biochemistry* 35, 373–382.

Gonod, L.V., Chadoeuf, J. and Chenu, C. (2006) Spatial distribution of microbial 2,4-dichlorophenoxy acetic acid mineralization from field to microhabitat scales. *Soil Science Society of America Journal* 70, 64–71.

Grammatika, M. and Zimmerman, W.B. (2001) Microhydrodynamics of flotation processes in the sea surface layer. *Dynamics of Atmospheres and Oceans* 34, 327–348.

Gray, T.R.G., Baxby, P., Hill, I.R. and Goodfellow, M. (1968) Direct observation of bacteria in soil. In: Gray, T.R.G. and Parkinson, D. (eds) *The Ecology of Soil Bacteria*. University of Toronto Press, Toronto, Canada, pp. 171–192.

Grundmann, G.L. (2004) Spatial scales of soil bacterial diversity – the size of a clone. *FEMS Microbiology Ecology* 48, 119–127.

Grundmann, G.L. and Debouzie, D. (2000) Geostatistical analysis of the distribution of NH_4^+ and NO_2^--oxidizing bacteria and serotypes at the millimeter scale along a soil transect. *FEMS Microbiology Ecology* 34, 57–62.

Grundmann, G.L., Dechesne, A., Bartoli, F., Flandrois, J.P., Chasse, J.L. and Kizungu, R. (2001) Spatial modeling of nitrifier microhabitats in soil. *Soil Science Society of America Journal* 65, 1709–1716.

Harris, R.F. (1981) Effect of water potential on microbial growth and activity. In: Parr, J.F., Gardner, W.R. and Elliott, L.F. (eds) *Water Potential Relations in Soil Microbiology, SSSA Special Publication Number 9*. Soil Science Society of America, Madison, Wisconsin, pp. 23–95.

Hassink, J., Bouwman, L.A., Zwart, K.B. and Brussaard, L. (1993) Relationships between habitable pore-space, soil biota and mineralization rates in grassland soils. *Soil Biology and Biochemistry* 25, 47–55.

Hense, B.A., Kuttler, C., Mueller, J., Rothballer, M., Hartmann, A. and Kreft, J.U. (2007) Opinion – Does efficiency sensing unify diffusion and quorum sensing? *Nature Reviews Microbiology* 5, 230–239.

Hentzer, M., Teitzel, G.M., Balzer, G.J., Heydorn, A., Molin, S., Givskov, M. *et al.* (2001) Alginate overproduction affects *Pseudomonas aeruginosa* biofilm structure and function. *Journal of Bacteriology* 183, 5395–5401.

Hilger, H.A., Liehr, S.K. and Barlaz, M.A. (1999) Exopolysaccharide control of methane oxidation in landfill cover soil. *Journal of Environmental Engineering-ASCE* 125, 1113–1123.

Hilger, H.A., Cranford, D.F. and Barlaz, M.A. (2000) Methane oxidation and microbial exopolymer production in landfill cover soil. *Soil Biology and Biochemistry* 32, 457–467.

Hillel, D. (1980) *Fundamentals of Soil Physics*. Academic Press, New York.

Holden, P.A. (2001) Biofilms in unsaturated environments. In: Doyle, R.J. (ed.) *Methods of Enzymology*. Academic Press, San Diego, California, pp. 125–143.

Holden, P.A. and Fierer, N. (2004) Vadose zone: microbial ecology. In: Hillel, D. (ed.) *Encyclopedia of Soils in the Environment*. Academic Press, London.

Holden, P.A. and Fierer, N. (2005) Microbial processes in the vadose zone. *Vadose Zone Journal* 4, 1–21.

Holden, P.A. and Pierce, D. (2001) New environmental scanning electron microscopic (ESEM) observations of bacteria on simulated soil substrates. *Microscopy and Microanalysis* 7, 736–737.

Holden, P.A., Hunt, J.R. and Firestone, M.K. (1995) Unsaturated zone gas-phase VOC biodegradation: the importance of water potential. In: *50th Purdue Industrial Waste Conference*. Lewis Publishers, CRC Press, Inc., West Lafayette, Indiana.

Holden, P.A., Hunt, J.R. and Firestone, M.K. (1997) Toluene diffusion and reaction in unsaturated *Pseudomonas putida* mt-2 biofilms. *Biotechnology and Bioengineering* 56, 656–670.

Holden, P.A., Hersman, L.E. and Firestone, M.K. (2001) Water content mediated microaerophilic toluene biodegradation in arid vadose zone materials. *Microbial Ecology* 42, 256–266.

Holden, P.A., LaMontagne, M.G., Bruce, A.K., Miller, W.G. and Lindow, S.E. (2002) Assessing the role of *Pseudomonas aeruginosa* surface-active gene expression in hexadecane biodegradation in sand. *Applied and Environmental Microbiology* 68, 2509–2518.

Hu, S., Chapin, F.S., Firestone, M.K., Field, C.B. and Chiariello, N.R. (2001) Nitrogen limitation of microbial decomposition in a grassland under elevated CO_2. *Nature (London)* 409, 188–191.

Hungate, B.A., Holland, R.B., Jackson, F.S., Chapin, F.S.I., Mooney, H.A. and Field, C.B. (1997) The fate of carbon in grasslands under carbon dioxide enrichment. *Nature* 388, 576–579.

Ineson, P., Cotrufo, M.F., Bol, R., Harkness, D.D. and Blum, H. (1996) Quantification of soil carbon inputs under elevated CO_2: C_3 plants in a C_4 soil. *Plant and Soil* 187, 345–350.

Israelachvili, J.N. (1992) *Intermolecular and Surface Forces*. Academic Press, San Diego, California.

Jesus, E.D., Marsh, T.L., Tiedje, J.M. and Moreira, F.M.D. (2009) Changes in land use alter the structure of bacterial communities in Western Amazon soils. *ISME Journal* 3, 1004–1011.

Joubert, L.M., Wolfaardt, G.M. and Botha, A. (2006) Microbial exopolymers link predator and prey in a model yeast biofilm system. *Microbial Ecology* 52, 187–197.

Jury, W.A., Gardner, W.R. and Gardner, W.H. (1991) *Soil Physics*. John Wiley and Sons, New York.

Kachlany, S.C., Levery, S.B., Kim, J.S., Reuhs, B.L., Lion, L.W. and Ghiorse, W.C. (2001) Structure and carbohydrate analysis of the exopolysaccharide capsule of *Pseudomonas putida* G7. *Environmental Microbiology* 3, 774–784.

Kaplan, J.B., Velliyagounder, K., Ragunath, C., Rohde, H., Mack, D., Knobloch, J.K.M. *et al.* (2004) Genes involved in the synthesis and degradation of matrix polysaccharide in *Actinobacillus actinomycetemcomitans* and *Actinobacillus pleuropneumoniae* biofilms. *Journal of Bacteriology* 186, 8213–8220.

Keeling, C.D., Bacastow, R.B., Carter, A.F., Piper, S.C., Whorf, T.P., Heimann, M. *et al.* (1989) A three-dimensional model of atmospheric CO_2 transport based on observed winds: 1. Analysis of observational data. In: Peterson, D.H. (ed.) *Aspects of Climate Variability in the Pacific and Western Americas*. American Geophysical Union, Washington, DC, pp. 165–236.

Kemmling, A., Kamper, M., Flies, C., Schieweck, O. and Hoppert, M. (2004) Biofilms and extracellular matrices on geomaterials. *Environmental Geology* 46, 429–435.

Kenzaka, T., Ishidoshiro, A., Yamaguchi, N., Tani, K. and Nasu, M. (2005) rRNA sequence-based scanning electron microscopic detection of bacteria. *Applied and Environmental Microbiology* 71, 5523–5531.

Kieft, T.L., Soroker, E. and Firestone, M.K. (1987) Microbial biomass response to a rapid increase in water potential when dry soil is wetted. *Soil Biology and Biochemistry* 19, 119–126.

Killham, K., Amato, M. and Ladd, J.N. (1993) Effect of substrate location in soil and soil pore-water regime on carbon turnover. *Soil Biology and Biochemistry* 25, 57–62.

Kiraly, Z., ElZahaby, H.M. and Klement, Z. (1997) Role of extracellular polysaccharide (EPS) slime of plant pathogenic bacteria in protecting cells to reactive oxygen species. *Journal of Phytopathology-Phytopathologische Zeitschrift* 145, 59–68.

Lamelas, C., Benedetti, M., Wilkinson, K.J. and Slaveykova, V.I. (2006) Characterization of H^+ and Cd^{2+} binding properties of the bacterial exopolysaccharides. *Chemosphere* 65, 1362–1370.

LaMontagne, M.G., Schimel, J.P. and Holden, P.A. (2003) Comparison of subsurface and surface soil bacterial communities in California grassland as assessed by terminal restriction fragment length polymorphisms of PCR-amplified 16S rRNA genes. *Microbial Ecology* 46, 216–227.

Lawrence, J.R., Swerhone, G.D.W., Leppard, G.G., Araki, T., Zhang, X., West, M.M. and Hitchcock, A.P. (2003) Scanning transmission X-ray, laser scanning, and transmission electron microscopy mapping of the exopolymeric matrix of microbial biofilms. *Applied and Environmental Microbiology* 69, 5543–5554.

Leadbetter, J.R. and Greenberg, E.P. (2000) Metabolism of acyl-homoserine lactone quorum-sensing signals by *Variovorax paradoxus*. *Journal of Bacteriology* 182, 6921–6926.

Lehmann, J., Solomon, D., Kinyangi, J., Dathe, L., Wirick, S. and Jacobsen, C. (2008) Spatial complexity of soil organic matter forms at nanometre scales. *Nature Geoscience* 1, 238–242.

Lindahl, V. and Bakken, L.R. (1995) Evaluation of methods for extraction of bacteria from soil. *FEMS Microbiology Ecology* 16, 135–142.

Long, T. and Or, D. (2005) Aquatic habitats and diffusion constraints affecting microbial coexistence in unsaturated porous media. *Water Resources Research* 41.

Lunsdorf, H., Erb, R.W., Abraham, W.R. and Timmis, K.N. (2000) 'Clay hutches': a novel interaction between bacteria and clay minerals. *Environmental Microbiology* 2, 161–168.

Mancinelli, R.L. (1995) The regulation of methane oxidation in soil. *Annual Review of Microbiology* 49, 581–605.

Marshall, K.C. (1999) Theoretical and practical significance of bacteria at interfaces. *Journal of Industrial Microbiology and Biotechnology* 22, 400–406.

Martens, D.A. and Frankenberger, W.T. (1991a) Determination of saccharides in biological materials by high-performance anion-exchange chromatography with pulsed amperometric detection. *Journal of Chromatography* 546, 297–309.

Martens, D.A. and Frankenberger, W.T. (1991b) Saccharide composition of extracellular polymers produced by soil microorganisms. *Soil Biology and Biochemistry* 23, 731–736.

Martin, J.P. (1946) Microorganisms and soil aggregation. 2. Influence of bacterial polysaccharides on soil structure. *Soil Science* 61, 157–166.

Marx, M.C., Kandeler, E., Wood, M., Wermbter, N. and Jarvis, S.C. (2005) Exploring the enzymatic landscape: distribution and kinetics of hydrolytic enzymes in soil particle-size fractions. *Soil Biology and Biochemistry* 37, 35–48.

Ménez, B., Rommevaux-Jestin, C., Salomé, M., Wang, Y., Philippot, P., Bonneville, A. *et al.* (2007) Detection and phylogenetic identification of labeled prokaryotic cells on mineral surfaces using scanning X-ray microscopy. *Chemical Geology* 240, 182–192.

Miller, A.E., Schimel, J.P., Meixner, T., Sickman, J.O. and Melack, J.M. (2005) Episodic re-wetting enhances carbon and nitrogen release from chaparral soils. *Soil Biology and Biochemistry* 37, 2195–2204.

Monier, J. and Lindow, S. (2003) Differential survival of solitary and aggregated bacterial cells promotes aggregate formation on leaf surfaces. *Proceedings of the National Academy of Sciences* 100, 15977–15982.

Morris, C.E., Barnes, M.B. and McLean, R.J.C. (2002) Biofilms on leaf surfaces: implications for the biology, ecology and management of populations of epiphytic bacteria. In: Lindow, S.E., Hecht-Poinar, E.I. and Elliot, V.J. (eds) *Phyllosphere Microbiology*. SAPS Press, St Paul, Minnesota, pp. 138–154.

Mummey, D., Holben, W., Six, J. and Stahl, P. (2006) Spatial stratification of soil bacterial populations in aggregates of diverse soils. *Microbial Ecology* 51, 404–411.

Myrold, D.D. and Tiedje, J.M. (1985) Diffusional constraints on denitrification in soil. *Soil Science Society of America Journal* 49, 651–657.

Neal, A.L., Dublin, S.N., Taylor, J., Bates, D.J., Burns, L., Apkarian, R. *et al.* (2007) Terminal electron acceptors influence the quantity and chemical composition of capsular exopolymers produced by anaerobically growing *Shewanella* spp. *Biomacromolecules* 8, 166–174.

Nicol, G.W., Glover, L.A. and Prosser, J.I. (2003) Spatial analysis of archaeal community structure in grassland soil. *Applied and Environmental Microbiology* 69, 7420–7429.

Nicolardot, B., Bouziri, L., Bastian, F. and Ranjard, L. (2007) A microcosm experiment to evaluate the influence of location and quality of plant residues on residue decomposition and genetic structure of soil microbial communities. *Soil Biology and Biochemistry* 39, 1631–1644.

Norby, R.J., O'Neill, E.G. and Luxmoore, R.J. (1986) Effects of CO_2 enrichment on the growth and mineral nutrition of *Quercus alba* seedlings in nutrient-poor soil. *Plant Physiology* 82, 83–89.

Nunan, N., Wu, K., Young, I.M., Crawford, J.W. and Ritz, K. (2002) In situ spatial patterns of soil bacterial populations, mapped at multiple scales, in an arable soil. *Microbial Ecology* 44, 296–305.

Nunan, N., Wu, K.J., Young, I.M., Crawford, J.W. and Ritz, K. (2003) Spatial distribution of bacterial communities and their relationships with the micro-architecture of soil. *FEMS Microbiology Ecology* 44, 203–215.

Oades, J.M. (1984) Soil organic matter and structural stability: mechanisms and implications for management. *Plant and Soil* 76, 319–337.

O'Donnell, A.G., Young, I.M., Rushton, S.P., Shirley, M.D. and Crawford, J.W. (2007) Visualization, modelling and prediction in soil microbiology. *Nature Reviews Microbiology* 5, 689–699.

Oldak, E. and Trafny, E.A. (2005) Secretion of proteases by *Pseudomonas aeruginosa* biofilms exposed to ciprofloxacin. *Antimicrobial Agents and Chemotherapy* 49, 3281–3288.

Ophir, T. and Gutnick, D.L. (1994) A role for exopolysaccharides in the protection of microorganisms from desiccation. *Applied and Environmental Microbiology* 60, 740–745.

Or, D., Phutane, S. and Dechesne, A. (2007a) Extracellular polymeric substances affecting pore-scale hydrologic conditions for bacterial activity in unsaturated soils. *Vadose Zone Journal* 6, 298–305.

Or, D., Smets, B.F., Wraith, J.M., Dechesne, A. and Friedman, S.P. (2007b) Physical constraints affecting bacterial habitats and activity in unsaturated porous media – a review. *Advances in Water Resources* 30, 1505–1527.

Pallud, C., Dechesne, A., Gaudet, J.P., Debouzie, D. and Grundmann, G.L. (2004) Modification of spatial distribution of 2,4-dichloro-phenoxyacetic acid degrader microhabitats during growth in soil columns. *Applied and Environmental Microbiology* 70, 2709–2716.

Papendick, R.I. and Campbell, G.S. (1981) Theory and measurement of water potential. In: Parr, J.F., Gardner, W.R. and Elliot, L.F. (eds) *Water Potential Relations in Soil Microbiology, SSSA Special Publication Number 9*. Soil Science Society of America, Madison, Wisconsin, pp. 1–22.

Park, E.-J., Sul, W.J. and Smucker, A.J.M. (2007) Glucose additions to aggregates subjected to drying/wetting cycles promote carbon sequestration and aggregate stability. *Soil Biology and Biochemistry* 39, 2758–2768.

Parkin, T.B. (1987) Soil microsites as a source of denitrification variability. *Soil Science Society of America Journal* 51, 1194–1195.

Passador, L., Cook, J.M., Gambello, M.J., Rust, L. and Iglewski, B.H. (1993) Expression of *Pseudomonas aeruginosa* virulence genes requires cell-to-cell communication. *Science* 260, 1127–1130.

Paul, E. and Clark, F. (1989) *Soil Microbiology and Biochemistry*. Academic Press, San Diego, California.

Pickett, S.T.A. and Cadenasso, M.L. (1995) Landscape ecology – spatial heterogeneity in ecological-systems. *Science* 269, 331–334.

Pinchuk, G.E., Ammons, C., Culley, D.E., Li, S.M.W., McLean, J.S., Romine, M.F. *et al.* (2008) Utilization of DNA as a sole source of phosphorus, carbon, and energy by *Shewanella* spp.: Ecological and physiological implications for dissimilatory metal reduction. *Applied and Environmental Microbiology* 74, 1198–1208.

Plusquellec, A., Beucher, M., Lelay, C., Legal, Y. and Cleret, J.J. (1991) Quantitative and qualitative bacteriology of the marine water-surface microlayer in a sewage-polluted area. *Marine Environmental Research* 31, 227–239.

Pollock, M.R. (1962) Exoenzymes. In: Gunsalus, I.C. and Stanier, R.Y. (eds) *The Bacteria*. Academic Press, New York.

Poynter, S. and Mead, G. (1964). *Symposium on Fresh Water Bacteria in Relation to Industrial Problems: Paper IV*, London.

Priester, J.H., Olson, S.G., Webb, S.M., Neu, M.P., Hersman, L.E. and Holden, P.A. (2006) Enhanced exopolymer production and chromium stabilization in *Pseudomonas putida* unsaturated biofilms. *Applied and Environmental Microbiology* 72, 1988–1996.

Priester, J.H., Horst, A.M., Van De Werfhorst, L.C., Saleta, J.L., Mertes, L.A.K. and Holden, P.A. (2007) Enhanced visualization of microbial biofilms by staining and environmental scanning electron microscopy. *Journal of Microbiological Methods* 68, 577–587.

Purevdorj, B., Costerton, J.W. and Stoodley, P. (2002) Influence of hydrodynamics and cell signaling on the structure and behavior of *Pseudomonas aeruginosa* biofilms. *Applied and Environmental Microbiology* 68, 4457–4464.

Ramasubbu, N., Thomas, L.M., Ragunath, C. and Kaplan, J.B. (2005) Structural analysis of dispersin B, a biofilm-releasing glycoside hydrolase from the periodontopathogen *Actinobacillus actinomycetemcomitans*. *Journal of Molecular Biology* 349, 475–486.

Ranjard, L. and Richaume, A.S. (2001) Quantitative and qualitative micro scale distribution of bacteria in soil. *Research in Microbiology* 152, 707–716.

Redfield, R.J. (2002) Is quorum sensing a side effect of diffusion sensing? *Trends in Microbiology* 10, 365–370.

Rich, J.J., Heichen, R.S., Bottomley, P.J., Cromack, K. and Myrold, D.D. (2003) Community composition and functioning of denitrifying bacteria from adjacent meadow and forest soils. *Applied and Environmental Microbiology* 69, 5974–5982.

Ritz, K., Dighton, J. and Giller, K.E. (eds) (1994) *Beyond the Biomass: Compositional and Functional Analysis of Soil Microbial Communities*. John Wiley and Sons, Chichester, UK.

Roberson, E.B. and Firestone, M.K. (1992) Relationship between desiccation and exopolysaccharide production in a soil *Pseudomonas* sp. *Applied and Environmental Microbiology* 58, 1284–1291.

Ryu, J.-H. and Beuchat, L.R. (2004) Factors affecting production of extracellular carbohydrate complexes by *Escherichia coli* O157:H7. *International Journal of Food Microbiology* 95, 189–204.

Schafer, A., Harms, H. and Zehnder, A.J.B. (1998) Bacterial accumulation at the air–water interface. *Environmental Science and Technology* 32, 3704–3712.

Schimel, J.P. (2001) Biogeochemical models: implicit vs. explicit microbiology. In: Schulze, E.D., Harrison, S.P., Heimannet, M. *et al.* (eds) *Global Biogeochemical Cycles in the Climate System*. Academic Press, New York, pp. 177–183.

Schimel, J.P. and Gulledge, J. (1998) Microbial community structure and global trace gases. *Global Change Biology* 4, 745–758.

Schimel, J.P. and Weintraub, M.N. (2003) The implications of exoenzyme activity on microbial carbon and nitrogen limitation in soil: a theoretical model. *Soil Biology and Biochemistry* 35, 549–563.

Schöning, I. and Kögel-Knabner, I. (2006) Chemical composition of young and old carbon pools throughout Cambisol and Luvisol profiles under forests. *Soil Biology and Biochemistry* 38, 2411–2424.

Sinsabaugh, R.L., Antibus, R.K. and Linkins, A.E. (1991) An enzymic approach to the analysis of microbial activity during plant litter decomposition. *Agriculture, Ecosystems and Environment* 34, 43–54.

Skujins, J. (1976) Extracellular enzymes in soil. *CRC Critical Reviews* 4, 383–421.

Smucker, A.J.M., Park, E.J., Dorner, J. and Horn, R. (2007) Soil micropore development and contributions to soluble carbon transport within macroaggregates. *Vadose Zone Journal* 6, 282–290.

Soroker, E.F. (1990) *Low Water Content and Low Water Potential as Determinants of Microbial Fate in Soil*. University of California, Berkeley, California.

Stark, J.M. and Firestone, M.K. (1995) Mechanisms for soil moisture effects on activity of nitrifying bacteria. *Applied and Environmental Microbiology* 61, 218–221.

Steinberger, R.E. and Holden, P.A. (2004) Macromolecular composition of unsaturated *Pseudomonas aeruginosa* biofilms with time and carbon source. *Biofilms* 1, 37–47.

Steinberger, R.E. and Holden, P.A. (2005) Extracellular DNA in single- and multiple-species unsaturated biofilms. *Applied and Environmental Microbiology* 71, 5404–5410.

Steinberger, R.E., Allen, A.R., Hansma, H.G. and Holden, P.A. (2002) Elongation correlates with nutrient deprivation in unsaturated *Pseudomonas aeruginosa* biofilms. *Microbial Ecology* 43, 416–423.

Stewart, P.S. (2003) Diffusion in biofilms. *Journal of Bacteriology* 185, 1485–1491.

Sutherland, I.W. (1977) *Surface Carbohydrates of the Prokaryotic Cell.* Academic Press, New York.

Sutherland, I.W. (1999a) Polysaccharases for microbial exopolysaccharides. *Carbohydrate Polymers* 38, 319–328.

Sutherland, I.W. (1999b) Polysaccharases in biofilms – sources – action – consequences! In: Wingender, J., Neu, T.R. and Flemming, H.-C. (eds) *Microbial Extracellular Polymeric Substances.* Springer-Verlag, Berlin, pp. 201–216.

Sutherland, I.W. (2001a) Biofilm exopolysaccharides: A strong and sticky framework. *Microbiology (UK)* 147, 3–9.

Sutherland, I.W. (2001b) The biofilm matrix: an immobilized but dynamic microbial environment. *Trends in Microbiology* 9, 222–227.

Tamaru, Y., Takani, Y., Yoshida, T. and Sakamoto, T. (2005) Crucial role of extracellular polysaccharides in desiccation and freezing tolerance in the terrestrial cyanobacterium *Nostoc commune. Applied and Environmental Microbiology* 71, 7327–7333.

Tanford, C. (1980) *The Hydrophobic Effect: Formation of Micelles and Biological Membranes.* Wiley, New York.

Teitzel, G.M. and Parsek, M.R. (2003) Heavy metal resistance of biofilm and planktonic *Pseudomonas aeruginosa. Applied and Environmental Microbiology* 69, 2313–2320.

Thieme, J., Schneider, G. and Knochel, C. (2003) X-ray tomography of a microhabitat of bacteria and other soil colloids with sub-100 nm resolution. *Micron* 34, 339–344.

Tokunaga, T.K., Wan, J.M., Hazen, T.C., Schwartz, E., Firestone, M.K., Sutton, S.R. *et al.* (2003) Distribution of chromium contamination and microbial activity in soil aggregates. *Journal of Environmental Quality* 32, 541–549.

Trumbore, S.E., Davidson, E.A., Decamargo, P.B., Nepstad, D.C. and Martinelli, L.A. (1995) Belowground cycling of carbon in forests and pastures of eastern Amazonia. *Global Biogeochemical Cycles* 9, 515–528.

van de Mortel, M. and Halverson, L. (2004) Cell envelope components contributing to biofilm growth and survival of *Pseudomonas putida* in low-water-content habitats. *Molecular Microbiology* 52, 735–750.

Venugopalan, V.P., Kuehn, A., Hausner, M., Springael, D., Wilderer, P.A. and Wuertz, S. (2005) Architecture of a nascent *Sphingomonas* sp. biofilm under varied hydrodynamic conditions. *Applied and Environmental Microbiology* 71, 2677–2686.

Vogel, J., Normand, P., Thioulouse, J., Nesme, X. and Grundmann, G.L. (2003) Relationship between spatial and genetic distance in *Agrobacterium* spp. in 1 cubic centimeter of soil. *Applied and Environmental Microbiology* 69, 1482–1487.

Wallenstein, M.D., Myrold, D.D., Firestone, M. and Voytek, M. (2006) Environmental controls on denitrifying communities and denitrification rates: Insights from molecular methods. *Ecological Applications* 16, 2143–2152.

Walsby, A.E. and Dunton, P.G. (2006) Gas vesicles in actinomycetes? *Trends In Microbiology* 14, 99–100.

Wan, J.M. and Wilson, J.L. (1994) Visualization of the role of the gas-water interface on the fate and transport of colloids in porous-media. *Water Resources Research* 30, 11–23.

Wan, J., Wilson, J.L. and Kieft, T.L. (1994) Influence of the gas-water interface on transport of microorganisms through unsaturated porous media. *Applied and Environmental Microbiology* 60, 509–516.

Wang, W., Shor, L.M., LeBoeuf, E.J., Wikswo, J.P. and Kosson, D.S. (2005) Mobility of protozoa through narrow channels. *Applied and Environmental Microbiology* 71, 4628–4637.

Whistler, R.L. and Kirby, K.W. (1956) Composition and behavior of soil polysaccharides 1,2. *Journal of the American Chemical Society* 78, 1755–1759.

Whitchurch, C.B., Tolker-Nielsen, T., Ragas, P.C. and Mattick, J.S. (2002) Extracellular DNA required for bacterial biofilm formation. *Science (Washington DC)* 295, 1487.

Williamson, K. and McCarty, P.L. (1976a) A model of substrate utilization by bacterial films. *Journal of the Water Pollution Control Federation* 48, 9–24.

Williamson, K.J. and McCarty, P.L. (1976b) Verification studies of the biofilm model for bacterial substrate utilization. *Journal of the Water Pollution Control Federation* 48, 281–296.

Wingender, J., Jaeger, K.-E. and Flemming, H.-C. (1999a) Interaction between extracellular polysaccharides and enzymes. In: Wingender, J., Neu, T.R. and Flemming, H.-C. (eds) *Microbial Extracellular Polymeric Substances.* Springer-Verlag, Berlin, pp. 231–251.

Wingender, J., Neu, T.R. and Flemming, H.-C. (1999b) What are bacterial extracellular polymeric substances? In: Wingender, J., Neu, T. R. and Flemming, H.–C. (eds) *Microbial Extracellular Polymeric Substances.* Springer-Verlag, Berlin, pp. 1–19.

Winogradsky, M.S. (1925) Etudes sur la microbiologie du sol, 1. Sur la méthode. *Annales Institut Pasteur Paris* 39, 299.

Wolfaardt, G.M., Lawrence, J.R. and Korber, D.R. (1999) Function of EPS. In: Wingender, J., Neu, T.R. and Flemming, H.-C. (eds) *Microbial Extracellular Polymeric Substances.* Springer–Verlag, Berlin, pp. 171–200.

Wollum, A.G. (1982) Cultural methods for soil microorganisms. In: Page, A.L., Miller, R.H. and Keeney, D.R (eds) *Methods of Soil Analysis, Part 2. Chemical and Microbiological Properties.* American Society of Agronomy, Madison, Wisconsin, pp. 781–802.

Xavier, J.B. and Foster, K.R. (2007) From the cover: Cooperation and conflict in microbial biofilms. *Proceedings of the National Academy of Sciences* 104, 876–881.

Xiang, S.R., Doyle, A., Holden, P.A. and Schimel, J.P. (2008) Drying and re-wetting effects on C and N mineralization and microbial activity in surface and subsurface California grassland soils. *Soil Biology and Biochemistry* 40, 2281–2289.

Young, I.M. and Crawford, J.W. (2004) Interactions and self-organization in the soil-microbe complex. *Science* 304, 1634–1637.

Zhou, J., Xia, B., Treves, D.S., Wu, L.Y., Marsh, T.L., O'Neill, R.V. *et al.* (2002) Spatial and resource factors influencing high microbial diversity in soil. *Applied and Environmental Microbiology* 68, 326–334.

Zuev, B.K., Chudinova, V.V., Kovalenko, V.V. and Yagov, V.V. (2001) The conditions of formation of the chemical composition of the sea surface microlayer and techniques for studying organic matter in it. *Geochemistry International* 39, 702–710.

8   Fungal Growth in Soils

R. Michael Miller* and Michael S. Fitzsimons

---

1 Introduction

From the very onset of colonization of land by the first land plants fungi have played a crucial role not only in moulding the evolution of higher plants but also by influencing the soils in which they grow. Sequencing studies along with more targeted studies of the fossil record suggest that the filamentous thallus most likely evolved on land but did so prior to the appearance of the first land plants. Even so, the feedbacks between plants and fungi, and, in turn, the consequences of these feedbacks for pedogenic processes, are just now beginning to be appreciated, especially in shaping the function and structure of soil fungi.

The taxonomic, structural and metabolic diversity of fungi is vast (Zak and Visser, 1996; Hawksworth, 2001; Brundrett, 2002; Martin *et al.*, 2003). There are fungal species that are unicellular and yeast-like, while others grow as a network of filamentous hyphal threads (Martin *et al.*, 2003). Among these filamentous forms are still other species that have an ability to produce a vast assortment of mycelial exploration types (Boddy, 1999; Agerer, 2001; Ritz, 2007) and various-shaped macroscopic reproductive structures such as cup fungi, puff-balls, mushrooms and stink-

horns. In addition, fungi are metabolically complex, playing a pivotal role in the breakdown of the complex polymers associated with litter and other soil-borne organic debris (Hudson, 1968; Frankland, 1998; Bardgett *et al.*, 2005; Deacon *et al.*, 2006). What the vast majority of fungi have in common is that they are eukaryotes, mostly growing by producing a network of filamentous hyphae that secrete enzymes for the extracellular rendering of complex substrates and relying on an osmotic mode of nutrition uptake based on membrane-localized importers (Martin *et al.*, 2003).

2 The Fungal Hypha

The fundamental unit of a fungus is its hypha. Fungal growth and morphogenesis are uniquely different from those found in plants and animals in that a hypha extends only at its apex, with cross-walls produced at right angles to the long axis of the hypha. Proliferation and differentiation, then, are through branching of hyphal tips (Moore *et al.*, 2006). In a broad sense fungal growth consists of an indeterminate apically growing and branching hyphal network (Talbot, 1997; Klein and Paschke, 2004). The ability

*Corresponding author: rmmiller@anl.gov

of fungi to translocate carbon and nutrients to growing hyphal tips permits fungi to grow and forage beyond the voids found between soil particles, but further allows for linkages to regions within the soil having high and low resource quality. This ability enables fungi to forage for nutrients and also exploit existing resources (Fogel and Hunt, 1983; Jennings, 1991; Frey *et al.*, 2003). The ability to forage is achieved by fungi having evolved a very efficient translocation system within its network of mycelia (Lindahl and Olsson, 2004). Translocation of resources in hyphae having accumulated higher concentrations of a nutrient than those of the distal portions of a mycelial growing front allows fungi to advance and grow through regions of lower resource quality (Boswell *et al.*, 2002; Lindahl and Olsson, 2004). An important consequence of the fungal translocation mechanism is that it allows for the movement of nutrients within the soil profile in greater amounts than could be expected by other means.

The most important component of soil fungi is their biomass, with biomass being directly proportional to hyphal length (Joergensen and Wichern, 2008). The amount of fungal biomass in a soil represents a significant pool of available nutrients, with its turnover having important consequences for carbon and nutrient cycling (Kjoller and Struwe, 1982; Langley and Hungate, 2003; Zhu and Miller, 2003;

Six *et al.*, 2006). The availability of hyphal components in the form of cell walls, cytoplasm and extracellular polysaccharides represents a relatively labile organic pool in soils (Klein and Paschke, 2004). In a forest stand composed of spruce and pine the amount of ectomycorrhizal hyphal biomass has been estimated at 700–900 kg/ha, with the amount of N in hyphae estimated to be over 180 kg N/ha (Wallander *et al.*, 2001, 2004). Arbuscular mycorrhizal (AM) fungi found in grassland soils also represent a major portion of the fungal biomass pool, with reports of 20–30% of the microbial biomass carbon being composed of mycorrhizal fungal biomass (Miller *et al.*, 1995; Olsson *et al.*, 1999; Leake *et al.*, 2004). Annual production of extra-radical hyphae of AM fungi in prairie soils is estimated to be 28 m/cm^3 of soil, with a calculated annual hyphal turnover of 26% (Miller *et al.*, 1995). However, higher hyphal turnover rates have been reported, suggesting that some AM fungal hyphae may have a lifespan of only 5–6 days (Staddon *et al.*, 2003).

In soils, a major obstacle to microbial growth is the air-filled voids created by the spatial arrangements of solids in soil (Elliott *et al.*, 1980; Hassink *et al.*, 1993; Otten *et al.*, 2004b). These voids represent vast deserts to microbes, with only those organisms possessing a filamentous morphology being able to bridge them (Fig. 8.1). Consequently, we find a partitioning of microbes

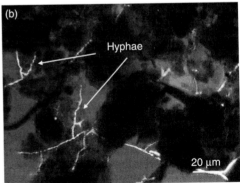

Fig. 8.1. The remarkable – and hugely consequential – ability of fungi to bridge the gaps between soil aggregates. (a) The arbuscular mycorrhizal fungus *Glomus etunicatum* growing in a highly aggregated silt loam (photograph by R.M. Miller); (b) the plant pathogen *Rhizoctonia solani* has a similar growth habit well adapted to growing through the soil pore network (from Harris *et al.*, 2003).

in soil between those whose growth is limited to growing along biofilm surfaces and those capable of producing aerial structures allowing for growth beyond surfaces of soil solids to traverse and grow within the air-filled chasms of soil pores.

A critical factor for fungal growth and exploitation in soil is the ability of the hyphal cell wall matrix to respond to an ever-changing soil environment. Fungal cell wall characteristics primarily comprise glycoproteins, glucans and chitin with exact composition being dependent on growing conditions (Wessels, 1993; Rast et al., 2003; Bowman and Free, 2006). Fungal cell walls are also fairly malleable when compared with the ridged cell walls of plants. This malleability of the fungal cell wall allows for exploration of hyphae to conform to the shapes of soil voids. The ability to alter cell wall surface properties is unique among filamentous fungi and is achieved through the secretion of hydrophobins (Wösten et al., 1993; Wessels, 1997; Rillig, 2005). Hydrophobins are small hydrophobic proteins that are produced at the growing tip of hyphae. These proteins have an ability to self-organize along the hyphal cell wall, creating hydrophobic surfaces that enable aerial growth (Wessels et al., 1991; Wösten, 2001). The hydrophobic nature of hydrophobins also acts as a desiccant protection mechanism and similarly aids in fungal tissue stabilization (Tagu et al., 1998; Richards et al., 2006).

For hyphae to effectively exploit nutrients from a substrate in a moist environment their cell wall surfaces need to have hydrophilic properties, whereas the efficient transport of nutrients elsewhere is best accomplished by producing hyphae or rhizomorphs having hydrophobic cell wall surfaces. This would suggest that a successful fungus should be able to produce both hydrophilic and hydrophobic hyphal wall surfaces, dependent on the conditions of the soil environment in which it is growing. It ends up that with ectomycorrhizal fungi hyphal cell wall surface characteristics and mode of hyphal exploitation are closely related. Those fungi that invest in a more distal foraging strategy and having cord-like rhizomorphs usually have hydrophobic surfaces, whereas those that proliferate in close proximity to a resource favour the production of hydrophilic cell wall surfaces (Unestam, 1991; Tagu et al., 1998; Olsson et al., 2003; Hobbie and Agerer, 2010).

3 Fungal Groups

Fungi are typically grouped into just a few categories based on their nutritional needs and ways in which they make their living in the spatially heterogeneous soil ecosystem. All fungi are heterotrophic, or derive their energy from other organisms. There are important divisions in this heterotrophic lifestyle, however. Saprobic fungi derive their energy from the degradation of senescent plant material. Another soil fungal group is composed primarily of plant pathogens, a group that consumes living plant cellular materials. The last major group of soil fungi is symbiotic with plant roots and is known as mycorrhizal fungi. These fungi have life histories shaped by their need to simultaneously infect plant roots to obtain carbohydrates and to explore the soil body for nutrients limiting to their own growth as well as limiting the growth of their plant hosts (Smith and Read, 2008). The relationship between mycorrhizal fungus and host is primarily an adaptation to resource limitation (Brundrett, 2002; Smith and Read, 2008). While each group of soil fungi differs in what it considers to be resources, they are all broadly similar in their mode of growth within the soil matrix and we will treat them as such throughout the chapter.

Saprobic fungi are critical components of soils due to their extensive control of nutrient cycling and their effect on nutrient availability and plant community dynamics (Bardgett et al., 2005; Six et al., 2006; Miller and Lodge, 2007). In a soil environment fungi are in large part responsible for degrading the majority of dead plant materials and are, with few exceptions, the only organisms capable of degrading lignin (Boddy and Watkinson, 1995; de Boer et al., 2005). The saprobic lifestyle is common throughout the kingdom of fungi and is found in four of the five fungal phyla (James et al., 2006).

Creating a living environment within the soil has led to the evolution of different life histories among saprobic fungi. With an estimated 1.5 million species and an astounding variety of morphologies this should not be surprising (Hawksworth, 2001). Some fungi, for instance, specialize on small, easily degradable compounds and others on more recalcitrant, large plant-derived molecules. These fungi can be coarsely, but informatively, grouped by successional stage (Hudson, 1968; Last et al., 1987; Frankland, 1998). Garrett (1951), in his classic study on the ecological groupings of soil fungi, was the first to visualize a substrate for decomposition as being either discrete or continuous. Soil fungi of continuous substrates inhabit a niche that continually receives new resources, whereas fungi of discrete substrates inhabit the substrate. The succession of fungi on a substrate results in a progressive deterioration in the capacity of the substrate to support further growth. In fact, the fungus eventually becomes the substrate. Garrett (1951) further states that the substrate comes directly to the soil microorganism, i.e. roots grow through soil and die in it; dead leaves fall upon the soil. The first example views a substrate as discrete patches, whereas the latter views a substrate as being continuous.

The first group to colonize newly available substrate is historically referred to as the 'sugar fungi' (Burges, 1939; Garrett, 1951). These fungi are opportunistic fast growers and produce enormous numbers of spores, which are frequently the most common spores found in the soil (Brown, 1958). While these fungi are taxonomically diverse, they are often members of the Zygomycota and Ascomycota. Sugar fungi generally do not produce macroscopic structures. These are the first fungi to colonize a substrate due to their ability to produce large numbers of spores, the ease with which the spores are disseminated and an ability to grow quickly, which are advantageous traits for colonization of easily decomposable substrates (de Boer et al., 2005; van der Wal et al.,

2006). For the most part, these opportunistic fungi can degrade simple soluble substrates, pectin and to some extent readily accessible cellulose and hemi-cellulose. Sugar fungi are incapable of producing lignin-degrading enzymes (van der Wal et al., 2006).

The next group of fungi to colonize substrate in the soil is known as the secondary colonizers. While they are genetically capable of living on the same simple sugars available earlier in succession, they are usually 'outgrown' by the faster-growing, early successional fungi or even suppressed by antibiotic production. Broadly, they are members of the Ascomycota and Basidiomycota phyla of which mushroom-forming fungi and many plant pathogens are also members. They are capable of producing enzymes that can degrade the cellulose and lignin that other fungi cannot (van der Wal et al., 2007). They tend to be slower growing and live on a different time scale from the fast pace of early successional fungi. They may also form extensive persistent structures within the soil (Thompson and Rayner, 1983). The terms 'early' and 'late' successional species should be used with some care as in some situations these fungi can exist concurrently, or the order in which they colonize a substrate may even be reversed (Hudson, 1968). The former may happen as these later-colonizing fungi break down cellulose and lignin into simple sugars; sugar fungi simply utilize some of the simple sugar by-products as they are produced. The latter situation may occur when fresh woody substrate falls onto the soil. Simple sugars are not a major component of a fresh woody substrate and it takes time for lignin and cellulose decomposers to produce sufficient simple sugar by-products to support a large sugar fungi community.

Another distinction between late and early successional fungi concerns how they forage for new substrates. Early successional fungi are said to be resource unit restricted and late successional fungi are non-resource unit restricted (Boddy, 1999). That is, early successional fungi colonize new resources via spores and cannot forage effectively much beyond where the substrate ends.

Fungi capable of degrading more complex organic compounds generally are not resource unit restricted, although within this group there remains a wide range of foraging capabilities (Boddy, 1999). So, while resource unit-restricted fungi grow as all fungi, via hyphae, non-resource unit-restricted fungi have taken it a step farther and can reach not only across a soil pore, but even metres away from large resource concentrations (Smith *et al.*, 1992; Boddy, 1993; Legrand *et al.*, 1996).

4 Biotic Interactions

As fungi grow and explore their habitats they interact with an enormous variety of organisms, including other fungi, bacteria, soil invertebrates and even plants. Interactions are certainly the rule and not the exception (Fitter *et al.*, 2005). These interactions include predation, competition and even mutualism. Understanding what underlies and determines the outcomes of these interactions is incredibly important for understanding the 'private life of fungi', including their success in the soil milieu.

As previously mentioned, fungi often account for the majority of soil biomass, while bacteria are numerically more diverse and abundant. One component of fungal competitive ability concerns the complicated and varied antibacterial and antifungal compounds they exude into their surrounds (Duffy *et al.*, 2003; de Boer *et al.*, 2005; Schüffler and Anke, 2009). Even so, soil and substrate structure is often the overriding determinant of whether fungi or bacteria will predominate in a particular environment (Maraun *et al.*, 2003). A greater challenge for fungi is their interactions with the soil faunal community. As fungi grow and proliferate within the soil environment they must vie with the soil microfauna and megafauna for resources. Although it may appear that soil fauna and fungi are competing for the same resources, it is unlikely that this competition is direct. Rather, what may be occurring is faunal grazing. Because a majority of soil invertebrates are mycophagous and possess chitinases in their gut they are well suited for grazing on hyphae (Siepel and Maaskamp, 1994; Berg *et al.*, 2004; McGonigle, 2007). Grazers often consume the hyphae associated with plant litter or detrital material rather than consuming litter directly. Because of the close proximity of hyphae and litter, it is difficult to group invertebrates into purely litter-consuming or fungi-consuming guilds. Even those fauna that apparently consume plant litter may be receiving more nutrients from the fungal and bacterial decomposers already colonizing and degrading litter. In fact, both supposed litter-consuming and fungal-consuming groups of soil fauna prefer fungal hyphae when given a choice between plant matter and fungi (Hedlund *et al.*, 1991; Kampichler *et al.*, 2004). Soil invertebrates also seem to prefer grazing on saprobic fungal hyphae rather than mycorrhizal hyphae (Klironomos *et al.*, 1999; Gormsen *et al.*, 2004).

Fungi respond dynamically to environmental conditions and grazing pressure. Fungi are able to switch from the exploitative growth pattern with plentiful diffuse hyphae to a more quickly growing exploratory strategy to allow escape when faced with heavy invertebrate grazing (Boddy, 1993; Tordoff *et al.*, 2006; Rotheray *et al.*, 2008). Some fungi when faced with high grazing pressure produce complex mycelial cords. Many of these cords consist of an outer rind encrusted with crystals that may be involved with protecting their resources from continued invertebrate grazing (Wicklow, 1979; Paterson *et al.*, 1987). In addition to dynamic responses, constitutive physical and chemical components may also deter invertebrate grazing (Shaw, 1992). Of course some fungi cannot effectively mount these defensive responses and so grazing pressure tends to select for unpalatable fungi. These feedbacks between fungi and grazers have important consequences for fungal and invertebrate community structure (Visser *et al.*, 1987; Maraun *et al.*, 2003; Tordoff *et al.*, 2008).

The activities of soil invertebrates offer many benefits to soil fungi as well. They can be especially important in terms of dispersal of fungal spores in the soil environment

(Reddell and Spain, 1991; Gange, 1993). In general, invertebrates do not actively disperse the spores although there are a few very interesting examples of insects obligately dependent on fungal species, including some beetle vectors of tree diseases and attine ants, which 'farm' particular species of fungi (Paine *et al.*, 1997; Seal and Tschinkel, 2007). Another, perhaps unexpected, interaction is that of soil fungi consuming soil invertebrates. Nematophagous fungi consume nematodes by trapping them in special noose-like hyphae for later consumption. These fungi can be either predatory or endoparasitic and appear to play an important role in controlling the population density of nematodes in native soils (Gray, 1987). Ectomycorrhizal fungi also appear to be able to immobilize and consume Collembola (Klironomos and Hart, 2001).

Soil fungi must also compete with plants for resources. While plants are very conspicuous members of the above-ground community they are the primary driver of below-ground processes. Plant roots after all are exploring the same soil milieu for mineral and organic resources and water. Soil saprobes, including fungi, are generally more capable of accessing available nutrients than are plants (Schimel *et al.*, 1989; Groffman *et al.*, 1993). It is likely only when the microbial population is under stress that increased mineralization occurs and allows the plants the opportunity to access nutrients previously held in microbial biomass. Another way in which plants can circumvent the intrinsically superior ability of saprobic fungi to gather nutrients is by rewarding fungi to do it for them. This is of course what plants do with mycorrhizal fungi (Smith and Read, 2008).

Competition among fungi, as within any group of related organisms, is most severe (Dowson *et al.*, 1988a, b). Interactions among fungal species undoubtedly occur throughout the soil body, but resources act as foci for mycelial effort (Donnelly and Boddy, 2001). Interactions between species are highly complex as fungi act to attain access to resources, defend territories and to gain access to new territories through replacement of resident fungi

(Dowson *et al.*, 1988c). The few studies that have been conducted suggest that a hierarchy of competitiveness exists among fungi (Chapela *et al.*, 1988; Holmer and Stenlid, 1997). In general, cord-forming decomposer fungi, as secondary colonizers, are more combative and replace non-cord-forming primary colonizers (Donnelly and Boddy, 2001). The types of interactions that occur are quite diverse, ranging from stalemate, to immediate conquest, to repeated and more forceful attempts to conquer a competitor if earlier attempts fail (Boddy and Abdalla, 1998; Harris and Boddy, 2005). Fungi are also capable of launching defensive manoeuvres such as mycelial ridges, but the function of such a structure in the soil context is unclear. While the above studies appear to form a rather rigid hierarchy, the outcomes are highly context dependent and can be swayed depending on local soil conditions and resource supply. For example, a 'less competitive' fungus controlling a large resource base may still be able to outcompete a more aggressive species given a large enough resource base. Fungi also appear to be capable of calculated decision making. They can 'decide', for instance, when given the option of colonizing multiple resources, to colonize preferentially the uncolonized resource, followed by resources with the weakest competitor.

5 Exploration of the Soil Body

The complexity of the soil body, with its water-filled pores, air-filled pores, anaerobic microsites and numerous microorganisms competing for substrate, is a critical feature to consider in order to understand how fungi live, reproduce or spread throughout soil. Historically, the growth of fungi has been modelled based on a physically and chemically homogeneous medium. While this captures some aspects of fungal vegetative reproduction, it is a gross oversimplification and a great deal of interesting biology and meaning have been lost by researching fungi in such a limited way. By mimicking the heterogeneity of the soil many interesting insights into fungal biology and ecology have been uncovered. Currently, we have only a limited notion of how fungi explore soil, but the available studies provide a wealth of knowledge and insight. For example, saprobic fungi have several traits that allow for efficient growth in N-limited habitats. These include being able to lyse and reassimilate nitrogen from degenerated hyphae; having directed growth to locally enriched nutrient sites; and having an ability to translocate cytoplasm to hyphal apices from mycelium in nitrogen-depleted regions (Moore, 1991; Klein and Paschke, 2004). Hyphae may also translocate mineral nitrogen to nitrogen-poor substrates where the absolute amount of nitrogen in a decomposing substrate increases during the early stages of decomposition (Melillo *et al.*, 1982; Holland and Coleman, 1987; Paustian and Schnurer, 1987; Unestam and Sun, 1995; Frey *et al.*, 2000).

6 Fungi and Scale

What follows is primarily a discussion of how fungi are influenced by their quest to find and exploit resources. Nevertheless, it is important to realize that these are not the only influences on fungal behaviour; fungi can perceive and respond to many aspects of their environment such as orientation and gravity (Moore, 1991; Brand and Gow, 2009), chemical cues from potential mates (Sutter *et al.*, 1996), heavy metals (Fomina *et al.*, 2000), sunlight (Carlile, 1970), radiation (Tugay *et al.*, 2006) and the topology of their environment (Read *et al.*, 1997). In addition soil fungi must respond to a great diversity of other soil organisms ranging through bacteria, invertebrates, other fungi and of course vegetation.

6.1 Small-scale heterogeneity

Beginning with the physical heterogeneity of soil at a very small scale, fungi perceive their world as very diverse with some paths for growth through the soil being favoured over others. Fungi appear to prefer to grow through pores in soil (Otten *et al.*, 2001,

2004b; Harris *et al.*, 2003). Moreover, they preferentially grow along the edges of pores and are restricted to pores larger than the width of their hyphae (Schack-Kirchner *et al.*, 2000). They probably hug the edge of the pores because soil aggregates are covered by a film of water in which the majority of nutrients are dissolved. While an even larger reservoir of nutrients is present in the aggregates themselves, this is generally protected from microbial decomposition including fungal penetration (Tisdall and Oades, 1982; Schack-Kirchner *et al.*, 2000). Even fungi, with their ability to cross pores, cannot directly unlock the nutrients hidden away within soil aggregates. These small-scale preferences in fungal growth can have large macroscopic effects on fungal expansion and even epidemics of plant-pathogenic fungi.

Another small-scale approach to fungal growth involves a concept in plant pathology known as the 'pathozone'. This is an area around the fungus beyond which no measurable (or below some threshold probability) infection of a host is detected (Gilligan, 1985). While pathozone explicitly refers to pathogens it can probably be extended or reduced to include saprobic growth as well. This conceptualization allows one to study the effect of changing one parameter or another on shrinking or expanding the pathozone of a fungus. These parameters could include nutritional status of host or pathogen, distance of inoculum from the host, the susceptibility of the host, and the biotic and abiotic properties of the soil (Bailey and Gilligan, 1997; Kleczkowski *et al.*, 1997; Bailey *et al.*, 2001; Otten *et al.*, 2001). The pathozone concept also links local-scale processes such as host infection with disease progress (Bailey and Gilligan, 1997; Stacey *et al.*, 2001).

6.2 Medium-scale heterogeneity

The next scale we will examine is where a fungus moves from one resource to another and how in some cases the organism integrates its response across its entire thallus. In order to discuss this literature, however, we must first take a detour into fractal geometry as much of the literature utilizes this concept. An object is said to be fractal if its spatial form appears essentially the same at whichever scale one views it. While natural systems are fractal over only a finite number of dimensions (Hastings and Sugihara, 1993), mycelial systems are sufficiently fractal to be described in this manner (Obert et al., 1990; Ritz and Crawford, 1990). While seemingly very abstract, this technique allows researchers to measure how intensively a fungus has explored or exploited a particular area or volume.

It is often calculated via the 'box-counting method', which consists of overlaying a mycelial image with grids of boxes of different sizes and counting the number of boxes intersected for each box size. Fractal images obey the power law relation over a length of scales, where D is the fractal dimension, N(s) is the total number of boxes of side length s that intersect the image, and c is a constant. D is estimated as the negative gradient of a regression line through the linear part of the plot of log N(s) against log s, for a sequence of scales (Hastings and Sugihara, 1993; Boddy, 1999). The interesting part from our perspective is that as D increases so does the intensity with which the fungus is exploring its resource.

$$N(s) = cs^{-D} \qquad (8.1)$$

When a fungus encounters a new resource it branches profusely, which is reflected in an increased D; this can be called the exploitative phase of growth. When a fungus is searching for additional resources it grows mostly linearly and is said to exhibit explorative growth (Ritz and Crawford, 1990). Fractal geometry allows for one to describe how fungal search pattern changes, but other measurements of fungal exploration of resources such as biomass, surface area or density may still be useful under different circumstances (Boddy, 1999).

In addition to measuring responses to resources, fractal geometry allows one to make interesting comparison among fungi. For example it allows one to distinguish between two types of non-resource unit-restricted fungi: short- and long-range

foragers (Boddy, 1999). Long-range foragers, which have well-defined, rapidly extending cords, such as Agrocybe giverosa, Coprinus picaceus, Phallus impudicus, Panerochaete velutina and Resinicium bicolor, are mass fractal, meaning they are fractal over their entire mass. Short-range foragers, such as Hypholoma fasciculare, Stropharia aeuruginosa and Stropharia caerulaea, are characterized by slowly extending search fronts and are said to be surface fractal as they are fractal only over the surface and not the entire mass. Short range-foragers are not mass fractal because they explore their environment so intensively that over time they are no longer fractal, but have formed a simple opaque sheet of colonization. Mass-fractal or long-range foragers grow outwards using rhizomorphs or cords and never explore their environment sufficiently to become non-fractal at their core. Interestingly, over time both types become more mass fractal because even short-range foragers regress some of their hyphae at their core to reallocate these resources to exploration (Boddy, 1999; Donnelly and Boddy, 2001).

The cord formers, as mentioned above, can translocate nutrients over a relatively large distance, which allows them to forage over very large areas (Smith et al., 1992). The thick cords or rhizomorphic mycelia underlie their ability to attain this size. The cords themselves are composed of aggregations of predominantly parallel mycelia, exhibit tissue differentiation and do not contribute significantly to nutrient uptake (Clipson et al., 1987; Wells and Boddy, 1995; Cairney, 2005). The formation of cords allows fungi great flexibility in finding resources and utilizing them simultaneously. One example from this group is Armillaria bulbosa, the almost mythical 'humongous fungus'; a single individual was found to extend over 15 ha (Smith et al., 1992). Generally, clones are much smaller than this, however. While the large extent of Armillaria does represent the ability of cord formers to extend successfully through a forest, it is probably unlikely that it is capable of transferring nutrients over such a distance.

6.3 Large-scale heterogeneity

At this largest of soil scales we will look at how fungi propagate among many localized nutrient sources. While there is an exceedingly large literature describing fungal pathogen spread, soil fungi have usually not been explicitly considered in these models, nor have existing studies used a spatially realistic approach. Network-based or spatially explicit epidemic models show great promise for providing much insight into how fungi grow through the soil environment (Park *et al.*, 2001; Jeger *et al.*, 2007). One successful approach to date has been the use of percolation models. These are models originally developed by physicists and chemists, but have also become a useful epidemiological tool. Here a network of plants is envisioned where some plants are linked to certain other plants, but not to others. In the case of plants this would represent a plant's neighbours, whereas for animal disease models it might represent family members or acquaintances. If one is interested in saprobic rather than pathogenic fungi the nodes could be considered resources such as plant litter quality. In this lattice of plants (or resources more generally) there exists a critical threshold probability for spread between adjacent cells where above this probability the disease will spread invasively (across the grid) or, if the probability is below this threshold, the epidemic will be restricted to relatively small patches (Grassberger, 1983; Bunde and Havlin, 1991).

Those studies that have been done with fungi in mind have looked at the impact of varying the distance among resources islands. Bailey *et al.* (2000) found this predicted threshold probability to be an accurate representation of how *Rhizoctonia solani* spreads in the environment. If resources were placed at distances larger than the threshold, colonization of a lattice of resource units was usually finite and would never fully colonize the system. One implication of this work is that the likelihood of a fungus propagating through the soil or via infected plants does not relate linearly to distance between soil resources; rather there is a threshold distance above which propagation is arrested. Another interesting conclusion one can draw here is that local-scale processes, distance from one resource to the next, can play an important role in larger-scale processes like a pathogen epidemic or saprophytic spread.

Also, percolation models have demonstrated that there needs to be a particularly dense arrangement of resource units in order for propagation to occur from one resource to another (Otten *et al.*, 2004a). In this case the authors held the distance constant among soil resources, but randomly removed resource islands. This has implication for both disease and propagation of antagonistic species throughout the soil, as having fewer hosts or less plant matter on the ground in the autumn may inhibit spread of the pathogen. These types of models work only for fungi that spread via vegetative growth, however, and spore production would require considerable retooling of the model structure.

Another useful large-scale approach to describing fungal spread in nature is through the use of cellular automata models; here, one models spatially discrete locations either colonized or not colonized by a particular organism with specific rules regarding dispersal among lattice points and interactions among organisms in the model space (Molofsky and Bever, 2004). Cellular automata models are especially useful for fungi, as mean field models (those that assume the population of organisms is well mixed spatially) are not well suited for sessile organisms. These types of models have resulted in some interesting conclusions. Bown *et al.* (1999) found evidence for emergent behaviour in community-scale fungal dynamics. Basically, one cannot summarize fungal interactions based solely on the local scale. This is because fungi translocate nutrients and the availability of resources to a fungus will impact its competition with other organisms. Some researchers have developed exceedingly complex approaches looking at fungal growth within a soil-like structure, even accounting for favoured growth along edges of soil pores and transitions from exploration to exploitation growth phases (Boswell *et al.*, 2007). Such approaches will probably provide tremendous insight in the future as to how local- and large-scale processes are controlled.

7 The Future

For the most part, this chapter has viewed soil fungi as decomposers. This view overlooks potentially far more important roles played by other groups of soil fungi: their contributions as symbionts with higher plant root systems; their role in the weathering of consolidated parent materials; their contributions to soil structure; and, taken together as a whole, an appreciation of the concept of mycelial networks (Boddy, 1999; Miller and Jastrow, 2000; Boswell *et al.*, 2002; Selosse *et al.*, 2006; Gadd, 2007). The complexity and scales of fungal contributions to ecosystem and community processes make quantifying their contributions difficult. Even so, what is becoming apparent is that the diversity on this earth is directly related to the ability of fungi to adapt and evolve in our ever-changing environment.

The challenges ahead are many. Although not discussed here, the rapid advancements in analytical and high-throughput methods by the molecular biology community offer new tools for improving our understanding of the contributions of fungi to soil systems, especially as they relate to the sustainable use of this critical resource (Jackson *et al.*, 2002). These tools may also give us an insight to better manage soils for carbon sequestration, remediation of heavy metal and degradation of hydrocarbons. Currently, though, fungi are the ugly step-sister to bacteria when it comes to funding for whole genome sequencing, a necessary step for integrating fungi with the new tools (Baker *et al.*, 2008).

The dearth of sequence information limits our ability optimally to take advantage of using metagenomic approaches for investigating soil fungi.

The past decade of fungal research has demonstrated that in many ways our crude ability to quantify mycorrhizal fungi is like that of the microbial ecologist's challenge of understanding the 'unculturable'. As we begin to factor in mycorrhizal fungi to ecosystem processes we are beginning to see a view that has been underappreciated, that this plant–fungus relationship may be the primary adaptation mechanism to limiting resources and a major mechanism driving the diversity we see both above and below ground (de Boer *et al.*, 2005). Phylogenetic approaches to the soil microbial community are important, but to really address the questions that are facing soil scientists and microbiologists today we need to be developing the tools to address function in a manner that gives a true representative view of fungi. Current estimates of fungal biomass, as well as our knowledge of how fungi grow and exploit the resources of the soil, e.g. their ability to produce hydrophobins, and the production of complex enzyme systems for the degradation of cellulose and lignin, emphasize the need for inclusion. What is overlooked in cutting-edge functional metagenomic approaches is that fungal systems are essentially absent.

The recent advancements in tomography are just now beginning to give us a view of fungi and soils and the habitats within them (De Gryze *et al.*, 2006; Lombi and Susini, 2009). Our challenge is to take these new tools and use them to make sense of how fungi grow. Also, along with the advancements in sequencing capabilities and actual sequencing of more branches of the fungal tree of life, we will be better able to integrate phylogeny with ecology and finally begin to address the questions of how fungi grow and function in soil.

References

Agerer, R. (2001) Exploration types of ectomycorrhizae - a proposal to classify ectomycorrhizal mycelial systems according to their patterns of differentiation and putative ecological importance. *Mycorrhiza* 11, 107–114.

Bailey, D.J. and Gilligan, C.A. (1997) Biological control of pathozone behaviour and disease dynamics of *Rhizoctonia solani* by *Trichoderma viride*. *New Phytologist* 136, 359–367.

Bailey, D.J., Otten, W. and Gilligan, C.A. (2000) Saprotrophic invasion by the soil-borne fungal plant pathogen *Rhizoctonia solani* and percolation thresholds. *New Phytologist* 146, 535–544.

Bailey, D.J., Thornton, C.R., Dewey, F.M. and Gilligan, C.A. (2001) A non-destructive immunoblotting technique for visualisation and analysis of the growth dynamics of *Rhizoctonia solani*. *Mycological Research* 105, 983–990.

Baker, S.E., Thykaer, J., Adney, W.S., Brettin, T.S., Brockman, F.J., D'Haeseleer, P. *et al.* (2008) Fungal genome sequencing and bioenergy. *Fungal Biology Reviews* 22, 1–5.

Bardgett, R.D., Bowman, W.D., Kaufmann, R. and Schmidt, S.K. (2005) A temporal approach to linking aboveground and belowground ecology. *Trends in Ecology and Evolution* 20, 634–641.

Berg, M.P., Stoffer, M. and van den Heuvel, H.H. (2004) Feeding guilds in collembola based on digestive enzymes. *Pedobiologia* 48, 589–601.

Boddy, L. (1993) Saprotrophic cord-forming fungi – warfare strategies and other ecological aspects. *Mycological Research* 97, 641–655.

Boddy, L. (1999) Saprotrophic cord-forming fungi: Meeting the challenge of heterogeneous environments. *Mycologia* 91, 13–32.

Boddy, L. and Abdalla, S.H.M. (1998) Development of *Phanerochaete velutina* mycelial cord systems: Effect of encounter of multiple colonised wood resources. *FEMS Microbiology Ecology* 25, 257–269.

Boddy, L. and Watkinson, S.C. (1995) Wood decomposition, higher fungi, and their role in nutrient redistribution. *Canadian Journal of Botany* 73, S1377–S1383.

Boddy, L., Wells, J.M., Culshaw, C. and Donnelly, D.P. (1999) Fractal analysis in studies of mycelium in soil. *Geoderma* 88, 301–328.

Boswell, G.P., Jacobs, H., Davidson, F.A., Gadd, G.M. and Ritz, K. (2002) Functional consequences of nutrient translocation in mycelial fungi. *Journal of Theoretical Biology* 217, 459–477.

Boswell, G.P., Jacobs, H., Ritz, K., Gadd, G.M. and Davidson, F.A. (2007) The development of fungal networks in complex environments. *Bulletin of Mathematical Biology* 69, 605–634.

Bowman, S.M. and Free, S.J. (2006) The structure and synthesis of the fungal cell wall. *Bioessays* 28, 799–808.

Bown, J.L., Sturrock, C.J., Samson, W.B., Staines, H.J., Palfreyman, J.W., White, N.A., Ritz, K. and Crawford, J.W. (1999) Evidence for emergent behaviour in the community-scale dynamics of a fungal microcosm. *Proceedings of the Royal Society B* 266, 1947–1952.

Brand, A. and Gow, A.R. (2009) Mechanisms of hypha orientation of fungi. *Current Opinion in Microbiology* 12, 350–357.

Brown, J.C. (1958) Soil fungi of some British sand dunes in relation to soil type and succession. *Journal of Ecology* 46, 641–664.

Brundrett, M.C. (2002) Coevolution of roots and mycorrhizas of land plants. *New Phytologist* 154, 275–304.

Bunde, A. and Havlin, S. (1991) *Percolation*. Springer-Verlag, Berlin.

Burges, A. (1939) Soil fungi and root infection – a review. *Broteria* 8, 64–81.

Cairney, J.W.G. (2005) Basidiomycete mycelia in forest soils: Dimensions, dynamics and roles in nutrient distribution. *Mycological Research* 109, 7–20.

Carlile, M.J. (1970) *The Photoresponses of Fungi*. Wiley-Interscience, London.

Chapela, I.H., Boddy, L. and Rayner, A.D.M. (1988) Structure and development of fungal communities in beech logs 4½ years after felling. *FEMS Microbiology Ecology* 53, 59–69.

Clipson, N.J.W., Cairney, J.W.G. and Jennings, D.H. (1987) The physiology of basidiomycete linear organs. 1. Phosphate-uptake by cords and mycelium in the laboratory and the field. *New Phytologist* 105, 449–457.

Deacon, L.J., Pryce-Miller, E.J., Frankland, J.C., Bainbridge, B.W., Moore, P.D. and Robinson, C.H. (2006) Diversity and function of decomposer fungi from a grassland soil. *Soil Biology and Biochemistry* 38, 7–20.

de Boer, W., Folman, L.B., Summerbell, R.C. and Boddy, L. (2005) Living in a fungal world: Impact of fungi on soil bacterial niche development. *FEMS Microbiology Reviews* 29, 795–811.

De Gryze, S., Jassogne, L., Six, J., Bossuyt, H., Wevers, M. and Merckx, R. (2006) Pore structure changes during decomposition of fresh residue: X-ray tomography analyses. *Geoderma* 134, 82–96.

Donnelly, D.P. and Boddy, L. (2001) Mycelial dynamics during interactions between *Stropharia caerulea* and other cord-forming, saprotrophic basidiomycetes. *New Phytologist* 151, 691–704.

Dowson, C.G., Rayner, A.D.M. and Boddy, L. (1988a) Foraging patterns of *Phallus impudicus*, *Phanerochaete laevis* and *Steccherinum fimbriatum* between discontinuous resource units in soil. *FEMS Microbiology Ecology* 53, 291–298.

Dowson, C.G., Rayner, A.D.M. and Boddy, L. (1988b) Inoculation of mycelial cord-forming basidiomycetes into woodland soil and litter. 2. Resource capture and persistence. *New Phytologist* 109, 343–349.

Dowson, C.G., Rayner, A.D.M. and Boddy, L. (1988c) The form and outcome of mycelial interactions involving cord-forming decomposer basidiomycetes in homogeneous and heterogeneous environments. *New Phytologist* 109, 423–432.

Duffy, B., Schouten, A. and Raaijmakers, J.M. (2003) Pathogen self-defense: Mechanisms to counteract microbial antagonism. *Annual Review of Phytopathology* 41, 501–538.

Elliott, E.T., Anderson, R.V., Coleman, D.C. and Cole, C.V. (1980) Habitable pore-space and microbial trophic interactions. *Oikos* 35, 327–335.

Fitter, A.H., Gilligan, C.A., Hollingworth, K., Kleczkowski, A., Twyman, R.M., Pitchford, J.W. *et al.* (2005) Biodiversity and ecosystem function in soil. *Functional Ecology* 19, 369–377.

Fogel, R. and Hunt, G. (1983) Contribution of mycorrhizae and soil fungi to nutrient cycling in a douglass-fir ecosystem. *Canadian Journal of Forest Research* 13, 219–232.

Fomina, M., Ritz, K. and Gadd, G.M. (2000) Negative fungal chemotropism to toxic metals. *FEMS Microbiology Letters* 193, 207–211.

Frankland, J.C. (1998) Fungal succession – unravelling the unpredictable. *Mycological Research* 102, 1–15.

Frey, S.D., Elliott, E.T., Paustian, K. and Peterson, G.A. (2000) Fungal translocation as a mechanism for soil nitrogen inputs to surface residue decomposition in a no-tillage agroecosystem. *Soil Biology and Biochemistry* 32, 689–698.

Frey, S.D., Six, J. and Elliott, E.T. (2003) Reciprocal transfer of carbon and nitrogen by decomposer fungi at the soil-litter interface. *Soil Biology and Biochemistry* 35, 1001–1004.

Gadd, G.M. (2007) Geomycology: biogeochemical transformations of rocks, minerals, metals and radionuclides by fungi, bioweathering and bioremediation. *Mycological Research* 111, 3–49.

Gange, A.C. (1993) Translocation of mycorrhizal fungi by earthworms during early succession. *Soil Biology and Biochemistry* 25, 1021–1026.

Garrett, S.D. (1951) Ecological groups of soil fungi: A survey of substrate relationships. *New Phytologist* 50, 149–156.

Gilligan, C.A. (1985) Probability-models for host infection by soilborne fungi. *Phytopathology* 75, 61–67.

Gormsen, D., Olsson, P.A. and Hedlund, K. (2004) The influence of collembolans and earthworms on a fungal mycelium. *Applied Soil Ecology* 27, 211–220.

Grassberger, P. (1983) Asymmetric directed percolation on the square lattice. *Journal of Physics A – Mathematical and General* 16, 591–598.

Gray, N.F. (1987) Nematophagous fungi with particular reference to their ecology. *Biological Reviews* 62, 245–304.

Groffman, P.M., Zak, D.R., Christensen, S., Mosier, A. and Tiedje, J.M. (1993) Early spring nitrogen dynamics in a temperate forest landscape. *Ecology* 74, 1579–1585.

Harris, K., Young, I.M., Gilligan, C.A., Otten, W. and Ritz, K. (2003) Effect of bulk density on the spatial organisation of the fungus *Rhizoctonia solani* in soil. *FEMS Microbiology Ecology* 44, 45–56.

Harris, M.J. and Boddy, L. (2005) Nutrient movement and mycelial reorganization in established systems of *Phanerochaete velutina*, following arrival of colonized wood resources. *Microbial Ecology* 50, 141–151.

Hassink, J., Bouwman, L.A., Zwart, K.B. and Brussaard, L. (1993) Relationships between habitable pore space, soil biota and mineralization rates in grassland soils. *Soil Biology and Biochemistry* 25, 47–55.

Hastings, H.M. and Sugihara, G. (1993) *Fractals: a User's Guide for the Natural Sciences*. Oxford University Press, New York.

Hawksworth, D.L. (2001) The magnitude of fungal diversity: The 1.5 million species estimate revisited. *Mycological Research* 105, 1422–1432.

Hedlund, K., Boddy, L. and Preston, C.M. (1991) Mycelial responses of the soil fungus, *Mortierella isabellina*, to grazing by *Onychiurus armatus* (collembola). *Soil Biology and Biochemistry* 23, 361–366.

Hobbie, E.A. and Agerer, R. (2010) Nitrogen isotopes in ectomycorrhizal sporocarps correspond to belowground exploration types. *Plant and Soil* 327, 71–83.

Holland, E.A. and Coleman, D.C. (1987) Litter placement effects on microbial and organic-matter dynamics in an agroecosystem. *Ecology* 68, 425–433.

Holmer, L. and Stenlid, J. (1997) Competitive hierarchies of wood decomposing basidiomycetes in artificial systems based on variable inoculum sizes. *Oikos* 79, 77–84.

Hudson, H.J. (1968) Ecology of fungi on plant remains above soil. *New Phytologist* 67, 837–874.

Jackson, R.B., Linder, C.R., Lynch, M., Purugganan, M., Somerville, S. and Thayer, S.S. (2002) Linking molecular insight and ecological research. *Trends in Ecology and Evolution* 17, 409–414.

James, T.Y., Kauff, F., Schoch, C.L., Matheny, P.B., Hofstetter, V., Cox, C.J. *et al.* (2006) Reconstructing the early evolution of fungi using a six-gene phylogeny. *Nature* 443, 818–822.

Jeger, M.J., Pautasso, M., Holdenrieder, O. and Shaw, M.W. (2007) Modelling disease spread and control in networks: Implications for plant sciences. *New Phytologist* 174, 279–297.

Jennings, D.H. (1991) The spatial-aspects of fungal growth. *Science Progress Oxford* 75, 141–156.

Joergensen, R.G. and Wichern, F. (2008) Quantitative assessment of the fungal contribution to microbial tissue in soil. *Soil Biology and Biochemistry* 40, 2977–2991.

Kampichler, C., Rolschewski, J., Donnelly, D.P. and Boddy, L. (2004) Collembolan grazing affects the growth strategy of the cord-forming fungus *Hypholoma fasciculare*. *Soil Biology and Biochemistry* 36, 591–599.

Kjoller, A. and Struwe, S. (1982) Microfungi in ecosystems – fungal occurrence and activity in litter and soil. *Oikos* 39, 389–422.

Kleczkowski, A., Gilligan, C.A. and Bailey, D.J. (1997) Scaling and spatial dynamics in plant-pathogen systems: From individuals to populations. *Proceedings of the Royal Society of London Series B – Biological Sciences* 264, 979–984.

Klein, D.A. and Paschke, M.W. (2004) Filamentous fungi: The indeterminate lifestyle and microbial ecology. *Microbial Ecology* 47, 224–235.

Klironomos, J.N. and Hart, M.M. (2001) Food-web dynamics – animal nitrogen swap for plant carbon. *Nature* 410, 651–652.

Klironomos, J.N., Bednarczuk, E.M. and Neville, J. (1999) Reproductive significance of feeding on saprobic and arbuscular mycorrhizal fungi by the collembolan, *Folsomia candida*. *Functional Ecology* 13, 756–761.

Langley, J.A. and Hungate, B.A. (2003) Mycorrhizal controls on belowground litter quality. *Ecology* 84, 2302–2312.

Last, F.T., Dighton, J. and Mason, P.A. (1987) Successions of sheathing mycorrhizal fungi. *Trends in Ecology and Evolution* 2, 157–161.

Leake, J.R., Johnson, D., Donnelly, D.P., Muckle, G.E., Boddy, L. and Read, D.J. (2004) Networks of power and influence: The role of mycorrhizal mycelium in controlling plant communities and agroecosystem functioning. *Canadian Journal of Botany* 82, 1016–1045.

Legrand, P., Ghahari, S. and Guillaumin, J.J. (1996) Occurrence of genets of *Armillaria* spp. in four mountain forests in central France: The colonization strategy of *Armillaria ostoyae*. *New Phytologist* 133, 321–332.

Lindahl, B.D. and Olsson S. (2004) Fungal translocation – creating and responding to environmental heterogeneity. *Mycologist* 18, 79–88.

Lombi, E. and Susini, J. (2009) Synchrotron-based techniques for plant and soil science: Opportunities, challenges and future perspectives. *Plant and Soil* 320, 1–35.

Maraun, M., Martens, H., Migge, S., Theenhaus, A. and Scheu, S. (2003) Adding to 'the enigma of soil animal diversity': Fungal feeders and saprophagous soil invertebrates prefer similar food substrates. *European Journal of Soil Biology* 39, 85–95.

Martin, W., Rotte, C., Hoffmeister, M., Theissen, U., Gelius-Dietrich, G., Ahr, S. *et al.* (2003) Early cell evolution, eukaryotes, anoxia, sulfide, oxygen, fungi first (?), and a tree of genomes revisited. *IUBMB Life* 55, 193–204.

McGonigle, T.P. (2007) Effects of animals grazing on fungi. In: Kubicek, C.P. and Druzhinina, I.S. (eds) *The Mycota. IV. Environmental and Microbial Relationships*. Springer-Verlag, Berlin, pp. 201–212.

Melillo, J.M., Aber, J.D. and Muratore, J.F. (1982) Nitrogen and lignin control of hardwood leaf litter decomposition dynamics. *Ecology* 63, 621–626.

Miller, R.M. and Jastrow, J.D. (2000) Mycorrhizal fungi influence soil structure. In: Kapulnik, Y. and Douds, D.D. (eds) *Arbuscular Mycorrhizas: Physiology and Function*. Kluwer Academic Publishers, Dordrecht, The Netherlands, pp. 3–18.

Miller, R.M. and Lodge, D.J. (2007) Fungal responses to disturbance: Agriculture and forestry. In: Kubicek, C.P. and Druzhinina, I.S. (eds) *The Mycota. IV. Environmental and Microbial Relationships*. Springer-Verlag, Berlin, pp. 47–68.

Miller, R.M., Reinhardt, D.R. and Jastrow, J.D. (1995) External hyphal production of vesicular-arbuscular mycorrhizal fungi in pasture and tallgrass prairie communities. *Oecologia* 103, 17–23.

Molofsky, J. and Bever, J.D. (2004) A new kind of ecology? *Bioscience* 54, 440–446.

Moore, D. (1991) Perception and response to gravity in higher fungi – a critical-appraisal. *New Phytologist* 117, 3–23.

Moore, D., McNulty, L.J. and Meskauskas, A. (2006) Branching in fungal hyphae and fungal tissue: Growing mycelia in a desktop computer. In: Davies, J.A. (ed.) *Branching Morphogenesis*. Springer, Berlin. pp. 75–90.

Obert, M., Pfeifer, P. and Sernetz, M. (1990) Microbial-growth patterns described by fractal geometry. *Journal of Bacteriology* 172, 1180–1185.

Olsson, P.A., Thingstrup, I., Jakobsen, I. and Baath, F. (1999) Estimation of the biomass of arbuscular mycorrhizal fungi in a linseed field. *Soil Biology and Biochemistry* 31, 1879–1887.

Olsson, P.A., Jakobsen, I. and Wallander, H. (2003) Foraging and resource allocation strategies of mycorrhizal fungi in a patchy environment. *Ecological Studies*, 93–116.

Otten, W., Hall, D., Harris, K., Ritz, K., Young, I.M. and Gilligan, C.A. (2001) Soil physics, fungal epidemiology and the spread of *Rhizoctonia solani*. *New Phytologist* 151, 459–468.

Otten, W., Bailey, D.J. and Gilligan, C.A. (2004a) Empirical evidence of spatial thresholds to control invasion of fungal parasites and saprotrophs. *New Phytologist* 163, 125–132.

Otten, W., Harris, K., Young, I.M., Ritz, K. and Gilligan, C.A. (2004b) Preferential spread of the pathogenic fungus *Rhizoctonia solani* through structured soil. *Soil Biology and Biochemistry* 36, 203–210.

Paine, T.D., Raffa, K.F. and Harrington, T.C. (1997) Interactions among scolytid bark beetles, their associated fungi, and live host conifers. *Annual Review of Entomology* 42, 170–206.

Park, A.W., Gubbins, S. and Gilligan, C.A. (2001) Invasion and persistence of plant parasites in a spatially structured host population. *Oikos* 94, 162–174.

Paterson, R.R.M., Simmonds, M.S.J. and Blaney, W.M. (1987) Mycopesticidal effects of characterized extracts of *Penicillium* isolates and purified secondary metabolites (including mycotoxins) on *Drosophila melanogaster* and *Spodoptora littoralis*. *Journal of Invertebrate Pathology* 50, 124–133.

Paustian, K. and Schnurer, J. (1987) Fungal growth-response to carbon and nitrogen limitation – a theoretical-model. *Soil Biology and Biochemistry* 19, 613–620.

Rast, D.M., Baumgartner, D., Mayer, C. and Hollenstein, G.O. (2003) Cell wall-associated enzymes in fungi. *Phytochemistry* 64, 339–366.

Read, N.D., Kellock, L.J., Collins, T.J. and Gundlach, A.M. (1997) Role of topography sensing for infection-structure differentiation in cereal rust fungi. *Planta* 202, 163–170.

Reddell, P. and Spain, A.V. (1991) Earthworms as vectors of viable propagules of mycorrhizal fungi. *Soil Biology and Biochemistry* 23, 767–774.

Richards, T.A., Dacks, J.B., Jenkinson, J.M., Thornton, C.R. and Talbot, N.J. (2006) Evolution of filamentous plant pathogens: Gene exchange across eukaryotic kingdoms. *Current Biology* 16, 1857–1864.

Rillig, M.C. (2005) A connection between fungal hydrophobins and soil water repellency? *Pedobiologia* 49, 395–399.

Ritz, K. (2007) *Spatial Organisation of Soil Fungi in Soils*. Kluwer, Dordrecht, The Netherlands.

Ritz, K. and Crawford, J.W. (1990) Quantification of the fractal nature of colonies of *Trichoderma viride*. *Mycological Research* 94, 1138–1141.

Rotheray, T.D., Jones, T.H., Fricker, M.D. and Boddy, L. (2008) Grazing alters network architecture during interspecific mycelial interactions. *Fungal Ecology* 1, 124–132.

Schack-Kirchner, H., Wilpert, K.V. and Hildebrand, E.E. (2000) The spatial distribution of soil hyphae in structured spruce-forest soils. *Plant and Soil* 224, 195–205.

Schimel, J.P., Jackson, L.E. and Firestone, M.K. (1989) Spatial and temporal effects on plant microbial competition for inorganic nitrogen in a California annual grassland. *Soil Biology and Biochemistry* 21, 1059–1066.

Schüffler, A. and Anke, T. (2009) Secondary metabolites of basidiomycetes. In: Anke, T. and Weber, D. (eds) *The Mycota. XV. Physiology and Genetics*. Springer-Verlag, Berlin, pp. 209–231.

Seal, J.N. and Tschinkel, W.R. (2007) Co-evolution and the superorganism: switching cultivars does not alter the performance of fungus-gardening ant colonies. *Functional Ecology* 21, 988–997.

Selosse, M.-A., Richard, F., He, X. and Simard, S. (2006) Mycorrhizal networks: les liaisons dangereuses. *Trends in Ecology and Evolution* 11, 621–628.

Shaw, P.J.A. (1992) *Fungi, Fungivores, and Fungal Food Webs*. Marcel Dekker, New York.

Siepel, H. and Maaskamp, F. (1994) Mites of different feeding guilds affect decomposition of organic-matter. *Soil Biology and Biochemistry* 26, 1389–1394.

Six, J., Frey, S.D., Thiet, R.K. and Batten, K.M. (2006) Bacterial and fungal contributions to carbon sequestration in agroecosystems. *Soil Science Society of America Journal* 70, 555–569.

Smith, M.L., Bruhn, J.N. and Anderson, J.B. (1992) The fungus *Armillaria bulbosa* is among the largest and oldest living organisms. *Nature* 356, 428–431.

Smith, S.E. and Read, D.J. (2008) *Mycorrhizal Symbiosis*. Academic Press, New York.

Stacey, A.J., Truscott, J.E. and Gilligan, C.A. (2001) Soil-borne fungal pathogens: Scaling-up from hyphal to colony behaviour and the probability of disease transmission. *New Phytologist* 150, 169–177.

Staddon, P.L., Ramsey, C.B., Ostle, N., Ineson, P. and Fitter, A.H. (2003) Rapid turnover of hyphae of mycorrhizal fungi determined by AMS microanalysis of C-14. *Science* 300, 1138–1140.

Sutter, R.P., Grandin, A.B., Dye, B.D. and Moore, W.R. (1996) Mating type-specific mutants of *Phycomyces* defective in sex pheromone biosynthesis. *Fungal Genetics and Biology* 20, 268–279.

Tagu, D., Kottke, I. and Martin, F. (1998) Hydrophobins in ectomycorrhizal symbiosis: Hypothesis. *Symbiosis* 25, 5–18.

Talbot, N.J. (1997) Fungal biology: Growing into the air. *Current Biology* 7, R78–R81.

Thompson, W. and Rayner, A.D.M. (1983) Extent, development and function of mycelial cord systems in soil. *Transactions of the British Mycological Society* 81, 333–345.

Tisdall, J.M. and Oades, J.M. (1982) Organic-matter and water-stable aggregates in soil. *Journal of Soil Science* 33, 141–163.

Tordoff, G.M., Boddy, L. and Jones, T.H. (2006) Grazing by *Folsomia candida* (collembola) differentially affects mycelial morphology of the cord-forming basidiomycetes *Hypholoma fasciculare*, *Phanerochaete uelutina* and *Resinicium bicolor*. *Mycological Research* 110, 335–345.

Tordoff, G.M., Boddy, L. and Jones, T.H. (2008) Species-specific impacts of collembola grazing on fungal foraging ecology. *Soil Biology and Biochemistry* 40, 434–442.

Tugay, T., Zhdanova, N.N., Zheltonozhsky, V., Sadovnikov, L. and Dighton, J. (2006) The influence of ionizing radiation on spore germination and emergent hyphal growth response reactions of microfungi. *Mycologia* 98, 521–527.

Unestam, T. (1991) Water repellency, mat formation, and leaf-stimulated growth of some ectomycorrhizal fungi. *Mycorrhiza* 1, 13–20.

Unestam, T. and Sun, Y.P. (1995) Extramatrical structures of hydrophobic and hydrophilic ectomycorrhizal fungi. *Mycorrhiza* 5, 301–311.

van der Wal, A., van Veen, J.A., Pijl, A.S., Summerbell, R.C. and de Boer, W. (2006) Constraints on development of fungal biomass and decomposition processes during restoration of arable sandy soils. *Soil Biology and Biochemistry* 38, 2890–2902.

van der Wal, A., de Boer, W., Smant, W. and van Veen, J.A. (2007) Initial decay of woody fragments in soil is influenced by size, vertical position, nitrogen availability and soil origin. *Plant and Soil* 301, 189–201.

Visser, S., Parkinson, D. and Hassall, M. (1987) Fungi associated with *Onychiurus subtenuis* (collembola) in an aspen woodland. *Canadian Journal of Botany* 65, 635–642.

Wallander, H., Nilsson, L.O., Hagerberg, D. and Baath, E. (2001) Estimation of the biomass and seasonal growth of external mycelium of ectomycorrhizal fungi in the field. *New Phytologist* 151, 753–760.

Wallander, H., Goransson, H. and Rosengren, U. (2004) Production, standing biomass and natural abundance of n-15 and c-13 in ectomycorrhizal mycelia collected at different soil depths in two forest types. *Oecologia* 139, 89–97.

Wells, J.M. and Boddy, L. (1995) Phosphorus translocation by saprotrophic basidiomycete mycelial cord systems on the floor of a mixed deciduous woodland. *Mycological Research* 99, 977–980.

Wessels, J.G.H. (1993) Wall growth, protein excretion and morphogenesis in fungi. *New Phytologist* 123, 397–413.

Wessels, J.G.H. (1997) Hydrophobins: Proteins that change the nature of the fungal surface. *Advances in Microbial Physiology* 38, 1–45.

Wessels, J.G.H., Devries, O.M.H., Asgeirsdottir, S.A. and Schuren, F.H.J. (1991) Hydrophobin genes involved in formation of aerial hyphae and fruit bodies in *Schizophyllum*. *Plant Cell* 3, 793–799.

Wicklow, D.T. (1979) Hair ornamentation and predator defense in *Chaetomium*. *Transactions of the British Mycological Society* 72, 107–110.

Wösten, H.A.B. (2001) Hydrophobins: Multipurpose proteins. *Annual Review of Microbiology* 55, 625–646.

Wösten, H.A.B., Devries, O.M.H. and Wessels, J.G.H. (1993) Interfacial self-assembly of a fungal hydrophobin into a hydrophobic rodlet layer. *Plant Cell* 5, 1567–1574.

Zak, J.C. and Visser, S. (1996) An appraisal of soil fungal biodiversity: The crossroads between taxonomic and functional biodiversity. *Biodiversity and Conservation* 5, 169–183.

Zhu, Y.-G. and Miller, R.M. (2003) Carbon cycling by arbuscular mycorrhizal fungi in soil-plant systems. *Trends in Plant Science* 8, 407–409.

9   Sensory Ecology in Soil Space

Iain M. Young* and Dmitry Grinev

1 Introduction

Knowing where you are helps getting you to where you need to be. Consider being dumped into an environment of which you had no prior knowledge. Add to this the fact that you are blind, and react to surfaces, electrical fields, gas concentrations and temperature. You also know what is up and what is down, so you are aware of gravity, which is good as you are spending your life underground. Also, most of the time you need to move in water, and your life cycle is at best a couple of days at optimum temperature, and the temperature is rarely optimal. Now, find your way to a food source and avoid obstacles and predators. What strategy would you adopt to keep yourself alive and functional?

This scenario roughly summarizes the challenges facing archaea, bacteria, protozoa, nematodes and, to a lesser extent, fungi in soil systems. These organisms control to a great extent our ability to utilize the Earth as a resource: growing food, clean water, etc.

These organisms reside and are active in a dark, dank, complex physical environment that exhibits extreme changes in moisture, temperature and architecture over very short spatial and temporal scales. Thus, they have

evolved perhaps unrivalled abilities to adapt rapidly to constantly changing environmental conditions. Alexander *et al.* (2006), in a summary of energy taxis in microorganisms, highlight the fact that genome sequencing of marine and soil microbes has shown that they possess large numbers of chemoreceptors, whereas microbes in environments that experience relatively constant environmental conditions are more likely to have significantly fewer chemoreceptors. The question is: How do they sense and respond to these conditions in soil?

2 The Habitat

In order to understand the strategies employed by soil organisms to enable them to negotiate through the most complex porous media system on the planet, we must first come to terms with this complexity. Various chapters in this book highlight several characteristics of soil. The salient points related to organism movement are that soils have (i) large internal surface areas (>20 m^2/g); (ii) complex organic and inorganic building blocks (and thus microhabitats); (iii) habitats that change geometry (dependent on clay content and water

*Corresponding author: iyoung4@une.edu.au

regimes); (iv) a hierarchy of pore throats that define the permeability and connectivity of gas and liquid phases; and (v) the greatest reservoir of biodiversity on the planet.

Within soil where light is absent or heavily restricted, except at the surface, organisms have developed strategies that do not rely on photosensitivity to negotiate the dark 3D matrix. This review will highlight certain strategies of soil biota in relation to specific functions. The soil will be presented as a fragmented habitat, where patchiness is a key characteristic. For the majority of soil-borne organisms, the patchy habitats that they experience impact directly on their behaviour, at both the individual and community level (Schooley and Wiens, 2003). How they perceive/detect a suitable habitat and environment are important determinants of what Schooley and Wiens (2003) term the 'Functional connectivity of landscapes'.

3 Sound Soil

An example of how soil and the geometry of its pore spaces impact directly on organism development and activity comes from an unlikely source. Lange et al. (2007) report on the cause of suboptimal hearing of subterranean rodents. Generally, mammals that reside in soil rely on audition for communication. Lange et al. (2007) measured the auditory signals (200–800 Hz) within the burrows created by rodents. Not surprisingly, attenuation of sound was far less within the burrows compared with surrounding soil. More surprisingly, however, the burrows also amplified the amplitude of the low-frequency sounds by a factor of 2 over 1 m. The authors suggest that such amplification selected for rodents had decreased hearing sensitivity to avoid overstimulation of the ear in soil, leading to the perceived suboptimal hearing of the rodents.

Sound or vibrations play key roles for many organisms in porous media. Drywood termites are able to assess the quality of wood using vibratory signals (Inta et al., 2007). Nocturnal scorpions can interpret vibrations through sand to determine the location of prey (Brownwell, 1984). To alert

the community to an attack by predators, head-banging termites hit their heads on the substrate (soil/wood) to produce substrate-borne vibrations. Parasitic nematodes utilize vibrations in soil to locate hosts (Torr et al., 2004), and their ability to do so depends on the organic matter component of the soil system.

Hill (2001) provides a sound review of vibration and animal communication, demonstrating that vibration is a potential attractant for many organisms, small and large. In soil, the efficacy of sound/vibrations as an attractant depends on the soil composition and the water regime, both of which will significantly modulate the impact of any vibration from source through the soil architecture.

4 Chemoattractants

Smell is a key facility in the life of microorganisms and nematodes in soil that allows them to sense potential attractants and/or repellents. Many soil organisms can detect and follow gaseous gradients (in air or in water) using a variety of mechanisms. Nematodes have chemosensors at their heads and tails. Attractants/repellents bind to receptors on the cell surface of bacteria (Thar and Kühl, 2003).

Anderson et al. (1997a, b) provide theoretical and experimental proof of how a complex soil architecture impacts on the movement of nematodes in both the presence and absence of an attractant. The pore space is seen significantly to mediate the diffusion of volatile attractants to nematodes as well as act as a significant barrier to nematode movement. Thus, at a wide range of scales the architecture acts as a modulator for nematode movement (mm+) and gas diffusion at small (nm+) and large (m+) scales. The extent of the impact of soil architecture on nematode activity is in itself modulated by the matric potential of the soil that determines the volume and distribution of water in three dimensions (3D) and thus modulates the gas diffusion and hydraulic connectivity of the system. Rönkko and Wong (2008), using a modelling approach termed 'particle system', suggest that the movement of nematodes can be attributed to its physical interactions with the virtual environments that they imposed. However, this is really a revisiting of much previous experimental work which has time and time again emphasized the importance of understanding nematode–habitat interactions.

Torr *et al.* (2004) suggest that organic matter can absorb specific chemoattractants, reducing their impact in helping nematodes to locate substrate. However, it is also clear that under decomposition conditions they emanate from a wide range of potential attractants/repellents and this, rather than any absorptive properties, impacts on the efficacy of specific chemoattractants in guiding pathogenic nematodes.

One important issue regarding the bacterial-feeding *Caenorhabditis elegans* nematode is that, despite it being one of the most powerful tools in neurobiology and despite the sensory ecology of the organism being much studied, little work has been carried out in relation to habitat–nematode interactions. The typical environment for *C. elegans* is an agar-filled Petri dish exposed to *Escherichia coli*, which clearly does not mirror its natural environment (soil, leaves, etc.). Rodger *et al.* (2004) highlight this discrepancy in relation to the nutrition available to *C. elegans*. Recently, this has again been emphasized by Lockery *et al.* (2008), who have developed what they term 'artificial dirt', which is a class of microfluidic devices that aims to mimic a moist soil matrix and allow highly precise delivery of fluid-borne stimuli. The Lockery *et al.* (2008) approach is a clever initiative in relation to current methodological deficiencies in *C. elegans* research. However, using real soil would also be a useful approach to habitat–nematode interactions, as it is seems impossible to reproduce the complexity of the physical architecture that is inherent in all soil systems.

Hall and Hedlund (1999) highlight the role of chemoattractants in collembolan activity. Collembola find fungi by moving within a diffusion gradient emanating from the fungi. The sting in the tail here is that the main predator of the collembola also uses signals emanating from the fungi to find the collembola.

An obvious candidate to evolve some significant chemoattractant response would be the symbiotic relationship between arbuscular mycorrhizae and roots. If no such biased foraging strategy existed then such interactions would be left to chance occurrences between the fungal tip and root membranes. Considering that 80% of all vascular plants benefit from such symbiotic relations, this seems unlikely. However, according to Sbrana and Giovannetti (2005), the ability of AM fungi to locate a host remains unclear. Whilst such attraction has been achieved in an *in vitro* system, this was for aerial hyphae and not soil-borne fungi (Gemma and Koske, 1988). The main issue, like all reports on potential chemoattractant effects, is to decouple attraction from growth enhancement. For true chemotaxis responses an organism, once it senses an attractant at a sufficient concentration, must change direction towards the attractant gradient to locate the source. Other responses, such as kinesis, do not require a change in the movement of an organism, but a change in response (faster speed, change in turning angles). Such responses are also seen in chemoattractant studies but accompanied by alterations in organism direction.

Sbrana and Giovannetti (2005) showed that the AM (arbuscular mycorrhizal) fungus *Glomus mosseae* reoriented its hyphae in the presence of a host root and was able to perceive attractant signals at 910 µm from the root surface. Such attractant has been evidenced in ectomycorrhizal fungi (Horan and Chilvers, 1990), and such mechanisms have clearly important consequences for the success of the fungus–root symbiotic relation.

A key chemical thought to be responsible for chemotaxic responses was, and remains in some quarters, CO_2. However, CO_2 is an unlikely candidate for anything other than large-scale attraction as, in any fertile soil, it is constantly being produced by a wide array of biotic components (Young *et al.*, 1988). Johnson *et al.* (2006) show that the movement of the root-feeding clover weevil is not significantly affected by

CO_2. Thus, CO_2 may be able to inform an organism that there is, generally, life (or decay) somewhere, but in a fertile soil this is less likely to provide the spatial coordinates of that activity.

5 Electrical Signals

Within soil systems, food sources reside in volume occupied by organic matter (either live cells or detritus). Anderson et al. (1997a, b) explain the importance of the root–bacteria–chemotaxis matrix in the ability of nematodes to seek and locate the rhizosphere of soil. Morris and Ward (1992) show that isoflavones, which occur in soybean root exudates, are effective in attracting zoospores. Also, when isoflavones were added to suspensions of active zoospores, rapid encystment and germination ensued. However, life in soil is not all about smell. van West et al. (2003), in an elegant set of experiments, highlight how zoospores use electrical fields emanating from root surfaces in locating roots. It would seem that electrical fields can work in tandem with other attractant processes, or in some cases override these processes in altering short-range tactic responses of zoospores.

Liu et al. (2002), using a simple capillary system, examined the importance of electrical signals in the movement of soil bacteria. At low voltages (<0.2 V/cm) typical bacterial movement was observed (swimming, tumbling), with a biased movement towards the anode and cathode. When the electrical field was removed swimming speed remained the same, and at higher voltages all bacterial cells moved only towards the cathode.

6 Oxygen

Soil-borne nematodes – as opposed to the typical 'lab-rat' C. elegans – have to manoeuvre across an incredibly complex 3D environment, relying on a range of attractants (and repulsions) to succeed in finding food. In soil, the half-life of any environment, or indeed physical habitat, can be short (seconds, minutes), dependent on, say, variations in matric potential. Thus, any soil-borne organism must not only be able to determine and respond to a range of chemical or environmental cues, but also be able to do so rapidly. In a short summary related to the memory of nematodes, Rankin (2005) highlights the response of nematodes to oxygen emanating from food sources. Rankin (2005) places emphasis on the apparent plasticity of the nematodes' response to environmental cues. It would seem that the preferred oxygen concentration of C. elegans may be lowered by cultivating the nematodes at lower levels of oxygen in the presence of food (Cheung et al., 2004). A similar response has been observed in relation to temperature thresholds (Mohri et al., 2005).

7 How Do We Observe Them in 3D Soil?

Monitoring or predicting any process within soil is rendered difficult, primarily due to the opacity of soils and the fact they are complex systems operating in 3D. Thus the majority of work on sensory ecology in soil has been limited to relatively simple constructs: quasi-soil systems that do not take account of the spatio-temporal heterogeneity in most real soils, or simple 2D systems that whilst permitting observations do not replicate the conditions in real soils. The advent of non-destructive 3D imaging techniques using a variety of radiation types (e.g. X-rays, MRI (magnetic resonance imaging) or neutron sources) offers real possibilities, not only to see inside soil and measure relevant pore characteristics that map directly on to the movement of biota in soils, but also to observe moisture in soils and ultimately the biota themselves. Using high-resolution X-ray tomography we have managed to examine the spatial position of biota in soils (Fig. 9.1). At present it is not possible to see microbes, but imaging technology is developing fast and it may be, in the not too distant future, that real-time

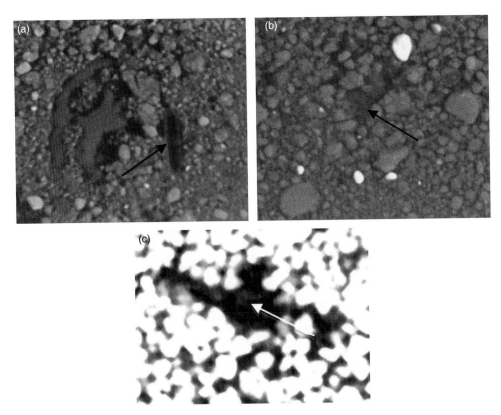

Fig. 9.1. (a) Wireworm (*Agriotes* spp.) near a piece of potato substrate; (b) larvae of *Sitona lepidus* in soil; (c) nematode (*Caenorhabditis elegans*) in a drop of water in packed sand. All organisms are arrowed in each image.

imaging of many biophysical processes, including those important for sensory ecology, can be captured non-invasively in real soils.

8 Conclusions

We know that the soil beneath our feet contains more individual organisms and a greater diversity of organisms than anywhere else on the planet. They happen to live in the most complex porous medium on the planet, which makes it extraordinarily difficult to observe and quantify their interactions with the soil matrix.

This complexity, both in the geometry of the soil architecture and the large variability of substrates that are accessible to them, means that soil organisms have developed unrivalled ways in which to navigate the soil maze, hide from predators and seek food. We have much to learn about our soil neighbours, and perhaps solutions to our problems above the soil will become evident as we invest more effort in what goes on beneath our feet.

References

Alexander, G., Greer-Pillips, S. and Zhulin, I.B. (2006) Ecological role of energy taxis in microorganisms. *FEMS Microbiology Reviews* 28, 113–126.

Anderson, R.A., Young, I.M., Sleeman, B.D., Griffiths, B.S. and Robertson, W.M. (1997a) Nematode movement along a chemical gradient in a structurally heterogenous environment. 1. Experiment. *Fundamental and Applied Nematology* 20, 157–163.

Anderson, R.A., Sleeman, B.D., Young, I.M. and Griffiths, B.S. (1997b) Nematode movement along a chemical gradient in a structurally heterogenous environment. 1. Theory. *Fundamental and Applied Nematology* 20, 165–172.

Brownwell, P.H. (1984) Prey detection of the sand scorpion. *Scientific American* 251, 86–97.

Cheung, H.H., Arellano-Carbajal, F., Rybicki, I. and de Bono, M. (2004) Soluble guanylate cyclasese act in neurons exposed to the body fluid to promote *C. elegans* aggregation behaviour. *Current Biology* 14, 1105–1111.

Gemma, J.N. and Koske, R.E. (1988) Pre-infection interactions between roots and the mycorrhizal fungus *Gigaspora gigantean*: chemotropism of germ-tubes and root growth response. *Transactions of the British Mycological Society* 91, 123–132.

Hall, M. and Hedlund, K. (1999) The predatory mite *Hyoaspis aculeifer* is attracted to food of its fungivorous prey. *Pedobiologia* 43, 119–131.

Hill, P.S.M. (2001) Vibration and animal communication: a review. *American Zoologist* 41, 1135–1142.

Horan, D.P. and Chilvers, G.A. (1990) Chemotropism – the key to ectomycorrhizal formation? *New Phytologist* 116, 297–301.

Inta, R., Lai, J.C.S., Fu, E.W. and Evans, T.A. (2007) Termites live in a material world. *Journal of the Royal Society Interface* 4, 735–744.

Johnson, S.N., Zhang, X., Crawford, J.W., Gregory, P.J., Hix, N.J., Jarvis, S.C. *et al.* (2006) Effects of CO_2 on the searching behaviour of the root-feeding clover weevil. *Bulletin of Entomological Research* 96, 361–366.

Lange, S., Burda, H., Wegner, R.E., Dammann, P., Begall, S. and Kawalika, M. (2007) Living in a 'stethoscope': burrow-acoustics promote auditory specializations in subterranean rodents. *Naturwissenschaften* 94, 134–138.

Liu, Z., Chen, W. and Papadopoulos, K.D. (2002) Electrokinetic movement of *Escherichia coli* in capillaries. *Environmental Microbiology* 1, 99–102.

Lockery, S.R., Lawton, K.J., Doll, J.C., Faumont, S., Coulthard, S.M., Thiele, T.R. *et al.* (2008) Artificial dirt: microfluidic substrates for nematode neurobiology. *Journal of Neurobiology* 99, 3136–3143.

Mohri, A., Kodama, E., Kimura, K.D., Koike, M., Mizuno, T. and Mori, I. (2005) Genetic control of temperature preference in the nematode *Caenorhabditis elegans*. *Genetics* 169, 1436–1450.

Rankin, C.H. (2005) Nematode memory: now where was I? *Current Biology* 15, R374–R375.

Morris, P.F. and Ward, E.W.R. (1992) Chemoattraction of zoospores of the Soybean pathogen, Phytophthorasojae, by isoflavones. *Physiological and Molecular Plant Pathology* 40, 17–22.

Rodger, R., Griffiths, B.S., McNicol, J.W., Wheatley, R. and Young, I.M. (2004) The impact of bacterial diet on the migration and navigation of *Caenorhabditis elegans*. *Microbial Ecology* 48, 358–365.

Rönkko, M. and Wong, G. (2008) Modelling *C. elegans* nematode and its environment using a particle system. *Journal of Theoretical Biology* 253, 316–322.

Sbrana, C. and Giovannetti, M. (2005) Chemotropism in the arbuscular mycorrhizal fungus *Glomus mosseae*. *Mycorrhiza* 15, 539–545.

Schooley, R.L. and Wiens, J.A. (2003) Finding habitat patches and directional connectivity. *Oikos* 102, 559–570.

Thar, R. and Kühl, M. (2003) Bacteria are not too small for spatial sensing of chemical gradients: an experimental evidence. *Proceedings of the National Academy of Sciences* 100, 5748–5753.

Torr, P., Heritage, S. and Wilson, M.J. (2004) Vibrations as a novel signal for host location by parasitic nematodes. *International Journal for Parasitology* 34, 997–999.

van West, P., Appiah, A.A. and Gow, N.A.R. (2003) Advances in research on oomycete root pathogens. *Physiological and Molecular Plant Pathology* 62, 99–113.

Young, I.M. (1998) Biophysical interactions at the root-soil interface: a review. *The Journal of Agricultural Science* 130, 1–7.

10 Managing the Interactions between Soil Biota and their Physical Habitat in Agroecosystems

Bev D. Kay[‡] and Lars J. Munkholm[*]

1 Introduction

Agroecosystems are, ideally, developed with a view to maximize harvestable yields of crops and livestock under conditions that maximize farm income and result in a net positive environmental impact. Swift and Anderson (1993) grouped biota in agroecosystems into (i) productive biota (plants and livestock); (ii) destructive biota (weeds, pathogens, pests); and (iii) resource biota (decomposer biota, cover crops, etc.). Soil biota residing in the soil include all three groups.

Soil biota serve many functions and provide a range of services that are critical to the sustainability of agriculture. These include the mineralization of organic materials, nutrient cycling, disease suppression, formation and stabilization of soil structure, fixation of atmospheric nitrogen in a form that can be utilized by plants and the transformation and detoxification of potential pollutants. However, agricultural management practices can also have a significant impact on the ability of biota to provide these services or functions. This occurs through management-induced changes in the physical habitat of soil biota and

alteration in their food sources and chemical environment. Management practices having the strongest impact on biota include choice of crops and crop sequence, tillage and residue management, traffic and the use of organic amendments. The effects of these practices are further influenced by water, pH and nutrient management. Management practices can have positive and negative impacts on habitat and food supply for biota and the balance that positive versus negative impacts of practices has on the long-term sustainable agricultural productivity of soil resources. These impacts can be manifested at scales ranging from the microscopic to the watershed.

Agricultural management practices are commonly combined in different ways to create farming systems. A large number of choices in management practices give rise to a multitude of different combinations or farming systems. Any given farming system reflects the resources on the farm (e.g. diversity of soils and field size, presence of hedgerows), access to finances, forces outside the farm (e.g. national or international agricultural policy, changes in climate) and the values and lifestyle choices of the farmer. Changes in any of these factors and

[‡]Deceased. [*]Corresponding author: lars.munkholm@agrsci.dk

the consequent alteration or replacement of one or more management practices create further challenges/opportunities for life in the soil.

Assessments of agricultural management practices have been based on yields, farm income and, more recently, on environmental impacts related to water and air quality. Less attention has been paid to soil biota and, until very recently, the research that has related management to biota has generally focused on the effects of management on the services or functions that biota perform. Little research has considered the impact of management practices on the physical habitat of biota or their food supply and concomitant effects on populations, activity and community structure.

This chapter will examine the impact of management practices on the services or functions that soil biota perform and, where possible, will relate that information to the impact of management practices on the physical habitat of biota and their food supply. Specific features of physical habitat that will be considered will be pore characteristics (total porosity, pore size distribution, pore continuity and tortuosity) and the balance between air- and water-filled pore space. Characteristics of food supply will include the total amount of crop residue returned to soil, its distribution with depth in the soil and its chemical composition. The chapter will primarily focus on current management practices employed in humid temperate environments. Preliminary assessments of practices that may be introduced in the future will be presented. Less consideration will be given to current or future practices in arid or tropical environments.

2 Choice of Crops and Crop Sequence

2.1 Crops and food sources for biota

Food quantity

In agroecosystems, photosynthates assimilated by crops serve as the main food source for soil biota – both the decomposer subsystem and the root-associated organisms (herbivores and pathogens and symbiotic biota). The net primary production (NPP) is usually high in agroecosystems in comparison with natural ecosystems but also highly variable depending on soil, crop and climate. This means that, potentially, there is an abundance of food available for soil biota in many agroecosystems. However, the net food supply will be markedly smaller when a substantial proportion of NPP is exported as plant and animal products. A relative large proportion of NPP is available for soil biota in systems such as grazed pastures where most resources are retained in the system, i.e. the only export of resources is meat. In contrast, lower proportions of NPP are available in systems with annual cash crops where a substantial part of NPP is exported in plant products. Monocultural cereal production systems with straw removal or burning are an extreme case, with a low proportion of NPP available for soil biota.

The photosynthate that is produced is distributed between above-ground and below-ground plant parts. Although there is an abundance of studies reporting crop yield and/or total above-ground biomass production in different farming systems, there is less information on below-ground inputs of photosynthates. This is partly due to the difficulties in sampling and measuring roots and root exudates. However, below-ground inputs are substantial (e.g. Jensen, 1993; Mutegi *et al.*, 2011). A large part of the assimilated C is root residue and the ratio of shoot to roots varies with species (Bolinder *et al.*, 1997). In addition, up to 20% of the assimilated C may be released into the soil during the growing season for different crops (Hütsch *et al.*, 2002). In a few studies the total production and distribution of photosynthates has been estimated. Kuzyakov and Domanski (2000) estimated the total C balance for a wheat crop yielding 6 t/ha and a pasture with 6 t/ha dry matter production (Table 10.1). They estimated a total C assimilation of 10 and 8 t/ha, respectively, for wheat and pasture. The proportion of total C assimilated

B.D. Kay and L.J. Munkholm

Table 10.1. Approximate total C input in the soil and root-derived CO_2 efflux from the soil under wheat with 6 t/ha grain yield[a] and in a pasture of about 6 t/ha dry matter production (Kuzyakov and Domanski, 2000).

	Total assimilated C (%)		Below-ground C (%)		C production (t/ha)	
	Wheat	Pasture	Wheat	Pasture	Wheat	Pasture
Shoot	50	30	–	–	4.8	2.4
Shoot CO_2[b]	25	30	–	–	2.4	2.4
Roots	13	20	52	50	1.2	1.6
Soil + OM[c]	3	5	12	13	0.3	0.4
Root CO_2[d]	9	15	36	38	0.9	1.2
Below-ground	25	40	100	100	2.4	3.2
Total assimilate	100	100			10.0	8.0

[a]It is assumed that total above-ground plant mass is twice that of grain yield and that the C content of dry roots and shoots is 40%; [b]shoot respiration; [c]remaining in soil and microorganisms; [d]root-derived CO_2: the sum of root respiration and rhizomicrobial respiration of rhizodeposits.

lost in shoot and rhizosphere respiration was estimated at 34% and 45%, respectively, for wheat and pasture. For wheat only 16% was assessed as remaining in below-ground inputs (soil, microorganisms and roots) in comparison with 25% for pasture. These authors considered that 50% of total assimilated C in wheat was accumulated in above-ground biomass. Grain yield would account for about 50% of total above-ground biomass, i.e. 25% of total assimilated C found in grain. This means that, for a wheat crop with straw removal, less than 20% of assimilated C is expected to end as food source for soil biota by the end of the growing season. In turn, much higher proportions of assimilated C will end as food source for soil biota in grazed pasture. Kuzyakov and Domanski (2000) determined that 30% of assimilated C was accumulated in above-ground biomass. Grazing animals will remove only a portion of this and, in addition, they will return most of the biomass intake as manure.

Food distribution in space and time

Photosynthates assimilated by crops are distributed to soil biota in the form of crop residues (root and shoot remains) and root exudates. In addition, above-ground herbivores indirectly supply soil biota with photosynthates in the form of excreta and dead organisms. The supply of photosynthates for soil biota shows large spatial and temporal variability. Most above-ground crop residues of annual cash crops are supplied at harvest time, when they may be retained on the surface or mechanically incorporated. Below-ground inputs from root residue are primarily released at harvest time, i.e. crop maturity in most cases. The vegetative growth phase is the time with highest relative translocation of photosynthates below ground (e.g. Jensen, 1993) and highest release of root exudates (Hütsch et al., 2002; Fig. 10.1). This means that below-ground inputs are higher for crops kept in the vegetative phase (e.g. grazed pasture) (Warembourg and Estelrich, 2000).

In space, roots play the role of dominant carrier of photosynthates to soil biota. Root-associated mycorrhizal hyphae represent an extension of the root system and carry substantial amounts of photosynthates beyond the roots, i.e. up to as much as 10% of NPP (Leake et al., 2004). Mycorrhizal hyphae distribute photosynthates to areas distant from the rhizosphere and interact with other soil microorganisms to create their own mycorhizosphere (Leake et al., 2004 and references therein). The distribution of photosynthates to deeper layers via the action of roots, mycorrhizae and also earthworms may be crucial for soil quality. Deep

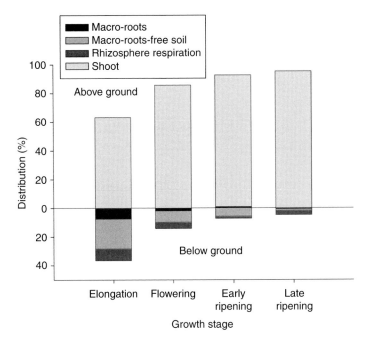

Fig. 10.1. Distribution of ^{14}C 6 days after start of labelling in a study with $^{14}CO_2$ pulse-labelled spring barley grown on a sandy loam (after Jensen, 1993).

rooting is essential for water and nutrient use efficiency and thereby productivity. In addition, developing and maintaining substantial decomposing potential at depths may be crucial for the decomposition of chemicals injurious to the environment.

Food quality

In general, food quality is high (i.e. easily decomposable) for most agroecosystems (i.e. low C:N ratio and low contents of recalcitrant crop material) (Table 10.2). Fertile ecosystems support decomposer food webs in which bacteria, microfauna and earthworms play important roles. In contrast, fungi and arthropods tend to dominate in infertile conditions (Wardle et al., 2004). Although food quality in general may be considered high in agroecosystems, there is an underlying large variation in food quality depending on plant species, growth stage and nutrient status (Swift et al., 1979; Rahn and Lillywhite, 2002). In general, crop residues high in N and low in lignin are readily decomposed

by soil biota (Melillo et al., 1982; Berendse et al., 1987). For instance, legume residue (rich in N) has been found to decompose faster than cereal residue (Cookson et al., 1998) or maize residue (Pare et al., 2000). In addition, Christensen (1986) found an increase in initial decomposition rate of barley straw with increased N content. The growth stage of the crop is also of importance as residue quality is higher in the premature stage (e.g. fresh vegetables and green manure) than in the mature stage (e.g. Seneviratne, 2000).

The quality of below-ground inputs also shows high variability. Root exudates are very labile and a large part of the input may decompose within a few days. Hütsch et al. (2002) showed that approximately 86% and 64% of root-borne C input was respired by microorganisms within 3 days for wheat and lucerne, respectively. For dead roots, as for above-ground material, availability depends on C:N ratio. In general, fine roots are more easily decomposed than coarse and more woody primary roots.

Table 10.2. Average content of major organic components in plant materials (from Haider, 1992).

Material	Component (%) dry weight			
	Cellulose	Hemicellulose	Lignin	Protein[a]
Ryegrass (mature)	19–26	16–23	4–6	12–20
Lucerne (stem)	13–33	8–11	6–16	15–18
Wheat straw	27–33	21–26	18–21	3
Pinus sylvestris (sawdust)	42–49	24–30	25–30	0.5–1.0

[a]N content $\times$ 6.25.

2.2 Crops and physical habitat

Soil biota live in highly urban societies. Young and Crawford (2004) estimated that only about 10^{-6}% of the soil surface area is covered by soil microbes. Soil biota colonize where food is available and abiotic factors favour activity. Crop residue and living roots constitute the dominant food sources for soil biota. Roots and associated mycorrhizal hyphae are the main carriers of photosynthates and soil microbes colonize preferentially along these food-source pathways. The microbial biomass is twofold or higher in the rhizosphere than in root-free soil (e.g. Cheng *et al.*, 1996). There is a direct effect of plant growth on physical habitat (i.e. pore space) as root proliferation creates new continuous coarse pores. However, root growth also induces soil compaction around roots (Bruand *et al.*, 1996; Clemente *et al.*, 2005). The compaction effect decreases more or less exponentially with distance from roots. Soil compaction around roots changes the physical and chemical conditions for microbial activity. Clemente *et al.* (2005) showed decreased water infiltration in the compacted zone around roots of *Eucalyptus grandis*.

Crops may also have an indirect effect as the primary supplier of organic matter. Organic matter and soil biota play an important role in soil structure formation and stabilization (Tisdall and Oades, 1982). Young and Crawford (2004) suggested a conceptual model for self-organization in the soil–microbe complex where availability of food source stimulates microbial activity, which subsequently causes a change towards a more open aggregated structure. This change may be beneficial for both productive and resource biota, i.e. it creates a better environment for both root growth (lower penetration resistance) and decomposition (increased gas exchange).

In the subsoil, tap-rooted crops in particular are able to modify the pore size distribution of dense soil by creating continuous macropores (Cresswell and Kirkegaard, 1995; Chen and Weil, 2010). These are, together with earthworm burrows, commonly labelled biopores. The vital importance of biopores as pathways for deep penetration of roots in dense soil has been shown in field studies (Ehlers *et al.*, 1983), model experiments (Dexter, 1986) and simulation studies (Jakobsen and Dexter, 1988). Deep rooting may increase nutrient uptake and retention in the root zone (Kristensen and Thorup-Kristensen, 2004) as well as increase available water content (Jakobsen and Dexter, 1988). Biopores may also improve infiltration and aeration (Ehlers *et al.*, 2000). In the subsoil, there is a much lower density of soil biota than in the topsoil (e.g. Jorgensen *et al.*, 2002) and a concentration along the biopores (Vinther *et al.*, 1999). The multiple benefits from biopores justify means to maintain an appropriate density of biopores. It is important to bear in mind that biopores may be produced rather slowly but may be long lasting (Jakobsen and Dexter, 1988).

2.3 Rotation and sequences of crops

Monoculture cropping will, as a rule, cause increased problems with destructive soil biota, i.e. increased incidence of soil-borne disease (Alabouvette *et al.*, 2004). For a few diseases, continuous cropping may result in disease decline over a number of years with take-all disease in cereals as the best-known example (Hornby, 1998). Crop rotations with non-host crops provide time for disease decrease due to decomposition of pathogens by resource biota. Monoculture cropping may also result in an unbalanced diet for soil biota. It has been assumed that low diversity in above-ground biota causes low food source diversity and subsequently low species and functional diversity in below-ground biota (Swift *et al.*, 1996). This suggests that management practices which increase above-ground biodiversity (e.g. crop rotations and cover crops) potentially would have positive effects on key soil functions such as disease suppression and decomposition. However, the causal link between above-ground and below-ground biota is not necessarily straightforward (Van der Putten *et al.*, 2009). A number of studies confirm the link between above-ground and below-ground diversity in agroecosystems (e.g. Lupwayi *et al.*, 1998; Alvey *et al.*, 2003), but experimental evidence for the link between diversity and function is lacking. For instance, a non-host crop should be more effective than fallows in disease suppression, but this has not always been confirmed in experimental studies (Alabouvette *et al.*, 2004). In addition, Degens (1998) found that an experimentally induced decrease in microbial diversity did not consistently result in reduced rate of wheat straw decomposition.

A number of studies have shown that crop rotations which include legumes and/ or grass increase microbial biomass in the soil (Biederbeck *et al.*, 1984; Collins *et al.*, 1992). This is supposedly strongly related to increased food input for soil biota. The inclusion of cover crops is expected to display a similar trend. A side effect of increased food input will in most cases be an increase in soil organic matter content (Dick and Gregorich, 2004). Soil organic matter (SOM) may serve as a buffer food source and thus help to maintain populations of soil biota in periods of low food inputs.

2.4 New management practices

Adaptation to climate change is one of the greatest challenges to agriculture. The topic has received considerable attention through the last decade and a number of reviews have been published on the consequences of climate change for agriculture (Olesen and Bindi, 2002; Lal, 2010). Farmers need to adapt new crops and crop rotations and learn to control new weeds and pests. They must also face new problems such as increased erosion risk in some areas (e.g. northern parts of the northern hemisphere) and increased risk of water shortage in others. The changes in climate and agricultural management will have important impacts on soil biota. Increased crop productivity from elevated CO_2 levels will result in enhanced food supply. However, this effect will be counteracted by decreased crop productivity due to water shortage in some areas. Soil habitat may be changed in a favourable direction in some agroecosystems: Rillig *et al.* (1999) showed increased aggregation for two grassland soils with elevated CO_2 levels. However, the net effect of climate change on food supply and habitat for soil biota will most likely display large variability. The response will depend on complex interactions between climate, soil and crop growth as well as climate triggered changes in management. Changes in the interactions between soil and soil biota itself may also affect food supply and habitat.

Organic farming has gained considerable interest in the last couple of decades. Mineral fertilizers and pesticides are avoided in organic farming. Nutrient supply and weed and pest control are achieved through the use of crop rotations (including legumes), cover crops, green manure crops and animal manure. Applying these management practices indicates a diverse food supply with implications for soil biota as described above. On the other hand, crop productivity is usually significantly lower in organically farmed systems, which indicates lower food supply. A large number of empirical studies have shown positive effects of organic farming on soil structure and biota (e.g. Reganold,

1988; Mader *et al.*, 2002; Schjønning *et al.*, 2002a; Elmholt *et al.*, 2008). Differences are in general largest when comparing organic farming systems with cereal-based conventional systems. For instance, Schjønning *et al.* (2002a) and Elmholt *et al.* (2008) found substantial differences when an organic system based on forages and animal manure was compared with a conventional cereal-based system without manure and with straw removal. Much smaller differences were found when comparing the organic system with a conventional system based on forages and animal manure. Also for grassland soil, small differences have been encountered in comparison with conventional forage-based cropping systems (Yeates *et al.*, 1997).

The incorporation of genetically modified crops into farming systems holds great promise for increasing agricultural output. These new crops include plants with resistance to specific herbicides or containing toxins for specific insects, plants with improved resistance to stress conditions, plants with better nutritional qualities and plants that produce industrial products or precursors. Among potential impacts of these crops are changes in the activity and population structure of microbial communities and consequently the rates at which the population of soil biota perform specific functions (e.g. crop residue decomposition and soil structure development). The effect of transgenic crops on soil biology has been intensively studied in recent years. In particular, concerns have been raised over the cultivation of *Bt*-modified crops as their crop residues remain in the soil. A microcosm study by Flores *et al.* (2005) indicated that plant residues from transgenic *Bt* plants decomposed more slowly than non-*Bt* plants. However, this has not been confirmed in more recent studies (Birch *et al.*, 2007). Also, Zurbrügg *et al.* (2010) did not find any difference in decomposition dynamics between *Bt* maize and non-*Bt* near-isolines. The concentration of *Bt* protein decreased rapidly within a few months after harvest of maize and therefore non-target soil organisms were exposed to relatively low concentrations of *Bt* proteins.

3 Tillage and Residue Management

Tillage causes a major disturbance of the soil ecosystem and thus affects most ot its components and functions. The reason for tillage is to improve the conditions for crop establishment and growth. Tillage may be required to produce a desirable seedbed, incorporate organic matter and to control weeds and pests. The latter may not necessarily be required in modern conventional agriculture where effective pesticides are available. In organic farming, however, tillage is needed for weed and pest control (Lampkin, 1990). There are a great variety of forms of tillage around the world, as tillage systems are adapted to different climates, soils and crops as well as to different management of residue, and weed and pest control. Conventional tillage (CT) consists in most parts of the world of mouldboard ploughing of the arable layer (c. 25 cm depth) and intensive seedbed preparation in the top 0–5 cm layer. This is a system with major disturbance (mixing and inversion) of the tilled layer and thus massive impact on food source distribution, as well as on physical habitat. Reduced-tillage systems involve in general less disturbance than conventional tillage. However, the term covers systems ranging from those involving a high degree of disturbance (harrowing or disking to more than 10 cm depth) to no-tillage (NT) systems with only the disturbance caused by the drill.

3.1 Tillage and physical habitat

Tillage drastically alters the physical habitat in the tilled layers and may also have significant effects on the physical habitat in adjacent layers (e.g. via tillage pan formation). In general, tillage operations result in an overall loosening of the soil, i.e. an increase in total porosity. The immediate increase in porosity after tillage may be very substantial. For instance, Andersson and Håkansson (1966) found that the depth of the ploughed layer was c. 7 cm higher immediately after autumn mouldboard ploughing than after harvest. This corresponds to an increase in total porosity from c. 50 m³/100 m³ to 63 m³/ 100 m³. In addition, deep-tine cultivation may cause a substantial increase in total porosity (e.g. Schjønning et al., 2002b). The pores produced by tillage are not stable and the soil becomes denser shortly after tillage due to natural densification or traffic. Schjønning et al. (2002b) showed a decline in total porosity from 62 to 56 m³/100 m³ in the 0–4 cm layer within 18 days after mouldboard ploughing for a sandy loam soil (Fig. 10.2). Five months after ploughing total porosity had decreased to the initial level.

Numerous studies have shown that continuous NT or shallow tillage produces lower total porosity in the non-cultivated topsoil (0–20 cm) (Douglas and Goss, 1987; Kay and VandenBygaart, 2002; Kadziene et al., 2011) and decreases the volume of macropores (Douglas et al., 1980; Schjønning, 1989). The effect of NT on volume of macropores appears to be soil type and time dependent. Schjønning and Rasmussen (2000) found that NT decreased the volume of macropores (>30 μm) in a sandy and a

silty soil but had the opposite effect for a sandy loam soil. VandenBygaart et al. (1999) showed that the number of large pores in the near-surface layer (0–3 cm) increased over time for an NT soil both in absolute terms and relative to a CT soil. However, at least 6 years of NT was needed before the number of irregular large pores was higher in the NT than in the CT soil. These authors related the increase in volume of large pores to accumulation of organic matter and increased earthworm activity in the surface layer. In addition, Voorhees and Lindstrom (1984) observed a recovery in porosity in the topsoil within a limited number of years (3–4) after conversion to shallow tillage. If NT is combined with cautious traffic, this stress release may in the long term cause the formation of strong but less dense aggregates due to repeated wetting and drying, as hypothesized by Horn (2004). This should increase pore continuity and improve air and water permeability. Horn (2004) showed an improved saturated hydraulic conductivity

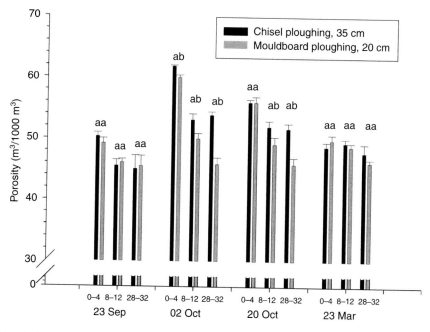

Fig. 10.2. Soil porosity at three depths (0–4, 8–12 and 28–32 cm) measured before tillage treatments (23 September) to the spring (23 March). The two intermediate dates are respectively a few days and three weeks following the tillage treatments. Columns with identical letters are not significantly different ($p > 0.05$).

over time for a sandy loam under shallow tillage in comparison with CT. The changes were most pronounced in the surface layers. but after >7 years of continuous shallow tillage the changes became detectable in the deeper layers as well.

The effect of shrinkage and swelling are most pronounced for clayey soils and in relative dry climates. Indeed, pore characteristics of smectite soils with high clay content appear to be determined largely by wetting and drying and have been found to be relatively insensitive to management practices (Eynard *et al.*, 2004). Better ability to conduct air and water for long- term shallow tillage may also be related to a larger number of persistent continuous biopores (i.e. root channels and earthworm channels), as found in many studies (Ball, 1981; Schjønning, 1989; Tebrügge and Düring, 1999). Schjønning (1989) also found a higher continuity of pores > 200 μm (e.g. biopores) for shallow tillage than for CT after 18 years of shallow tillage of a fine loam. In contrast, he showed lower continuity of smaller macropores (30–200 μm) in shallow-tilled soil. The influence of continuous large macropores was not enough to secure equal or better soil aeration in shallow-tilled than in CT soil as modelled by Schjønning (1989). He calculated substantially lower O_2 concentrations below c. 5 cm depth for the shallow-tilled as compared with the CT soil. The large decline in O_2 concentrations below 5 cm depth in shallow-tilled soil was related to the formation of a dense rotovator pan. Other studies have shown also that drills for NT may directly cause densification below and/or around the disturbed layer (e.g. Munkholm *et al.*, 2003).

An interesting characteristic often observed for NT soil is the development of a platy structure in the topsoil (Drees *et al.*, 1994; VandenBygaart *et al.*, 1999; Munkholm *et al.*, 2003). Platy structural units are separated by lateral interconnected voids. These voids are supposedly of key importance in relation to aeration, root profilation, etc. – especially of soil distant from vertical biopores. The study by Ball and Robertson (1994a) supports this hypothesis. They showed that air-filled porosity, diffusivity and permeability were significantly greater in horizontal than vertical samples taken in an NT soil.

Tillage-induced changes in pore characteristics, aeration, water content, soil temperature and depth of residue incorporation induce a wide range of responses among different species of soil biota. Kladivko (2001) concluded, in a comprehensive review, that most organism groups have greater abundance or biomass in NT than CT and that larger organisms in general appear to be more sensitive to tillage operations than smaller organisms. The diversity of microfauna was relatively unaffected by tillage, but macrofauna showed a wide range in response with diversity sometimes increasing and sometimes decreasing with increasing soil disturbance.

Little attention has been paid to the interaction of tillage and the nature of crop residue. Fox (2003) noted this deficiency and concluded that 'more research is needed to determine whether variations in organic residue composition and quantity in combination with different intensities of disturbance from tillage will lead to changes in species composition and on the capacity of biota to contribute to soil processes'.

3.2 Residue management

Residues as food source and habitat

Postharvest above-ground crop residues may constitute an important food source for soil biota. For a wheat crop, Kuzyakov and Domanski (2000) estimated that approximately 25% of total assimilated C was accumulated in above-ground straw and stubble residues. However, in many cropping systems crop residues are removed from the field (e.g. livestock feed or energy source) or burnt on the site. Removal or burning may cause a gradual depletion of SOM and a diminished food source for soil biota. Results from Danish long-term (36 years) field experiments show that straw burning or removal resulted in an increased depletion of SOM by 50% in comparison with straw incorporation (Schjønning *et al.*, 2009). The treatments were conducted in a monoculture spring barley cropping system with no external organic matter inputs. Therefore, SOM content decreased even in straw-incorporated soil. Model estimates indicated that half of the potential increase (or smaller decrease) in SOM content caused by straw incorporation was reached after only 5 years. The SOM content may increase linearly with the applied annual input of straw biomass, as observed by Thomsen and Christensen (2004). They reported results from another Danish long-term study (18 years) in which the annual addition of 0, 4, 8 and 16 t/ha barley straw was incorporated. The results indicate that the rate of straw decomposition was not reduced even with extremely high annual inputs (16 t/ha). However, prior removal or burning of crop residues may significantly reduce decomposition rate. In a New Zealand study, prior removal of wheat straw was found to reduce the decomposition rate of wheat straw by as much as 50% (Cookson et al., 1998).

Many studies have shown that soil biota respond strongly to food inputs from crop residues. For instance, Powlson et al. (1987) found that straw incorporation produced a c. 40% increase in microbial biomass in an 18-year Danish experiment on two sandy soils. For macrofauna the effect of straw retention may be even more pronounced. Haukka (1988) found that removal of straw reduced the earthworm population by 50% in a 12-year Finnish field experiment.

Crop residues may serve as a temporary habitat for a great number of pathogens. Inoculum from pathogens may survive on crop residues and subsequently infect the following crop (Alabouvette et al., 2004). In contrast, retention of crop residues may boost the level of general suppression in the soil. The role of crop residues as a bridge in the disease cycle is especially important in monoculture cropping systems.

Tillage effects

Incorporation of crop residues may be a principal reason for tillage and have a range of benefits in relation to crop production. Crop residue incorporation may accelerate decomposition, decrease the amount of pathogen-infected plant litter on the soil surface, facilitate better seedbed quality and result in the burial of newly shed weed seeds. A number of studies have shown that decomposition is increased by incorporation (Christensen, 1986; Curtin et al., 2008). Incorporation strongly improves the soil/residue contact and provides more favourable moisture and temperature conditions for decomposition. The incorporation of residues to different depths in the topsoil (0–20 cm) seems not to influence decomposition rate (Christensen, 1986). The decomposition of SOM in general (including previously applied crop residues) is usually also increased by tillage due to improved conditions for decomposition (moisture, aeration and temperature) and by exposing protected SOM (Dick and Gregorich, 2004). Despite this, the effect of shallow tillage or NT on SOM content is not so clear when the whole soil profile is taken into account. Some studies show that NT increases SOM content as compared with conventional tillage (Mishra et al., 2010), whereas other studies show no difference (Blanco-Canqui and Lal, 2008; Hermle et al., 2008, Christopher et al., 2009).

3.3 Decomposer food web

Resource soil biota responds directly to changes in physical habitat and food source input and distribution induced by tillage and residue management. Hendrix et al. (1986) found that CT favoured bacterial-based rather than fungal-based decomposition. This hypothesis has been confirmed in a number of later studies (Wardle et al., 1995). Thus, NT systems decomposer food webs have in some respects more in common with characteristic food webs from infertile ecosystems than from fertile ecosystems (Wardle et al., 2004). From other perspectives, for instance, the dominant role of earthworms in NT soil, the decomposer food web is more similar to the characteristics of fertile ecosystems. The slow decomposition rate may be beneficial in agroecosystems in relation to closed nutrient cycling, erosion control and carbon storage. Adversely, slow decomposition indicates low nutrient supply rate, which may be unfavourable for crop productivity in low-nutrient input or manure-based systems.

3.4 New management practices

A search for alternative energy sources is occurring in many developed countries in order to reduce reliance on imported fuels. Crop residues are being actively considered as an alternative source in many jurisdictions. Removal of crop residue must be balanced against potential increases in erosion, decrease in SOM content and subsequent deterioration in soil structure. Blanco-Canqui and Lal (2009) concluded in a major review showing that SOM content rapidly decreased with increased residue removal, and this may have severe consequences in terms of reduced soil quality. They called for more research on the establishment of site-specific threshold levels of residue removal rates. In Denmark, crop residues have been used as an energy source in combined heat and power stations for more than a decade as a consequence of a national plan passed by parliament in 1993. Christensen (2002) evaluated the plan and he estimated a decrease of up to 15% in organic matter content as a result of straw removal. The decrease in SOM content would be expected to have detrimental effects on the physical habitat of soil biota. These results reinforce the cautions expressed by Blanco-Canqui and Lal (2009) and suggest that continuous straw removal in cereal-based cropping systems in Denmark is incompatible with sustainable productivity. Christensen (2002) suggested compensatory changes in management (e.g. use of perennial grass–clover and cover crops) as a means to counteract a decline in organic matter content.

4 Traffic

Traffic with farm machinery may cause soil compaction and thereby significantly alter the physical habitat as well as the physical/chemical conditions for crops and soil biota. Recovery from soil compaction is faster in topsoil than in subsoil due to the greater prevalence in the topsoil of a combination of both natural processes (biological activity, freeze–thaw and drying–wetting cycles) and mechanical loosening. However, spatial variability in pore characteristics in the topsoil is increased by traffic. Fragments of the compacted topsoil in wheel tracks are mixed with less compacted soil by tillage, and these fragments often have no visual pores (Richard et al. 1999). The volume fraction of these fragments in the whole soil is a function of the aerial extent of trafficked zones, the extent of compaction in the wheel track (Richard et al., 1999), tillage (Roger-Estrade et al., 2009) and would be expected to vary with wetting–drying and freezing–thawing. In contrast, subsoil compaction is more persistent and, in combination with the increasing weight of farm machinery, has been recognized as one of the most severe threats to soil quality in modern agroecosystems.

4.1 Traffic impact on habitat

Traffic-induced soil compaction alters the size distribution and connectivity of the pore system. This topic was extensively reviewed by Kooistra and Tovey (1994). Large pores are first reduced in size and therefore macroporosity is normally affected more than total porosity. Traffic induces mainly vertical stresses near the surface and therefore horizontal and obliquely oriented pores are in general more affected by soil compaction than vertically oriented (Hartge and Sommer, 1980; Kremer et al., 2002; Schäffer et al., 2008). Soil compaction not only minimizes the space available for soil biota but may have an even stronger effect on the connectivity of the pore system. Ball and Robertson (1994b) observed that compaction caused a small reduction in air-filled porosity but a large decrease in air diffusivity and permeability when compacting two medium-textured Scottish soils.

In arable soil soil biota does not encounter uniformly compacted soil, as there is a large spatial and temporal variability. Spatially, the degree of compaction varies across the landscape due to the occurrence of wheel tracks, shifts in texture, etc. In the profile the degree of soil compaction varies as a result of, for example, tillage/traffic pans, increasing overburden pressure with depth and changes in texture. For tillage/traffic pans a platy structure is commonly found and this may strongly impede root growth and water and air transport. The platy structure is formed as a de-loading effect after wheel passage (Kooistra and Tovey 1994). Temporally, soil hardness varies with water content. Furthermore, compacted soil may over time be naturally alleviated through the effect of weather (drying–wetting and frost–thaw cycles) and soil biological activity (e.g. earthworm burrowing and root growth).

4.2 Response of soil biota

A multitude of studies have shown that soil compaction may strongly affect spatial and temporal root development. In heavily compacted soils, roots are normally concentrated in surface layers but the total root biomass is not necessarily affected (Whalley et al., 1995). Plant roots respond to compacted layers by thickening and/or deflection. When a root is deflected, it often grows horizontally until it finds a vertical pore (e.g. Munkholm, 2000). The greater the number of vertical pores, the greater the likelihood of bypassing the compacted layer. Extensive soil compaction often results in slow growth, which Passioura (2002) labelled 'the bonsai effect', as the slow growth might reflect a similar response to that of bonsai plants grown in small containers.

Excessive soil compaction has also been found to reduce earthworm abundance in many studies. For instance, there was a strong initial response to traffic in regard to earthworm numbers, i.e. 680 earthworms/m^2 and 160/m^2 in lightly and normally trafficked soil, respectively, 3–5 years after the establishment of a long-term Norwegian field experiment (Hansen and Engelstad, 1999). The abundance of earthworms decreased with time in the lightly trafficked treatment, which was attributed to acidification and low manure input. Interestingly, epigeic (surface-living species) and endogeic (within-soil living) displayed a similar response to soil traffic.

The concentration of food source distributors such as roots and earthworms in loose zones means that food distribution to soil microorganisms is restricted in severely compacted zones. Furthermore, the supply of oxygen is also reduced in heavily compacted zones due to reduced diffusivity of air. The result of a smaller food supply and reduced rate of oxygen transport will inevitably be a reduced population of soil microorganisms. This may itself cause an even poorer soil structure due to the interactions between soil and microbes, as hypothesized by Young and Crawford (2004). Apparently, a compaction event may trigger a vicious circle in the complex soil–soil biota system. A long duration with stress release may be needed for soil to recover.

4.3 New management practices

In principle there are two strategies that can be followed to minimize excessive compaction in agricultural soils – either to reduce soil stresses and increase soil strength or to minimize the compacted area. Increasing the contact area and lowering ground pressure by using wider, larger and more flexible tyres or tracks may reduce soil stress. The question is to what extent this strategy is valid. The weight of farm machinery is ever increasing, which accelerates the need for increasingly larger and better tyres or tracks. The use of lightweight automated vehicles may constitute a realistic alternative in the future, and no doubt this 'small is beautiful' approach is very promising (Alakukku et al., 1997). However, the challenge is to develop flexible and efficient systems. In addition to a reduction in stress, an increase in soil strength is a means of decreasing soil compaction. A well-known means of increasing soil strength is to improve drainage; this may cause a drier and thereby stronger soil at the time of traffic.

Controlled traffic farming (CTF) is an effective measure to decrease the compacted area. Traffic is avoided between wheel tracks within and across years. The aim is to attain a loose, well-structured soil between firm wheel tracks. Pioneering work (e.g. Chamen *et al.*, 1994) has demonstrated promising effects of CTF on soil structure and crop growth. Modern guidance technology has facilitated the use of CTF and, today, this technology has been widely adopted (Tullberg *et al.*, 2007; Qingjie *et al.*, 2009; Vermeulen and Mosquera, 2009). Other prevention strategies to minimize compaction have been discussed in recent reviews (van den Akker and Schjønning, 2004; Hamza and Anderson, 2005).

5 Water Management

Water is managed in agricultural soils to (i) minimize the stress on crops created by too little or too much water; (ii) improve the timeliness of different operations such as tillage, seeding and harvesting; and (iii) diminish the negative impacts of traffic. The two most common forms of water management are irrigation and drainage. Irrigation is used to minimize the occurrence of drought, particularly where high-value crops are grown or where arid and semi-arid conditions jeopardize production of most crops. Artificial drainage is used to remove excess water, particularly on medium- and fine-textured soils in humid temperate environments and on irrigated sites. Changes in water management practices can result in changes in (i) the amount of crop residue returned to soil; (ii) the distribution with depth in the soil profile of root residues; and (iii), where changes in water management result in selection of different crops, alteration of the composition of residues being returned. Changes in water management practices also imply changes in the temporal variation in water-filled pore space through the growing season. In the longer term,

changes in water management may result in changes to pore characteristics.

5.1 Irrigation

Irrigation commonly results in higher crop yields, with more surface and root residues returned to the soil. However, irrigation also tends to increase the rate of soil C mineralization (Kochsiek *et al.*, 2009). In a review, Jarecki and Lal (2003) reported soil organic carbon contents increasing from −50 to +1187 kg/ha/year after irrigation was begun. Increases were smallest under irrigated mouldboard-ploughed crops and greatest under pastures. While increases in plant residues being returned to the soil influence the food supply of biota and may indirectly lead to changes in the physical habitat, more immediate impacts of irrigation on soil structure occur as a consequence of structural collapse due to high water contents. Pagliai *et al.* (2004) found that porosity of surface soils decreased during the irrigation season and that the decrease was significantly greater when irrigation was by impounding rather than by sprinkler irrigation. The loss in porosity was greatest among pores >50 μm that were elongated in shape.

The influence of irrigation on soil structure is exacerbated when the irrigation water is saline or sodic. Water with a large concentration of Na relative to that of Ca and Mg (characterized by the sodium adsorption ratio, SAR) results in slaking and the development of an apedal surface layer. Continued application of water of this quality increases the swelling and dispersion of clay tactoids with reduced size of the conducting pores, increased blockage of pore necks and a progressive increase in the thickness of the apedal layer (So and Aylmore, 1993; Emdad *et al.*, 2004).

These changes in soil structure reduce infiltration rates and irrigation application efficiency and increase the incidence of surface waterlogging. The sensitivity of hydraulic conductivity to these changes in SAR varies with soil texture, and the limits on acceptable irrigation water SAR should generally be decreased as clay content increases (Curtin *et al.*, 1994). Although the effect of irrigation, particularly the quality of irrigation water, on soil structure has been studied extensively, less information is available on the effect of irrigation on soil biota. While we may speculate on the effects of irrigation on biota as a consequence of changes in total porosity, pore continuity and frequency that pores are air-filled, there are feedback processes that could not be predicted. For instance, Sarig *et al.* (1993) showed that the rate of C mineralization was decreased under saline water irrigation and the deterioration of soil structure was not due solely to chemical and physical interactions between soil particles and solution salts, but was also due to the fact that saline water irrigation caused a reduction in carbohydrates produced by microorganisms and thus a reduction in aggregate stability.

5.2 Drainage

Artificial drainage is intended to remove excess water from the root zone, and the rate of water removal increases with decreasing drain spacing and increasing depth (Kladivko *et al.*, 2004). Artificial drainage diminishes the length of time that the soil has high water contents and a low structural stability and enhances the extent of wetting and drying. The risk of compaction by traffic and smearing or structural deformation by tillage is reduced with improved drainage. Drainage increases the diversity of crops that can be grown, creates a better environment for the production of most arable crops and increases the amount of crop residue that may be added to the soil (Fausey and Lal, 1989). However, improved aeration also creates a more favourable environment for aerobic biota and would be expected to result in increased rate of C turnover. Although there are limited data on the changes in organic carbon content, studies in Ohio, USA, suggest small (Jacinthe *et al.*, 2001) or non-significant decreases (Sullivan *et al.*, 1998) in carbon content after drainage is introduced when conventional or ridge tillage is practised.

Similarly, drainage of grassland did not cause a change in the organic carbon content in south-west England after 15 years (Clegg *et al.*, 2003). Increased rates of mineralization appear, at least on these soils, to compensate for increased C inputs. There is a paucity of data on population dynamics of soil biota following installation of artificial drainage. Clegg *et al.* (2003) noted that drainage had a significant effect on the community structures of actinomycetes and pseudomonads.

While the purpose of drainage is to transfer excess water in the soil to surface water bodies, an unfortunate component of this process is the associated transfer of dissolved and suspended material from the soil surface or the soil matrix to surface water bodies. Some of these materials subsequently contribute to the contamination of water resources (Ross and Donnison, 2003; Kladivko *et al.*, 2004; Ulen *et al.*, 2007).

5.3 New management practices

The challenges related to water management that will face agriculture in the future include (i) increasing competition for high-quality irrigation water; (ii) increased use of low-quality water; (iii) sustaining the productivity of irrigated soils; and (iv) reducing the movement of water-borne agricultural chemicals and pathogens to water resources. Potential new strategies to address these challenges include (i) selection of crops with reduced irrigation requirements or tolerance to moderate salinity; (ii) use of organic materials in irrigation water to diminish destabilization of soil structure during irrigation; and (iii) soil and crop management practices to reduce loss of water-borne materials to surface water bodies. Development of genotypes for greater drought resistance would reduce irrigation requirements, but may not be achieved without reduced biomass accumulation (Tardieu, 2005). A similar cost may be associated with increased salt tolerance. Use of organic soil conditioners in irrigation, particularly furrow irrigation, increases the stability of aggregates that are first contacted by irrigation water by ensnaring and bonding mineral particles (Ross *et al.*, 2003). However, this effect is limited to the first few millimetres of soil being infiltrated by the amended irrigation water. Soil and crop management practices to reduce the loss of dissolved or suspended material from the root zone via artificial drainage are being actively investigated and include measures at the field scale (e.g. use of 'trap' crops, disruption of macropores contributing to bypass water flow) and at the watershed scale in the form of, for example, constructed wetlands. Although the impact of many of the new strategies to address challenges related to water management will influence the nature and quantity of crop residue returned to soil as well as the physical habitat of soil biota, the impact of these strategies on soil biota remains largely unexplored.

6 Soil Amendments

Materials are commonly added to soil in order to change soil chemical properties (pH or plant nutrient levels) or physical properties (structure), to control weeds, insects or diseases, or simply as a means of disposal.

6.1 Liming materials

Soil pH is strongly influenced by the composition of soil parent material and the extent of weathering of this material. In some parts of the world, acidification of soils by acidic atmospheric deposition is becoming increasingly important (Goulding and Blake, 1998). An additional cause of acidification on agricultural soils is the application of ammoniacal fertilizers without any neutralizing material. Soil pH is managed to maximize the availability of nutrients and minimize the availability of metals that are often naturally occurring in soils and can be toxic to many plants and soil biota. The optimum pH for most agricultural crops is from 6.5 to 7.0. Acidic conditions are the normal reasons for adjusting the pH, and calcitic and/or dolomitic

limestone-based materials are the most common amendments. Increasing the pH of acidic soil results in increases in the amount of crop residue returned to the soil, but also changes the chemical environment for soil biota. In the absence of plant growth, the addition of $CaCO_3$ at a rate of 1 t/ha/year for 70 years in plots at Versailles, France resulted in an increase in pH of 2.6 units and an increase in cation exchange capacity (CEC) of 50% (Pernes-Debuyser and Tessier, 2004). Soil cohesion was improved, aggregate stability increased and rates of wetting and drying enhanced. The classic experiments at Rothamsted in the UK show that acidification leads to the release of Al and potentially toxic metals (Cd, Pb, Zn and Cu) into soil water and plants (Goulding and Blake, 1998), but regular liming can prevent these problems (Goulding and Blake, 1998). Increasing soil pH from 5.1 to 6.9 with liming materials has also been found to increase the population of a range of soil biota, including those antagonistic to certain pathogens (e.g. cavity spot disease of carrots) (Eitarabily *et al.*, 1996).

6.2 Fertilizers

Nutrient levels in soils are generally inadequate to support high yields of agricultural crops, and therefore the natural levels are supplemented by addition of nutrients in organic or inorganic forms. Fertilization would be expected to increase the amount of crop residue produced and the turnover of the organic materials returned to the soil, although the impact of increased residue production would be expected to be diminished if the crop residue is removed. Secondary effects of some kinds of inorganic fertilizer include reduction in soil pH (following addition of ammonium-based fertilizers) and addition of Cd (with use of some phosphate-based fertilizers).

The impacts of nutrient management are most obvious on long-term research sites. In a review of the effects of manures (farmyard manure, slurry and green manure) and fertilizers on soil productivity in 14 long-term trials (20–120 years), Edmeades (2003) found that the input of nutrients, as either fertilizers or manures, had very large effects (150–1000%) on soil productivity, as measured by yields. Manured soils had higher contents of organic matter and numbers of bacteria and actinomycetes than fertilized soils. There were no differences in the population of fungi. Poll *et al.* (2003) also found addition of farmyard manure for 120 years on a soil in Germany increased total microbial biomass, and a shift of the microbial community towards a more bacterial-dominated community in the coarse sand fraction. However, microbial communities in the finer fractions were less affected by the manure. Edmeades (2003) found that the higher organic matter content coincided with lower bulk density, higher hydraulic conductivity and higher aggregate stability, relative to fertilized soils. However, with the exception of the Rothamsted classic long-term trials, there was no significant difference between fertilizers and manures in their long-term effects on yields. The inputs of manures and the accumulation of organic matter were larger in the Rothamsted trials than in the other trials, and Edmeades suggested that manures may only have a benefit on soil productivity, over and above their nutrient content, when large inputs are applied over many years. A comparison of the chemical characteristics of organic matter on three fertilizer treatments (nil, inorganic fertilizer and animal manure) on long-term plots at Rothamsted and Askov (Denmark) led Randall *et al.* (1995) to conclude that fertilizer regimes did not significantly affect the chemical composition of SOM. They concluded that the chemical composition was determined primarily by the interaction between the biota responsible for decomposition and the mineral matrix rather than the nature of substrate input.

Fertilizers can also have effects on soil properties that are independent of increased organic matter content. For instance, the addition of ammoniacal fertilizers for 70 years to plots at Versailles resulted in pH dropping as much as two units (Pernes-Debuyser and Tessier, 2004). Plots treated in this way exhibited a decrease in CEC and were more sensitive to degradation of their

hydraulic properties. Long-term application of phosphate fertilizer can result in increased Cd content of soils (Nicholson *et al.*, 1994). However, the accumulation of Cd in herbage can be minimized with liming (Nicholson *et al.*, 1994), and this effect would be expected to extend to soil biota.

6.3 Urban and industrial wastes

Waste materials of urban or industrial origin often have large contents of organic carbon. Application of these materials to agricultural land represents an inexpensive means of disposal and, depending on rates of addition and composition, can increase SOM contents and plant nutrient levels. Concomitant improvements in soil structure and changes in microbiological characteristics are observed. For example, the paper industry faces a challenge of economically sound and environmentally safe disposal of large amounts of residue; paper sludges are rich in cellulose and lignin (Bellamy *et al.*, 1995). Analyses show low levels of N, P and K and, with the exception of Cu, small concentrations of metals (Beauchamp *et al.*, 2002). Fatty and resin acids and polycyclic hydrocarbons were the organic compounds found to be present at highest concentrations (Beauchamp *et al.*, 2002), but the sludge was free of phytotoxic substances (Madejon *et al.*, 2001). Addition of de-inking paper sludge resulted in an increase in soil water content (Chantigny *et al.*, 2000) and an increase in water-stable aggregation, hydraulic conductivity, available water and air-filled porosity in a fine-textured soil but a decrease in water-stable aggregation and hydraulic conductivity in a coarse-textured soil (Nemati *et al.*, 2000). Microbial biomass and enzyme activity increased following addition of the sludge (Chantigny *et al.*, 2000; Madejon *et al.*, 2001), N_2 fixation by legumes was increased (Allahdadi *et al.*, 2004) and the population of lumbricid earthworms was increased (Piearce and Boone, 1998). All of these studies suggest that changes in both food supply and the physical habitat of biota resulting from the addition of this waste material have a positive effect on soil biota. However, the impacts of waste application on soil biota are not always positive, particularly when the wastes contain materials that are toxic to soil biota. Sewage sludge can sometimes illustrate this limitation. Sewage sludges often contain high contents of heavy metals and the deleterious effects of the addition of these metals on soil biota can persist for many years. For instance, measurements on the garden experimental plots at Rothamsted made nearly four decades after application of the sludges had ceased still exhibited an altered microbial community and reduced total microbial biomass, microbial biomass as a percentage of soil organic carbon as well as microbial numbers and activity (Abaye *et al.*, 2005).

6.4 Soil conditioners

The term 'soil conditioner' is commonly applied to any material that is added to soil for the specific purpose of improving its physical or biological properties. Many of the waste materials that are applied to soils can be considered as soil conditioners although they vary in effectiveness (e.g. Reynolds *et al.*, 2003). Commercially produced synthetic conditioners began appearing on the market in the 1950s and have been used to stabilize soil structure, thereby reducing seal formation, increasing infiltration and diminishing runoff and erosion. Their effectiveness varies with polymer characteristics (type and amount of charge, configuration and molecular weight), soil properties (clay content and mineralogy, electrical conductivity and pH), methods of application (spraying on soil surface or applying in irrigation water) and amounts applied (Stewart, 1975; Seygold, 1994). Few studies on the impact of these materials on soil biota have been completed, although unpredictable effects on microbial biomass have been observed (Steinberger *et al.*, 1993). The cost of commercially produced polymers and the need for

repeated applications, as a consequence of their susceptibility to biodegradation (Wolter *et al.*, 2002), continue to constrain their widespread use in agriculture. However, synthetic conditioners are an important amendment in a restricted range of management systems (e.g. that using furrow irrigation). Increasing demand for food in the future may well result in increased use of soils that are prone to seal formation, erosion, etc. and, if the value of the crops produced on these soils increases, use of synthetic conditioners on such soils may become more feasible. A futuristic alternative to synthetic organic conditioners may be the inoculation of soils with biota capable of producing organic materials *in situ* that are effective soil conditioners. Microbial inoculants have been developed and used for this purpose on a limited scale, and the search continues for superior strains, natural or engineered, that are effective in the field (Zimmerman, 1993).

6.5 Herbicides and pesticides

Weeds, insects and other biological pests are often controlled in farming systems using specific organic chemicals synthesized for that purpose. These materials are applied in small quantities using relatively lightweight equipment and their application would not be expected to have a direct effect on the physical habitat of soil biota. The application of these materials may have immediate effects on non-target biota (e.g. Heilmann *et al.*, 1995; Dunfield *et al.*, 2000). Although any generalizations about the long-term impact of these materials on non-target biota should be treated with caution, these materials are biodegradable and do not tend to accumulate in soils. Measurements made on a long-term study of five different pesticides (aldicarb, benomyl, chlorfenvinphos, glyphosate and chlorotoluron) at Rothamsted revealed no effects on microbial processes after nearly 20 years of repeated applications of these pesticides (Bromilow *et al.*, 1996). No pesticide residues could be detected in soil sampled

17 months after the last experimental treatment (Bromilow *et al.*, 1996). Although similar studies have not been conducted on all pesticides that are applied to soils, there is little evidence that the application of these organic materials would have a long-term effect on the population of soil biota or the functions they fulfil in soils. This generalization may not apply to materials containing metals as an active ingredient.

7 Methods of Assessing/Monitoring the Impact of Management on Soil Biota

Twenty years prior to the publication of this chapter, Van Veen and Kuikman (1990) asserted in a discussion of the decomposition of SOM that while 'soil architecture is the dominant control over microbially mediated decomposition processes in terrestrial ecosystems, ... suitable methodology to quantify the relation between soil structure and biological processes as a function of different types and conditions of soils is still lacking'. An unfortunate result of this deficiency is that assessments or monitoring of management practices do not routinely consider the effects of management practices on the physical habitat of soil biota. Recent advances in the characterization of the physical habitat of biota are considered elsewhere in this volume, and the role of structure in different biological processes is examined comprehensively. However, limitations are imposed on the methods that can be employed if information on the physical habitat of soil biota is to be used for assessing/monitoring purposes and for supporting optimal management decisions in agriculture.

For the purpose of assessing the impact of agricultural management practices on the physical habitat of biota, characterization methods that will be most informative must meet two criteria. First, they must provide data that can be interpreted in terms of agriculturally important functions or services that the biota is performing. The mineralization of organic N and the oxidation/reduction of inorganic N is one such group of processes that are of significant economic

and environmental importance. Secondly, the methods must be responsive to factors that account for the spatial and temporal variability in these processes under different management practices. Parkin (1993) has reviewed the factors controlling the variability in microbial processes at the scale of micro- and macro-aggregates, field plots, landscapes and regions. Although the factors having the dominant influence on the variability in microbial processes change with the different scales, a common important factor across all scales is soil water content. Soil water content varies in space (laterally and with depth) and time and can also be influenced by management. Therefore, from the perspective of assessing the impact of management on soil biota, two different pieces of information are required: the effect of management on the relation between the rate of the relevant processes and soil water content and the effect of management on soil water content. Both pieces of information would be expected to vary spatially, and therefore either direct measurement of their spatial variability is required or the spatial variability must be predicted from other, more readily measurable, variables. A comprehensive method to provide the first piece of information is determination of the least limiting water range (LLWR) and the non-limiting water range (NLWR) for the relevant processes.

The activity of many biota in soil with a given structure varies in a curvilinear manner with water content (Fig. 10.3). Inadequate aeration can affect reaction rates at high water contents, whereas diminished rates of substrate diffusion and low water potential (Sommers *et al.*, 1980; Skopp *et al.*, 1990; Stark and Firestone, 1995) and reduced mobility of nematodes and protozoan grazers (Killham *et al.*, 1993) are important at low water contents. The architecture of soil pore space controls all of these factors and consequently the shape of the curve. It is possible to identify maximum and minimum water contents at which the rates for processes such as net N mineralization approach zero, giving rise to upper- and lower-limiting water contents, respectively. The difference between the upper- and lower-limiting water

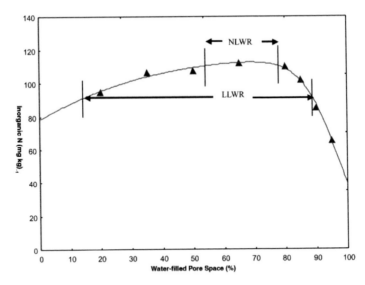

Fig. 10.3. Illustration of the variation in shape of the curve relating net N mineralization to water-filled pore space for a Conestogo silt loam, relative compaction 0.91, 3-month incubation (initial N content 91.9 mg N/kg soil) (adapted from Drury et al., 2003).

contents is referred to as the least-limiting water range, LLWR (Drury et al., 2003). Within the LLWR, a range of intervening water contents exist where biological processes are independent of water content, or exhibit limited variation with water content. This range of water contents represents the non-limiting water range, NLWR (Drury et al., 2003). The overall shape of the curve relating net mineralization to water content can then be represented by three parameters: the maximum reaction rate, the LLWR and the NLWR. The upper and lower limits of the limiting water ranges can then be related to either measured water contents (e.g. Dharmakeerthi et al., 2005) or critical water contents such as those at field capacity, permanent wilting point or penetration resistance or gas diffusivity at critical levels (e.g. Drury et al., 2003; Kadziene et al., 2011). Management practices that alter the physical habitat in ways that decrease the magnitude of the limiting water ranges, especially the NLWR, have a detrimental effect on the ability of soil biota to perform vital services. Similarly, detrimental effects arise when management practices change the soil water regime so that the water content falls outside the NLWR with increasing frequency.

Increasing awareness of the importance of the services or functions that soil biota perform in agricultural systems would be expected to lead to other new methods of characterizing the effects of management of the quality of the physical environment in which biota perform these functions.

8 Summary

The effects of agricultural management practices on crop yield and soil physical and chemical properties are more commonly documented than the effects of management on soil biota. Physical properties that are often measured include pore characteristics (total porosity and pore size distribution), water content and, to a lesser extent, pore continuity, aeration and soil temperature. Chemical characteristics have tended to focus on nutrient dynamics. The spatial and temporal variability in both physical and chemical characteristics has, until recently, generally been ignored. The promise of increased profitability and reduced negative environmental impact of site-specific or precision agriculture is resulting in improved documentation of the variability in some physical and chemical characteristics of soils. However, notwithstanding the benefits of improved understanding of the spatial and temporal variability in soil physical and chemical properties, the data on these characteristics are normally collected to help comprehend the effects of management on plant growth or environmental effects related to erosion or the transport of water and solutes, and few of the data are at a scale or in a form that can be readily related to soil biota.

Management practices can profoundly alter the nature of the food supply of soil biota, as well as their physical habitat. When the effects of management on soil biota are investigated, studies often focus on total biomass, the activity of biota, the populations of specific organisms or groups of organisms or the rate of a given process of significance to agriculture. The relative importance of management-induced effects on food supply and physical habitat on these characteristics is seldom investigated. This limitation may be due, in part, to the paucity of information on the most important aspects of the physical habitat to measure. In addition, management-induced changes in food supply may occur over a much shorter time scale than changes in the pore structure and water regime, introducing an additional complication in assessing their relative importance. The failure of physical and ecological sciences effectively to link their disciplinary solitudes in studies of soil biota may also contribute to our limited understanding of the relation between management-induced changes in habitat and the characteristics and functions of soil biota. Irrespective of the reasons for this lack of information, there are two obvious consequences. First, the greater amount of information on the effect of food supply on biota may lead to the superficial conclusion that management-induced changes in food supply have a larger effect on organisms than on physical habitat. Yet this is not necessarily the case. Secondly, lack of information on the physical habitat has resulted in little attention being paid to this variable in the assessment or monitoring of management practices.

While one might intuitively expect that agricultural management practices that provide a food supply and physical environment for soil biota that support greatest biological diversity would also be expected to result in long-term sustainable productivity, there have been relatively few studies formally evaluating this expectation.

References

Abaye, D.A., Lawlor, K., Hirsch, P.R. and Brookes, P.C. (2005) Changes in the microbial community of an arable soil caused by long-term metal contamination. *European Journal of Soil Science* 56, 93–102.

Alabouvette, C., Backhouse, D., Steinberg, C., Donovan, N.J., Edel Hermann, V. and Burgess, L.W. (2004) Microbial diversity in soil – effects on crop health. In: Schjønning, P., Elmholt, S. and Christensen, B.T. (eds) *Managing Soil Quality: Challenges in Modern Agriculture*. CABI Publishing, Wallingford, UK, pp. 121–138.

Alakukku, L., Pöyhönen, A. and Sampo, M. (1997) Soil compaction control with a light, unmanned tractor in two tillage systems. In: Fotyma, M., Jozefaciuk, M., Malicki, L. and Borowiecki, J. (eds) *Fragmenta Agronomica TOM 2A/97, Proceedings 14th ISTRO Conference*, 27 July–1 August 1997, Pulawy, Poland. Polish Society of Agrotechnical Sciences, Poland, pp. 19–22.

Allahdadi, I., Beauchamp, C.J. and Challifour, F.P. (2004) Symbiotic dinitrogen fixation in forage legumes amended with high rates of de-inking paper sludge. *Agronomy Journal* 96, 956–965.

Alvey, S., Yang, H., Buerkert, A. and Crowley, D.E. (2003) Cereal/legume rotation effects on rhizosphere bacterial community structure in West African soils. *Biology and Fertility of Soils* 37, 73–82.

Andersson, S. and Håkansson, I. (1966) Markfysiologiska undersögelser i odlad jord. XVI. Strukturdynamiken i matjorden. En fältstudie. *Grundförbätring* 19, 191–228.

Ball, B.C. (1981) Pore characteristics of soils from two cultivation experiments as shown by gas diffusivities and permeabilities and air-filled porosities. *Journal of Soil Science* 32, 483–498.

Ball, B.C. and Robertson, E.A.G. (1994a) Effects of soil water hysteresis and the direction of sampling aeration and pore function in relation to soil compaction and tillage. *Soil and Tillage Research* 32, 51–60.

Ball, B.C. and Robertson, E.A.G. (1994b) Effects of uniaxial compaction on aeration and structure of ploughed or direct drilled soils. *Soil and Tillage Research* 31, 135–148.

Beauchamp, C.J., Charest, M.H. and Gosselin, A. (2002) Examination of environmental quality of raw and composting de-inking paper sludge. *Chemosphere* 46, 887–895.

Bellamy, K.L., Chong, C. and Cline, R.A. (1995) Paper sludge utilization in agriculture and container nursery culture. *Journal of Environmental Quality* 24, 1074–1082.

Berendse, F., Berg, B. and Bosatta, E. (1987) The effect of lignin and nitrogen on the decomposition of litter in nutrient-poor ecosystems – a theoretical approach. *Canadian Journal of Botany-Revue Canadienne de Botanique* 65, 1116–1120.

Biederbeck, V.O., Campbell, C.A. and Zentner, R.P. (1984) Effect of crop-rotation and fertilization on some biological properties of a loam in southwestern Saskatchewan. *Canadian Journal of Soil Science* 64, 355–367.

Birch, A.N.E., Griffiths, B.S., Caul, S., Thompson, J., Heckmann, L.H., Krogh, P.H. *et al.* (2007) The role of laboratory, glasshouse and field scale experiments in understanding the interactions between genetically modified crops and soil ecosystems: A review of the ECOGEN project. *Pedobiologia* 51, 251–260.

Blanco-Canqui, H. and Lal, R. (2008) No-tillage and soil-profile carbon sequestration: An on-farm assessment. *Soil Science Society of America Journal* 72, 693–701.

Blanco-Canqui, H. and Lal, R. (2009) Crop residue removal impacts on soil productivity and environmental quality. *Critical Reviews in Plant Sciences* 28, 139–163.

Bolinder, M.A., Angers, D.A. and Dubuc, J.P. (1997) Estimating shoot to root ratios and annual carbon inputs in soils for cereal crops. *Agriculture, Ecosystems and Environment* 63, 61–66.

Bromilow, R.H., Evans, A.A., Nicholls, P.H., Todd, A.D. and Briggs, G.G. (1996) The effect on soil fertility of repeated applications of pesticides over 20 years. *Pesticide Science* 48, 63–72.

Bruand, A., Cousin, I., Nicoullaud, B., Duval, O. and Begon, J.C. (1996) Backscattered electron scanning images of soil porosity for analyzing soil compaction around roots. *Soil Science Society of America Journal* 60, 895–901.

Chamen, W.C.T., Dowler, D., Leede, P.R. and Longstaff, D.J. (1994) Design, operation and performance of a gantry system: experience in arable cropping. *Journal of Agricultural Engineering Research* 59, 45–60.

Chantigny, M.H., Angers, D.A. and Beauchamp, C.J. (2000) Active carbon pools and enzyme activities in soils amended with de-inking paper sludge. *Canadian Journal of Soil Science* 80, 99–105.

Chen, G. and Weil, R. (2010) Penetration of cover crop roots through compacted soils. *Plant and Soil* 331, 31–43.

Cheng, W., Zhang, Q., Coleman, D.C., Ronald Carroll, C. and Hoffman, C.A. (1996) Is available carbon limiting microbial respiration in the rhizosphere? *Soil Biology and Biochemistry* 28, 1283–1288.

Christensen, B.T. (1986) Barley straw decomposition under field conditions – effect of placement and initial nitrogen-content on weight-loss and nitrogen dynamics. *Soil Biology and Biochemistry* 18, 523–529.

Christensen, B.T. (2002) Biomass removal for energy purposes – consequences for soil carbon balance in agriculture and forestry. DJF Report, *Markbrug* 72, 1–75. 2002. Ministry of Food, Agriculture and Fisheries, Danish Institute of Agricultural Sciences, Tjele, Denmark.

Christopher, S.F., Lal, R. and Mishra, U. (2009) Regional study of no-till effects on carbon sequestration in Midwestern United States. *Soil Science Society of America Journal* 73, 207–216.

Clegg, C.D., Lovell, R.D. and Hobbs, P.J. (2003) The impact of grassland management regime on the community structure of selected bacterial groups in soils. *FEMS Microbiology Ecology* 43, 263–270.

Clemente, E.P., Schaefer, C.E.G.R., Novais, R.F., Viana, J.H. and Barros, N.F. (2005) Soil compaction around *Eucalyptus grandis* roots: a micromorphological study. *Australian Journal of Soil Research* 43, 139–146.

Collins, H.P., Rasmussen, P.E. and Douglas, C.L. (1992) Crop-rotation and residue management effects on soil carbon and microbial dynamics. *Soil Science Society of America Journal* 56, 783–788.

Cookson, W.R., Beare, M.H. and Wilson, P.E. (1998) Effects of prior crop residue management on microbial properties and crop residue decomposition. *Applied Soil Ecology* 7, 179–188.

Cresswell, H.P. and Kirkegaard, J.A. (1995) Subsoil amelioration by plant roots – the process and the evidence. *Australian Journal of Soil Research* 33, 221–239.

Curtin, D., Steppuhn, H. and Selles, F. (1994) Structural stability of Chernozemic soils as affected by exchangeable sodium and electrolyte concentration. *Canadian Journal of Soil Science* 74, 157–164.

Curtin, D., Francis, G.S. and McCallum, F.M. (2008) Decomposition rate of cereal straw as affected by soil placement. *Australian Journal of Soil Research* 46, 152–160.

Degens, B.P. (1998) Decreases in microbial functional diversity do not result in corresponding changes in decomposition under different moisture conditions. *Soil Biology and Biochemistry* 30, 1989–2000.

Dexter, A.R. (1986) Model experiments on the behaviour of roots at the interface between a tilled seed-bed and a compacted sub-soil. II. Entry of pea and wheat roots into sub-soil cracks. *Plant and Soil* 95, 135–147.

Dharmakeerthi, R.S., Kay, B.D. and Beauchamp, E.G. (2005) Factors contributing to changes in plant available nitrogen across a variable landscape. *Soil Science Society of America Journal* 69, 453–462.

Dick, W.A. and Gregorich, E.G. (2004) Developing and maintaining soil organic matter levels. In: Schjønning, P., Elmholt, S. and Christensen, B.T. (eds) *Managing Soil Quality: Challenges in Modern Agriculture.* CABI Publishing, Wallingford, UK, pp. 103–120.

Douglas, J.T. and Goss, M.J. (1987) Modification of pore space by tillage in two stagnogley soils with contrasting management histories. *Soil and Tillage Research* 10, 303–317.

Douglas, J.T., Goss, M.J. and Hill, D. (1980) Measurements of pore characteristics in a clay soil under ploughing and direct drilling, including use of a radioactive tracer (144Ce) technique. *Soil and Tillage Research* 1, 11–18.

Drees, L.R., Karathanasis, A.D., Wilding, L.P. and Blevins, R.L. (1994) Micromorphological characteristics of long-term no-till and conventionally tilled soils. *Soil Science Society of America Journal* 58, 508–517.

Drury, C.F., Zhang, T. and Kay, B.D. (2003) The non-limiting and least-limiting water ranges for soil N mineralization: Measurement and effects of soil texture, compaction and legume residue addition. *Soil Science Society of America Journal* 67, 1388–1404.

Dunfield, K.E., Siciliano, S.D. and Germida, J.J. (2000) The fungicides thiram and captan affect the phenotypic characteristics of *Rhizobium leguminosarum* strain C1 as determined by FAME and Biolog analyses. *Biology and Fertility of Soils* 31, 303–309.

Edmeades, D.C. (2003) The long-term effects of manures and fertilizers on soil productivity and quality: a review. *Nutrient Cycling in Agroecosystems* 66, 165–180.

Ehlers, W., Kopke, U., Hesse, F. and Bohm, W. (1983) Penetration resistance and root growth of oats in tilled and untilled loess soil. *Soil and Tillage Research* 3, 261–275.

Ehlers, W., Werner, D. and Mähner, T. (2000) Wirkung mechanischer Belastung auf Gefüge und Ertragsleistung einer Löss-Parabraunerde mit zwei Bearbeitungssystemen. *Journal of Plant Nutrition and Soil Science* 163, 321–333.

Eitarabily, K.A., Hardy, G.E.S.J., Sivasithamparam, K. and Kurtboke, I.D. (1996) Microbiological differences between limed and unlimed soils and their relationship with cavity spot disease of carrots (*Daucus carota* L) caused by Pythium in Western Australia. *Plant and Soil* 183, 279–290.

Elmholt, S., Schjønning, P., Munkholm, L.J. and Debosz, K. (2008) Soil management effects on aggregate stability and biological binding. *Geoderma* 144, 455–467.

Emdad, M.R., Raine, S.R., Smith, R.J. and Fardad, H. (2004) Effect of water quality on soil structure and infiltration under furrow irrigation. *Irrigation Science* 23, 55–60.

Eynard, A., Schmacher, T.E., Lindstrom, M.J. and Malo, D.D. (2004) Porosity and pore size distribution in cultivated Ustolls and Usterts. *Soil Science Society of America Journal* 68, 1927–1934.

Fausey, N.R. and Lal, R. (1989) Drainage-tillage effects on Crosby-Kokomo soil association in Ohio. *Soil Technology* 2, 359–370.

Flores, S., Saxena, D. and Stotzky, G. (2005) Transgenic Bt plants decompose less in soil than non-Bt plants. *Soil Biology and Biochemistry* 37, 1073–1082.

Fox, C.A. (2003) Characterizing soil biota in Canadian agroecosystems: State of knowledge in relation to soil organic matter. *Canadian Journal of Soil Science* 83, 245–257.

Goulding, K.W.T. and Blake, L. (1998) Land use, liming and the mobilization of potentially toxic metals. *Agriculture Ecosystems and Environment* 67, 135–144.

Haider, K. (1992) Problems related to the humification processes in soils of temperate climates. In: Stotzky, G. and Bollag, J.-M. (eds) *Soil Biochemistry*, vol. 7. Marcel Dekker, New York, pp. 55–94.

Hamza, M.A. and Anderson, W.K. (2005) Soil compaction in cropping systems: A review of the nature, causes and possible solutions. *Soil and Tillage Research* 82, 121–145.

Hansen, S. and Engelstad, F. (1999) Earthworm populations in a cool and wet district as affected by tractor traffic and fertilisation. *Applied Soil Ecology* 13, 237–250.

Hartge, K.H. and Sommer, C. (1980) The effect of geometric patterns of soil structure on compressibility. *Soil Science* 130, 180–185.

Haukka, J. (1988) Effect of various cultivation methods on earthworm biomasses and communities on different soil types. *Annales Agriculturae Fenniae* 27, 263–269.

Heilmann, B., Lebuhn, M. and Beese, F. (1995) Methods for investigation of metabolic activity and shifts in the microbial community in soil treated with a fungicide. *Biology and Fertility of Soils* 19, 186–192.

Hendrix, P.F., Parmelee, R.W., Crossley, D.A., Coleman, D.C., Odum, E.P. and Groffman, P.M. (1986) Detritus food webs in conventional and no-tillage agroecosystems. *Bioscience* 36, 374–380.

Hermle, S., Anken, T., Leifeld, J. and Weisskopf, P. (2008) The effect of the tillage system on soil organic carbon content under moist, cold-temperate conditions. *Soil and Tillage Research* 98, 94–105.

Horn, R. (2004) Time dependence of soil mechanical properties and pore functions for arable soils. *Soil Science Society of America Journal* 68, 1131–1137.

Hornby, D. (1998) *Take-all Disease of Cereals: a Regional Perspective*. CAB International, Wallingford, UK.

Hütsch, B.W., Augustin, J. and Merbach, W. (2002) Plant rhizodeposition – an important source for carbon turnover in soils. *Journal of Plant Nutrition and Soil Science* 165, 397–407.

Jacinthe, P.A., Lal, R. and Kimble, J.M. (2001) Organic carbon storage and dynamics in croplands and terrestrial deposits as influenced by subsurface drainage. *Soil Science* 166, 322–335.

Jakobsen, B.F. and Dexter, A.R. (1988) Influence of biopores on root growth, water uptake and grain yield of wheat (*Triticum aestivum*) based on predictions from a computer model. *Biology and Fertility of Soils* 6, 315–321.

Jarecki, M.K. and Lal, R. (2003) Crop management for soil carbon sequestration. *Critical Reviews in Plant Sciences* 22, 471–502.

Jensen, B. (1993) Rhizodeposition by $^{14}CO_2$-pulse-labeled spring barley grown in small-field plots on sandy loam. *Soil Biology and Biochemistry* 25, 1553–1559.

Jorgensen, R.G., Raubuch, M. and Brandt, M. (2002) Soil microbial properties down the profile of a black earth buried by colluvium. *Journal of Plant Nutrition and Soil Science-Zeitschrift für Pflanzenernahrung und Bodenkunde* 165, 274–280.

Kadziene, G., Munkholm, L.J. and Mutegi, J. (2011) Root growth conditions in the topsoil as effected by tillage intensity. *Geoderma*, doi:10.1016/J.geoderma.2011.07.013.

Kay, B.D. and VandenBygaart, A.J. (2002) Conservation tillage and depth stratification of porosity and soil organic matter. *Soil and Tillage Research* 66, 107–118.

Killham, K., Amato, M. and Ladd, J.N. (1993) Effect of substrate location in soil and soil-water regime on carbon turnover. *Soil Biology and Biochemistry* 25, 57–62.

Kladivko, E.J. (2001) Tillage systems and soil ecology. *Soil and Tillage Research* 61, 61–76.

Kladivko, E.J., Frankenberger, J.R., Jaynes, D.B., Meek, D.W., Jenkinson, B.J. and Fausey, N.R. (2004) Nitrate leaching to subsurface drains as affected by drain spacing and changes in crop production system. *Journal of Environmental Quality* 3, 1803–1813.

Kochsiek, A.E., Knops, J.M.H., Walters, D.T. and Arkebauer, T.J. (2009) Impacts of management on decomposition and the litter-carbon balance in irrigated and rainfed no-till agricultural systems. *Agricultural and Forest Meteorology* 149, 1983–1993.

Kooistra, M.J. and Tovey, N.K. (1994) Effect of compaction on soil microstructure. In: Soane, B.D. and Van Ouwerkerk, C. (eds) *Soil Compaction in Crop Production*. Elsevier, Amsterdam, pp. 91–112.

Kremer, J., Wolf, B. and Matthies, D. (2002) Structural deformation of artificial macropores under varying load and soil moisture. *Journal of Plant Nutrition and Soil Science-Zeitschrift für Pflanzenernahrung und Bodenkunde* 165, 627–633.

Kristensen, H.L. and Thorup-Kristensen, K. (2004) Root growth and nitrate uptake of three different catch crops in deep soil layers. *Soil Science Society of America Journal* 68, 529–537.

Kuzyakov, Y. and Domanski, G. (2000) Carbon input by plants into the soil. Review. *Journal of Plant Nutrition and Soil Science-Zeitschrift für Pflanzenernahrung und Bodenkunde* 163, 421–431.

Lal, R. (2010). Managing soils for a warming earth in a food-insecure and energy-starved world. *Journal of Plant Nutrition and Soil Science* 173, 4–15.

Lampkin, N. (1990) *Organic Farming*. Farming Press Books, Ipswich, UK.

Leake, J.R., Johnson, D., Donnelly, D., Muckle, G., Boddy, L. and Read, D. (2004) Networks of power and influence: the role of mycorrhizal mycelium in controlling plant communities and agroecosystem functioning. *Canadian Journal of Botany* 82, 1016–1045.

Lupwayi, N.Z., Rice, W.A. and Clayton, G.W. (1998) Soil microbial diversity and community structure under wheat as influenced by tillage and crop rotation. *Soil Biology and Biochemistry* 30, 1733–1741.

Madejon, E., Burgos, P., Murillo, J.M. and Cabrera, F. (2001) Phytotoxicity of organic amendments on activities of selected soil enzymes. *Communications in Soil Science and Plant Analysis* 32, 2227–2239.

Mader, P., Fliessbach, A., Dubois, D., Gunst, L., Fried, P. and Niggli, U. (2002) Soil fertility and biodiversity in organic farming. *Science* 296, 1694–1697.

Melillo, J.M., Aber, J.D. and Muratore, J.F. (1982) Nitrogen and lignin control of hardwood leaf litter decomposition dynamics. *Ecology* 63, 621–626.

Mishra, U., Ussiri, D.A.N. and Lal, R. (2010) Tillage effects on soil organic carbon storage and dynamics in Corn Belt of Ohio USA. *Soil and Tillage Research* 107, 88–96.

Munkholm, L.J. (2000) The spade analysis – a modification of the qualitative spade diagnosis for scientific use. DIAS Report No. 28. Plant production, 1–40. Danish Institute of Agricultural Sciences, Tjele, Denmark.

Munkholm, L.J., Schjønning, P., Rasmussen, K.J. and Tanderup, K. (2003) Spatial and temporal effects of direct drilling on soil structure in the seedling environment. *Soil and Tillage Research* 71, 163–173.

Mutegi, J.K., Petersen, B.M., Munkholm, L.J. and Hansen, E.M. (2011) Belowgound carbon input and translocation potential of fodder radish cover-crop. *Plant and Soil* 344, 159–175.

Nemati, M.R., Caron, J. and Gallichand, J. (2000) Using paper de-inking sludge to maintain soil structural form. *Soil Science Society of America Journal* 64, 275–285.

Nicholson, F.A., Jones, K.C. and Johnston, A.E. (1994) Effect of phosphate fertilizers and atmospheric deposition on long-term changes in the cadmium content of soils and crops. *Environmental Science and Technology* 28, 2170–2175.

Olesen, J.E. and Bindi, M. (2002) Consequences of climate change for European agricultural productivity, land use and policy. *European Journal of Agronomy* 16, 239–262.

Pagliai, M., Vignozzi, N. and Pellegrini, S. (2004) Soil structure and the effect of management practices. *Soil and Tillage Research* 79, 131–143.

Pare, T., Gregorich, E.G. and Nelson, S.D. (2000) Mineralization of nitrogen from crop residues and N recovery by maize inoculated with vesicular-arbuscular mycorrhizal fungi. *Plant and Soil* 218, 11–20.

Parkin, T.B. (1993) Spatial variability of microbial processes in soil – a review. *Journal of Environmental Quality* 22, 409–417.

Passioura, J.B. (2002) Soil conditions and plant growth. *Plant Cell and Environment* 25, 311–318.

Pernes-Debuyser, A. and Tessier, D. (2004) Soil physical properties affected by long-term fertilization. *European Journal of Soil Science* 55, 505–512.

Piearce, T.G. and Boone, G.C. (1998) Responses of invertebrates to paper sludge application to soil. *Applied Soil Ecology* 9, 393–397.

Poll, C., Thiede, A., Wermbter, N., Sessitsch, A. and Kanddeler, E. (2003) Micro-scale distribution of microorganisms and microbial enzyme activities in a soil with long-term organic amendment. *European Journal of Soil Science* 54, 715–724.

Powlson, D.S., Brookes, P.C. and Christensen, B.T. (1987) Measurement of soil microbial biomass provides an early indication of changes in total soil organic matter due to straw incorporation. *Soil Biology and Biochemistry* 19, 159–164.

Qingjie, W., Hao, C., Hongwen, L., Wenying, L., Xiaoyan, W., McHugh, A.D. *et al.* (2009) Controlled traffic farming with no tillage for improved fallow water storage and crop yield on the Chinese Loess Plateau. *Soil and Tillage Research* 104, 192–197.

Rahn, C.R. and Lillywhite, R.D. (2002) A study of the quality factors affecting the short-term decomposition of field vegetable residues. *Journal of the Science of Food and Agriculture* 82, 19–26.

Randall, E.W., Mahieu, N., Powlson, D.S. and Christensen, B.T. (1995) Fertilization effects on organic matter in physically fractionated soils as studied by C-13 NMR: Results from two long-term field experiments. *European Journal of Soil Science* 46, 557–565.

Reganold, J.P. (1988) Comparison of soil properties as influenced by organic and conventional farming systems. *American Journal of Alternative Agriculture* 3, 144–154.

Reynolds, W.D., Yang, X.M., Drury, C.F., Zhang, T.Q. and Tan, C.S. (2003) Effects of selected conditioners and tillage on the physical quality of a clay loam soil. *Canadian Journal of Soil Science* 83, 381–393.

Richard, G., Boizard, H., Roger-Estrade, J., Boiffin, J. and Guerif, J. (1999) Field study of soil compaction due to traffic in northern France: Pore space and morphological analysis of the compacted zones. *Soil and Tillage Research* 51, 151–160.

Rillig, M.C., Field, C.B. and Allen, M.F. (1999) Soil biota responses to long-term atmospheric CO_2 enrichment in two California annual grasslands. *Oecologia* 119, 572–577.

Roger-Estrade, J., Richard, G., Dexter, A.R., Boizard, H., De Tourdonnet, S., Bertrand, M. *et al.* (2009) Integration of soil structure variations with time and space into models for crop management. A review. *Agronomy for Sustainable Development* 29, 135–142.

Ross, C. and Donnison, A. (2003) Campylobacter and farm dairy effluent irrigation. *New Zealand Journal of Agricultural Research* 46, 255–262.

Ross, C.W., Sojka, R.E. and Foerster, J.A. (2003) Scanning electron micrographs of polyacrylamide-treated soil in irrigation furrows. *Journal of Soil and Water Conservation* 58, 327–331.

Sarig, S., Roberson, E.B. and Firestone, M.K. (1993) Microbial activity soil-structure-response to saline water irrigation. *Soil Biology and Biochemistry* 25, 693–697.

Schäffer, B., Mueller, T.L., Stauber, M., Müller, R., Keller, M. and Schulin, R. (2008) Soil and macro-pores under uniaxial compression. II. Morphometric analysis of macro-pore stability in undisturbed and repacked soil. *Geoderma* 146, 175–182.

Schjønning, P. (1989) Long-term reduced cultivation. II. Soil pore characteristics as shown by gas diffusivities and permeabilities and air-filled porosities. *Soil and Tillage Research* 15, 91–103.

Schjønning, P. and Rasmussen, K.J. (2000) Soil strength and soil pore characteristics for direct drilled and ploughed soils. *Soil and Tillage Research* 57, 69–82.

Schjønning, P., Elmholt, S., Munkholm, L.J. and Debosz, K. (2002a) Soil quality aspects of humid sandy loams as influenced by organic and conventional long-term management. *Agriculture Ecosystems and Environment* 88, 195–214.

Schjønning, P., Rasmussen, K.J., Munkholm, L.J. and Nielsen, P.S. (2002b) *Jordbearbejdning i økologisk jordbrug – pløjedybde og ikke-vendende jordløsning.* Danish Institute of Agricultural Sciences, Tjele, Denmark.

Schjønning, P., Heckrath, G. and Christensen, B.T. (2009) Threats to soil quality in Denmark. DJF Report on Plant Science 143, pp. 1–121.

Seneviratne, G. (2000) Litter quality and nitrogen release in tropical agriculture: a synthesis. *Biology and Fertility of Soils* 31, 60–64.

Seygold, C.A. (1994) Polyacrylamide review: Soil conditioning and environmental fate. *Communications in Soil Science and Plant Analysis* 25, 2171–2185.

Skopp, J., Jawson, M.D. and Doran, J.W. (1990) Steady-state aerobic microbial activity as a function of soil water content. *Soil Science Society of America Journal* 54, 1619–1625.

So, H.B. and Aylmore, L.A.G. (1993) How do sodic soils behave? The effects of sodicity on soil physical behavior. *Australian Journal of Soil Research* 31, 761–777.

Sommers L.E., Gilmour, C.M., Wildung, R.E. and Beck, S.M. (1980) The effect of water potential on decomposition processes in soils. In: Parr, J.F., Gardner, W.R. and Elliott, L.F. (eds) *Water Potential Relations in Soil Microbiology.* Soil Science Society of America, Madison, Wisconsin, pp. 97–117.

Stark, J.M. and Firestone, M.K. (1995) Mechanisms for soil-moisture effects on activity of nitrifying bacteria. *Applied and Environmental Microbiology* 61, 218–221.

Steinberger, Y., Sarig, S., Nadler, A. and Barnes, G. (1993) The effect of synthetic soil conditioners on microbial biomass. *Arid Soil Research and Rehabilitation* 7, 303–306.

Stewart, B.A. (1975) *Soil Conditioners.* Soil Science Society of America Special Publication 7, Soil Science Society of America, Madison, Wisconsin.

Sullivan, M.D., Fausey, N.R. and Lal, R. (1998) Long-term effects of subsurface drainage on soil organic carbon content and infiltration in the surface horizons of a lakebed soil in Northwest Ohio. In: Lal, R., Kimble, J.M., Folett, R.F. and Stewart, B.A. (eds) *Management of Carbon Sequestration in Soil.* CRC Press, Boca Raton, Florida.

Swift, M.J. and Anderson, J.M. (1993) Biodiversity and ecosystem function in agricultural systems. In: Schulze, E.D. and Mooney, H.A. (eds) *Biodiversity and Ecosystem Function.* Springer, Berlin, pp. 15–41.

Swift, M.J., Heal, O.W. and Anderson, J.M. (1979) *Decomposition in Terrestrial Ecosystems Studies in Ecology,* vol. 5. Blackwell Scientific Publications, Oxford, UK.

Swift, M.J., Vandermeer, J., Ramakrishnan, P.S., Anderson, J.M., Ong, C.K. and Hawkins, B.M. (1996) Biodiversity and agroecosystem function. In: Mooney, H.A., Cushman, J.H., Medina, E., Sala, O.E. and Schulze, E.D. (eds) *Functional Roles of Biodiversity – a Global Perspective.* John Wiley and Sons, Chichester, UK, pp. 261–298.

Tardieu, F. (2005) Plant tolerance to water deficit: physical limits and possibilities for progress. *Comptes Rendus Géoscience* 337, 57–67.

Tebrügge, F. and Düring, R.-A. (1999) Reducing tillage intensity – a review of results from a long-term study in Germany. *Soil and Tillage Research* 53, 15–28.

Thomsen, I.K. and Christensen, B.T. (2004) Yields of wheat and soil carbon and nitrogen contents following long-term incorporation of barley straw and ryegrass catch crops. *Soil Use and Management* 20, 432–438.

Tisdall, J.M. and Oades, J.M. (1982) Organic matter and water-stable aggregates in soils. *Journal of Soil Science* 33, 141–163.

Tullberg, J.N., Yule, D.F. and McGarry, D. (2007) Controlled traffic farming – From research to adoption in Australia. *Soil and Tillage Research* 97, 272–281.

Ulen, B., Bechmann, M., Folster, J., Jarvie, H.P. and Tunney, H. (2007) Agriculture as a phosphorus source for eutrophication in the north-west European countries, Norway, Sweden, United Kingdom and Ireland: a review. *Soil Use and Management* 23, 5–15.

Van den Akker, J. and Schjønning, P. (2004) Subsoil compaction and ways to prevent it. In: Schjønning, P., Elmholt, S. and Christensen, B.T. (eds) *Managing Soil Quality: Challenges in Modern Agriculture*. CAB International, Wallingford, UK, pp. 163–184.

VandenBygaart, A.J., Protz, R., Tomlin, A.D. and Miller, J.J. (1999) Tillage system effects on near-surface soil morphology: Observations from the landscape to micro-scale in silt loam soils of southwestern Ontario. *Soil and Tillage Research* 51, 139–149.

Van der Putten, W.H., Bardgett, R.D., de Ruiter, P.C., Hol, W.H.G., Meyer, K.M., Bezemer, T.M. *et al.* (2009) Empirical and theoretical challenges in aboveground–belowground ecology. *Oecologia* 161, 1–14.

Van Veen, J.A. and Kuikman, P.J. (1990) Soil structural aspects of decomposition of organic-matter by micro-organisms. *Biogeochemistry* 11, 213–233.

Vermeulen, G.D. and Mosquera, J. (2009) Soil, crop and emission responses to seasonal-controlled traffic in organic vegetable farming on loam soil. *Soil and Tillage Research* 102, 126–134.

Vinther, F.P., Eiland, F., Lind, A.-M. and Elsgaard, L. (1999) Microbial biomass and numbers of denitrifiers related to macropore channels in agricultural and forest soils. *Soil Biology and Biochemistry* 31, 603–611.

Voorhees, W.B. and Lindstrom, M.J. (1984) Long-term effects of tillage method on soil tilth independent of wheel traffic compaction. *Soil Science Society of America Journal* 48, 152–156.

Wardle, D.A., Yeates, G.W., Watson, R.N. and Nicholson, K.S. (1995) The detritus food-web and the diversity of soil fauna as indicators of disturbance regimes in agroecosystems. *Plant and Soil* 170, 35–43.

Wardle, D.A., Bardgett, R.D., Klironomos, J.N., Setala, H., van der Putten, W.H. and Wall, D.H. (2004) Ecological linkages between aboveground and belowground biota. *Science* 304, 1629–1633.

Warembourg, F.R. and Estelrich, H.D. (2000) Towards a better understanding of carbon flow in the rhizosphere: a time-dependent approach using carbon-14. *Biology and Fertility of Soils* 30, 528–534.

Whalley, W.R., Dumitru, E. and Dexter, A.R. (1995) Biological effects of soil compaction. *Soil and Tillage Research* 35, 53–68.

Wolter, M., Weisch, C.I.D., Zadraziul, F., Hey, S., Haselbach, J. and Schnug, E. (2002) Biological degradation of synthetic superabsorbent soil conditioners. *Landbauforsch Volkenrode* 52, 43–52.

Yeates, G.W., Bardgett, R.D., Cook, R., Hobbs, P.J., Bowling, P.J. and Potter, J.F. (1997) Faunal and microbial diversity in three Welsh grassland soils under conventional and organic management regimes. *Journal of Applied Ecology* 34, 453–470.

Young, I.M. and Crawford, J.W. (2004) Interactions and self-organization in the soil-microbe complex. *Science* 304, 1634–1637.

Zimmerman, W.J. (1993) Microalgal biotechnology and applications in agriculture. In: Metting, F.B. (ed.) *Soil Microbial Ecology. Applications in Agricultural and Environmental Management*. Marcel Dekker, New York, pp. 457–479.

Zurbrügg, C., Hönemann, L., Meissle, M., Romeis, J. and Nentwig, W. (2010) Decomposition dynamics and structural plant components of genetically modified Bt maize leaves do not differ from leaves of conventional hybrids. *Transgenic Research* 19, 257–267.

11 Contaminated Soils and Bioremediation: Creation and Maintenance of Inner Space

Richard J.F. Bewley* and Simon Hockin

1 Introduction

This chapter explores the importance of spatial distribution of soil components in terms of the treatment of contaminated soil using bioremediation. We shall first describe the spatial distribution of contaminants within typical contaminated sites and then examine how a successful bioremedial strategy seeks to overcome problems imposed by heterogeneity and engineer conditions that facilitate contaminant treatment.

Most developed countries face growing pressures to remediate land that has been contaminated as a result of industrial activities. Such pressures originate from the need to develop 'brownfield' sites especially for residential use, but also from the requirement to protect water resources. Over the last 20 years there has been an increasing application of technologies for the treatment of contaminated soil in preference to 'dig and dispose' (Evans *et al.*, 2001). The latter is currently becoming more expensive as restrictions on landfill disposal have begun to take effect, especially in Europe as a result of the Landfill Directive.

The definition of 'contaminated land' varies according to legislative terminology. In England, for example, it embodies the concept of 'significant harm' to specific receptors or 'pollution of controlled waters' being caused, or potentially being caused, as a result of contamination (DETR, 2000). For land to be contaminated there has to be a 'pollutant linkage' that includes the contaminant, the receptor and a viable pathway between the two. Remedial technologies therefore seek to address one or more components of this pollutant linkage. Bioremediation is one such treatment technology. It can be defined as the optimization of physical, chemical and biological factors to achieve conditions under which microorganisms may degrade contaminants to harmless substances or otherwise change their physico-chemical properties so as to obviate the pollutant linkage.

Amongst the emerging technologies for treatment of contaminated soil and groundwater, bioremediation has achieved a significant track record of application both *ex situ* and *in situ*. A successful bioremediation strategy will depend upon an appropriate level of understanding of the broader 'conceptual site model' in terms of the relationship of the contaminant, pathway and receptor, but also the factors that currently determine why the contaminants in question persist in soil and groundwater. The

*Corresponding author: richard.bewley@urs.com

latter include factors relating to the contaminant itself and those that relate to the environment (Bewley, 1996a). At both a macro and micro scale, the spatial distribution of the contaminant in relation to the physico-chemical environment and the microflora is therefore of critical importance. We shall examine this in the following section.

2 Nature of Soil Contamination at Industrial and Urban Sites

2.1 Sources of contamination at industrial sites and factors affecting their distribution

Soil contamination may be incidental or accidental in nature (Finnecy, 1987). The former arises as a consequence of a deliberate act such as the storage of uncontained bulk material on soil, or a failure to act, for example to repair a leaking valve or improve the integrity of a drainage system. At an industrial site incidental contamination commonly results from the delivery, storage and intra-factory movement of materials, from the production processes, storage of products or from waste disposal. Accidental soil contamination occurs as a result of an unplanned event such as a sudden failure of containment, collision during transportation or an explosion or fire. Incidental contamination is typified by a non-uniform distribution across the site, differing according to the properties and origin of the contaminant concerned. Accidental contamination typically results in more highly contaminated soil, the volume of which is dependent upon the nature of the incident concerned, as well as the contaminant.

Both across and below the site, therefore, the distribution of contaminants will be influenced by the origin and nature of release, the physico-chemical properties of both the compound in question and the recipient environment. The below-ground infrastructure of industrial facilities is likely to have an important influence, particularly on the migration of organic contaminants. Drainage runs and service lines will create preferential pathways through more permeable surrounding fill material, promoting lateral migration. Foundations that intercept soil horizons of varying permeability will promote vertical migration. Typical migration pathways for contamination arising at an operational site are illustrated in Fig. 11.1.

Organic contaminants are generally more mobile than inorganic forms and, if released as free product (for example, from a leakage of solvent or fuel oil), will migrate vertically downwards, towards the water table as a non-aqueous-phase liquid (NAPL). If the spill is relatively small and from a non-continuous source, downward percolation in the unsaturated zone will not continue once the total spill volume has reached residual saturation. However, continuing dissolution of the NAPL from infiltration may continue to make it a threat to underlying groundwater. If the source of NAPL is continuous, however, or occurs as a series of successive pulses, then further downward migration will take place. Variations in the permeability of soil horizons will have a major influence on migration pathways: lenses of low-permeability clay, for example, will cause the product to move laterally along its surface until a breach occurs and more permeable material is encountered. Further vertical migration will then take place, although capillary forces will also accentuate horizontal spreading (Domenico and Schwartz, 1998).

Once the water table is encountered then, depending upon its relative density, free product will tend to float on the water table as a light non-aqueous-phase liquid (LNAPL), a category that includes most hydrocarbons, especially those from petrol and diesel spillages. On the other hand, for organic compounds that are denser than water, dense non-aqueous-phase liquids (DNAPL) will sink and move downward to the base of the aquifer. Chlorinated solvents are the most commonly encountered contaminants falling into this category, especially perchloroethylene (PCE), trichloroethylene (TCE), dichloroethane (DCA) and dichloromethane (DCM). As with their behaviour in the unsaturated zone these will form zones of residual saturation as they migrate to the base of the aquifer, where they

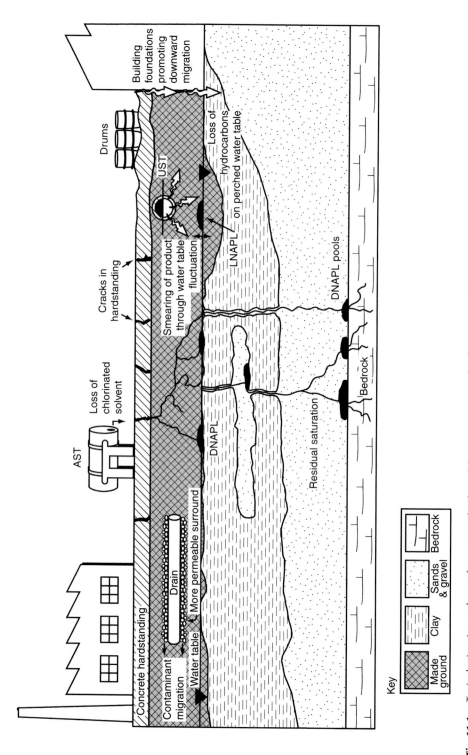

Fig. 11.1. Typical migration pathways from contamination arising at an operational industrial site. AST, above-ground storage tank; DNAPL, dense non-aqueous phase liquid; LNAPL, light non-aqueous-phase liquid; UST, underground storage tank.

may form DNAPL pools. In the case of both LNAPL and DNAPL, therefore, such product will provide an ongoing source of contamination that will continue to dissolve into the aqueous phase.

Most of the key heavy metals occur in cationic form and, as such, their mobility in soil will be strongly retarded by the presence of clays and organic matter, which can mitigate their influence on microbial activity (e.g. Bewley and Stotzky, 1983). Notable exceptions to these are arsenic and hexavalent chromium, which may occur in the anionic form as arsenate and chromate, respectively. As such they tend to be highly mobile in the sub-surface environment (Bewley *et al.*, 2000).

2.2 Contamination in urban soils and heterogeneity issues

In urban areas, however, soil contamination may have arisen from multiple occupancy of a given site over many decades of use. Moreover, the material that constitutes 'made ground' in industrialized areas may have originated from highly contaminated sources or even represent a source of contamination in its own right. An example of this concerns chromium ore processing residue originating from a chromium works, which was used as fill material at a significant number of sites in south-east Glasgow during the 19th and 20th centuries. This has resulted in highly elevated concentrations of hexavalent chromium (Cr (VI)) in urban soils within the area, which remain a significant local environmental issue (Farmer *et al.*, 1999; Bewley *et al.*, 2001a).

Urban–industrial soils therefore can be expected to demonstrate significant differences in comparison with soils of natural origin, and such differences will have implications for contaminant distribution.

Intense human activity is both a defining characteristic of urban–industrial soils and the dominant active agent in urban soil development. Frequent cycles of disturbance disrupt the soil profile and tend to impede the development of features that

characterize mature soils. Such soils are further distinguished by the presence of solid materials originating from prior human uses. Both endogenous and exogenous anthropogenic materials may be present and commonly include brick, masonry, concrete, tarmac, glass, metal, plastic, timber, organics, ash, clinker and production wastes. In addition, exogenous subsoils and dredgings are frequently encountered in made ground. These materials may be present at particle sizes ranging from silt to large boulders and are often unevenly mixed in the soil profile, where they substantially affect the soil's physical, chemical and biological properties. The influences of climate and *in situ* parent material that dominate natural soil formation are therefore of reduced relative importance in the genesis of urban–industrial soils, making it extremely difficult to predict the characteristics of these soils without both a knowledge of site history and the use of detailed ground investigations.

Many of these anthropogenic components inherently contain significant levels of contamination, and their broad-scale distribution is likely to be very uneven across a site area (Lethbridge and Wenham, 2004). Natural and anthropogenic materials are frequently mixed and are used to fill depressions and excavations and to make ground level, often resulting in burial of the former soil surface and formation of relict horizons. Cutting to level is also common, truncating the former soil profile. Urban–industrial soils therefore frequently contain abrupt lithologic discontinuities with markedly differing physical, chemical and biological characteristics, and with associated distinct zonal heterogeneity in the concentration of chemical contaminants (Craul, 1999). Point-source contamination arising from on-site activities generates distinct and spatially heterogeneous soil contamination patterns that both cut across those intrinsic to the made ground and are further influenced by differential transport-fate characteristics of the highly spatially variable physico-chemical environment of underlying soils. Such patterns may be established over several cycles of human land use and soil manipulation.

The relative immaturity of urban–industrial soils, the lack of surface vegetation and the frequent presence of an impermeable covering mean that true organic horizons rarely develop. Soil invertebrates are lacking, or much reduced, thus curtailing comminution and soil mixing and reducing the rates of aggregation, structure formation and nutrient cycling. A combination of low organic matter and high silt content from anthropogenic components also tends towards natural compaction under the moisture conditions and physical loads associated with construction/demolition cycles (Craul and Klein, 1982). Crushing of materials to smaller grades and their addition, with surface compaction, to the evolving soil surface is a requirement for the operation of heavy construction/demolition plant and is common practice in improving the structural engineering characteristics of the soil for future use. The bulk densities of urban–industrial soils are consequently often higher than those commonly encountered in 'natural' soils, further reducing permeability, oxygenation and biological activity and tending to favour the preservation of initially heterogeneous contaminant distribution patterns. The preservation of micro-/meso-scale heterogeneity and the associated spatial separation of contaminants and soil microorganisms may therefore be important in limiting rates of contaminant degradation, and intrinsic rates of natural biological attenuation of soil contaminants may be severely curtailed in comparison with those theoretically achievable. Bioremedial strategies are, as discussed, essentially targeted at enhancing such rate-limiting factors.

The typical spatial heterogeneity of such soils and the associated patterns of contamination have important implications for the management and remediation of contaminated land. High heterogeneity transmits to a low signal:noise ratio and into high coefficients of variance from soil samples. In practice it is very difficult to generate single baseline and endpoint measures from the data that such environments generate.

An example of the variation of contamination with depth and the relationship to the materials sampled at a former gasworks site is provided in Fig. 11.2.

High sample variance and time/cost constraints on realistic sampling frames mean that standard inferential statistical approaches have frequently proved inadequate. This is exacerbated by the non-normally distributed nature of sample contaminant concentrations. Risk-based approaches are therefore increasingly adopted in assessing the need for remedial action and in evaluating the outcomes of intervention programmes. The US Environmental Protection Agency has identified soil and contaminant spatial heterogeneity as a fundamental limiting factor in achieving effective and cost-efficient site clean-up and has been identified as the key driver in the development of new instruments for site characterization and programme evaluation (Crumbling et al., 2003).

Lethbridge and Wenham (2004) examined this problem in the context of proposed Integrated Pollution Prevention and Control (IPPC) guidance seeking to establish baseline and endpoint criteria on industrial site operations, contrasting this with the risk-based approach to historical contamination under Part IIA of the 1990 Environmental Protection Act. The report concludes that

> The key challenge to determining whether new contamination of a site has taken place under the lifetime of a PPC permit is that of the highly heterogeneous distribution of chemicals, of either natural or anthropogenic origin, in the ground.

Data from contaminated sites in the UK illustrate the typically heterogeneous distribution of contaminants in urban–industrial soils and how this constrains our ability precisely to quantify the degree of contamination and to evaluate (bio)remedial outcomes (Fig. 11.3 and Table 11.1). Theoretically, the closer together two samples are taken, the lower the variability between them would be expected to be. To test this hypothesis an area from former tank farm sites was sampled at 15 cm centres and analysed for total petroleum hydrocarbons (TPH) (Fig. 11.4 and Table 11.2). The between-sample variability was less for

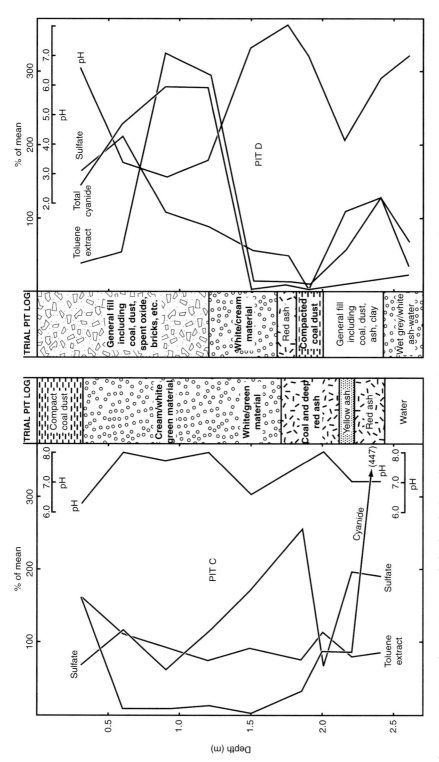

Fig. 11.2.   Variation of contamination with depth and relationship to materials sampled for two trial pits, excavated on a former gasworks site (redrawn from Smith and Ellis, 1986).

Table 11.1. Heterogeneity of total petroleum hydrocarbons (TPH) in a medium-fine sandy soil beneath a former tank farm storing crude oil.

	Depth (cm)		
	0–25	25–50	50–75
n	34	25	25
Minimum value (mg/kg)	124	58.0	50
Maximum value (mg/kg)	17,091	16,462	2,353
Mean (mg/kg)	6,221	2,392	479
Standard deviation (mg/kg)	5,960	3,844	626
Coefficient of variation (%)	96	161	131

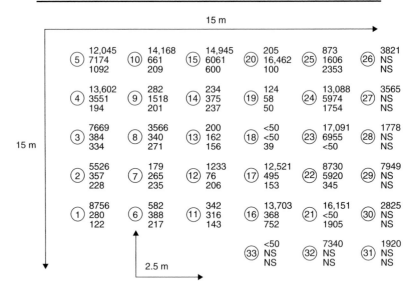

Fig. 11.3. Distribution of total petroleum hydrocarbons (TPH, mg/kg) and associated statistics (see Table 11.1) in a medium-fine sandy soil beneath a former tank farm storing crude oil (redrawn from Lethbridge and Wenham, 2004).

these samples when compared with those collected at lower density over the whole 15 × 15 m area and the coefficient of variation was reduced from >100% to <50% of the mean. Nevertheless, variability remained high, whilst sampling at such high densities under operational conditions would be prohibitively expensive.

3 Contaminant Distribution and the Microbial Environment

The complexity of contaminant distribution, particularly within made ground,

therefore represents a significant challenge in the design, implementation and validation of bioremediation systems. At the micro scale too, contaminant distribution in relation to the microflora is of significance in terms of the rate of biodegradation and will be mediated by the physico-chemical properties of the environment.

Whilst patchy distributions of contaminants and active soil biota may overlap at coarser scales, it is at the micrometre–millimetre scale that highly heterogeneous, overlapping and mismatched spatial and phase distributions of contaminant molecules and the soil microbiota most constrain both the degree of intrinsic degradation and the

Table 11.2. Example of small-scale variability in total petroleum hydrocarbons (TPH) in a medium-fine sandy soil beneath a former tank farm storing crude oil.

	Depth (cm)		
	0–25	25–50	50–75
n	7	7	7
Minimum value (mg/kg)	7,060	3,158	313
Maximum value (mg/kg)	12,104	6,215	697
Mean (mg/kg)	9,536	4,622	558
Standard deviation (mg/kg)	1,873	1,281	131
Coefficient of variation (%)	20	28	23

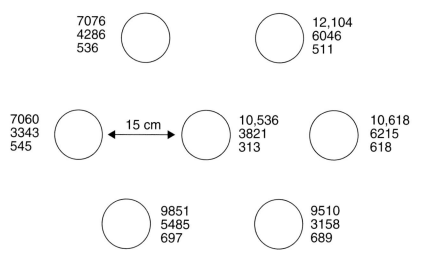

Fig. 11.4. Small-scale variability in total petroleum hydrocarbons (TPH, mg/kg) and associated statistics (see Table 11.2) in a medium-fine sandy soil beneath a former tank farm storing crude oil (redrawn from Lethbridge and Wenham, 2004).

efficacy of remedial interventions. The aggregate effect of these interacting nanomicroscale factors is implicit in the concept of bioavailability, which is widely recognized within the research scientific community as key to the understanding of substrate turnover in complex matrices such as soil, and is fundamental not only to understanding and manipulating biodegradative activity but also in setting appropriate remedial endpoints. The concept of bioavailability is intrinsic to the source–pathway–receptor model typified, for example, in the UK by the Environmental Protection Act (1990), yet regulatory closure is invariably determined by total extractable concentrations, based on

small numbers of widely dispersed samples and on non-speciating extraction protocols that may poorly represent the risks to humans and other organisms.

At the micrometre–millimetre scale, matrix attributes control the distribution and mobility of both contaminants and soil microbiota. Soil organic matter, clay components, fine structure, porosity and surface chemistry interact with contaminant parameters (hydrophobicity, lipophilicity, ion exchange and surface charge characteristics, viscosity, surface tension) in controlling the degree to which organic contaminant molecules and inorganic ions are distributed, adsorbed and sequestered among mineral and organic soil components (Reid *et al.*, 2000; Semple *et al.*, 2003). Initial adsorption of a contaminant fraction by hydrogen bonding and Van der Waals forces is generally accepted to be very rapid, whilst over the longer term stronger bonds may form. The sequestration of hydrophobic organic contaminants therefore tends to increase over time and, for most compounds, is expected to reach an equilibrium that will be controlled by the interaction of contaminant chemistry and local soil characteristics. Common and widespread organic contaminants, including many polyaromatic hydrocarbons (PAHs) and polychlorinated biphenyls (PCBs), share these common characteristics of structural recalcitrance to enzyme attack and poor bioavailability, whilst for some contaminants, including PCBs, covalent bonding to soil organic components can lead to near-irreversible sequestration that is not equilibrium dependent (Dec and Bollag, 1997). Soils can therefore be seen as comprising an important environmental sink for certain classes of organic contaminants (Semple *et al.*, 2003).

The distribution and activity of the microbial biomass are constrained by chemical and physical limitations, primarily soil particle surface chemistry, pore size and moisture availability. Cell surface chemistry and sorption characteristics are also important, as is nutrient availability, which may control overall metabolic activity, microbial growth habit and enzyme production. In the low-dissolved nutrient environment of most soils (and particularly urban–industrial soils) most soil bacteria are present as surface-attached populations comprising mixed functional group biofilms associated with both soil particles and plant rootlets where surface vegetation is present. Most soil bacteria, however, appear unable to penetrate the tortuous, porous matrix of soil organic matter, or to occupy the nanometre-scale pores common in the inorganic fraction. In contrast, organic contaminant molecules are able to penetrate soil organic matter by diffusion that is limited by both the soil matric potential and surface interactions that constrain only the

rate of molecular movement. The low organic matter content of industrial soils is, in this context, something of an advantage in maintaining the mobility and bioavailability of xenobiotics, whilst organic contaminants themselves may constitute an important substrate for soil enzyme activity, particularly the more labile hydrocarbon fractions. At coarser scales this would be expected to generate a positive correlation between the distributions of the contaminant and of competent degrading organisms, and so would be regarded as highly desirable for bioremediation. However, direct toxicity, recalcitrance as a growth substrate, high C:N ratios and poorly co-distributed minor nutrients often act to constrain this.

The distribution of hygroscopic, capillary and free water is critical to the presence and activity of microorganisms, and to their mobility and capacity for chemotactic redistribution along dissolved chemical gradients as local conditions change. Whilst soil fungi are able to redistribute patchy substrates across a localized hyphal network, and in this respect are less constrained than bacteria by local soil heterogeneity, dissolved chemical gradients are also important for the location of organic substrates by fungi (Boswell *et al.*, 2002). Soil moisture is also important in controlling the availability of nutrients and terminal electron acceptors that constrain microbial distribution and activity. At the low matric potentials characteristic of many urban–industrial soils, larger pores (2–5 µm) will contain only capillary water and thus be accessible to aerobic bacteria. Smaller pores, even those that are apparently large enough to support bacterial metabolism, will be limited by the low diffusion rates of dissolved gases and will be limited to lower-energy, lower-activity metabolisms (Fig. 11.5). This diffusion limitation at a range of scales may be a significant factor in controlling degradation rates, particularly where the target molecule is degraded by enzymes commonly produced by bacterial metabolic groups associated with particular ranges of pH and Eh.

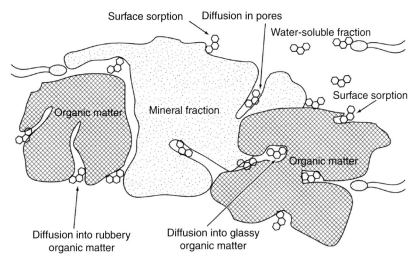

Fig. 11.5. A summary of the physical behaviour of a contaminant within the soil (redrawn from Semple *et al.*, 2003).

4 Application of Bioremediation

4.1 Introduction

Bioremediation systems vary in the degree of engineered complexity, but can broadly be divided into *ex situ* and *in situ* systems, the former involving removal of the soil, the latter without any displacement. The differences between these two approaches have profound implications for the spatial configuration of microorganisms, contaminants, electron acceptors and other substrates. These systems are discussed below. Finally, some consideration is given to the potential for passive approaches to contaminant treatment, particularly in conjunction with phytoremediation.

4.2 *Ex situ* systems and pre-processing requirements

It is the *ex situ* processes that hold the greatest potential for the manipulation of the physico-chemical factors relating to the soil environment, particularly in terms of the spatial distribution of microorganisms, contaminants, electron acceptors and nutrients. Such systems involve the excavation of soil

from the area of concern, mechanical processing and screening to achieve suitable particle size and placement in a purposefully designed treatment cell. This incorporates a drainage layer constructed over an impermeable membrane to confer protection to the underlying soil from contaminated leachate. The soil is typically arranged in a mound shaped in trapezoidal or pyramidical cross-section, referred to as a treatment bed. *Ex situ* bioremediation can be subdivided into mechanically aerated systems and static systems. The spatial distribution of the key elements of the bioremediation process is of profound importance for the operation of both types of approach.

Mechanically aerated systems have the advantage over static systems of accelerating the mass transfer of oxygen into the soil treatment bed, and periodic working of the soil allows for the ongoing redistribution of microorganisms, nutrients and the contaminant, exposing fresh surfaces to microbial attack. This process may already have been initiated during the pre-processing stage during which the soil is sorted prior to placement in the treatment bed. In some cases bricks and concrete rubble, which often comprise key components of made ground, have been mechanically removed from the soil during the pre-processing stage, crushed and then returned to the soil undergoing treatment prior to placement. This may help to create a more granular structure within the material undergoing treatment and facilitate the passage of air through such macropores into the core of the soil treatment bed. Specific equipment employed during the pre-processing phase may also facilitate the break-up of clay material through a shearing action (e.g. through the employment of an 'alu' bucket or similar).

The incorporation of such processes may be particularly important where a static system is to be employed under which there is limited opportunity for further mixing once placed in the treatment bed.

4.3 Mechanically aerated systems

The simplest form of bioremediation using mechanical aeration is referred to as land farming. The soil treatment bed is constructed typically to a height of 500–600 mm, which is the maximum depth at which most agricultural equipment can operate. The equipment used to turn the soil over is generally some form of rotovator.

Some of the more advanced forms of rotovation equipment include spades on the blades to achieve a more efficient mixing (Fig. 11.6).

During the process a significant degree of homogenization takes place, as reflected in a reduction of the variability of contaminant concentrations. This has been reported in a number of full-scale bioremediation projects, including soil undergoing *ex situ* treatment for hydrocarbons at oil refinery sites (Bewley *et al.*, 1990; Balba and Bewley, 1991; Bewley, 1996b) and phenols at a coal gasification site (Bewley *et al.*, 1989).

As discussed above, the mixing of microorganisms, nutrients, electron acceptors and contaminant will accelerate the overall treatment process. However, the redistribution of contamination 'hotspots' within the material does itself have at least two beneficial effects on degradation. First, localized areas of high concentration of organics where toxicity may be associated with, for example, action on cell membranes (Bossert and Bartha, 1984) will become diluted with adjacent, less contaminated, areas. Secondly, the overall surface area of the contamination may be increased, particularly where localized product has accumulated.

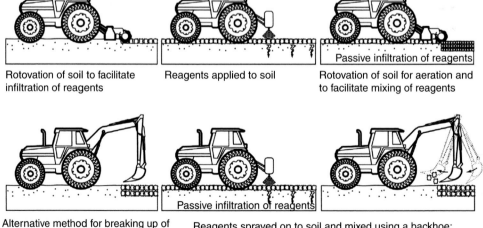

Rotovation of soil to facilitate infiltration of reagents

Reagents applied to soil

Rotovation of soil for aeration and to facilitate mixing of reagents

Alternative method for breaking up of soil to facilitate infiltration of reagents

Reagents sprayed on to soil and mixed using a backhoe: periodic turning to enhance aeration

Fig. 11.6. Simplest forms of land farming involving nutrient application and mechanical aeration.

Whilst land farming constituted some of the earliest successful bioremediation projects, it has rather unfairly been viewed as a 'low-tech' approach. None the less, given the overriding importance of facilitating mass transfer of oxygen in hydrocarbon degradation, a suitably powerful rotovator may be sufficient to achieve appropriate remedial targets, assuming that other limiting factors such as nutrients and moisture content have been accounted for.

The second form of *ex situ* treatment involves the use of windrow systems, where soil is loosely placed in a pyramid-shaped mound. Classic windrowing involves the incorporation of organic matter as a bulking agent, similar to that in a static biopile, so that the system essentially becomes one of composting. Depending upon the nature and quantity of the organic bulking agent used, a significant temperature increase is achievable that will accelerate the rate of degradation. As in land farming, mechanical aeration is used to achieve the appropriate degree of oxygenation, optimize structural conditions (through the break-up of clods of soil) and also to dissipate excess heat. Apart from the addition of organic material, one of the key differences between land farming and windrowing is the application of specialist plant to achieve rapid turning of soil, an example of which is shown in Fig. 11.7. This enables a much greater volume of soil to be treated during a particular cycle, vastly improves the efficiency of oxygen transfer and continually promotes structural amelioration. Such improvements have a significant effect on the overall treatment costs, through the greater volumes treated and reduced time for treatment. Additionally, the application of specialist plant with such a capacity for structural improvement may obviate the requirement or application of an organic bulking agent altogether, or at least reduce the overall requirement. This is of particular benefit from a geotechnical standpoint, as discussed below.

4.4 Static systems

Static *ex situ* systems, referred to as biopiles, rely on passive or forced aeration via

Fig. 11.7.   Windrow turning machine (from LandClean).

perforated pipes placed within the soil treatment mound to supply the oxygen required for microbial activity (Figs 11.8 and 11.9). As such, they typically incorporate an organic bulking agent for structural amelioration to promote bulk flow of air into the soil mound. Typical bulking agents include wood mulch, mushroom compost and bark chippings. The organic bulking agent may confer additional advantages in terms of moisture conservation, providing an additional nutrient source or increasing the effective surface area for microbial activity. This may be of particular concern where free product is present within the soil, which will then become sorbed to the organic matter where microbial activity is likely to be most concentrated. To what extent this actually confers an advantage will depend very much on both the nature of the organic material and the contaminant concerned.

In particular, hydrophobic organic molecules will readily become sorbed to organic

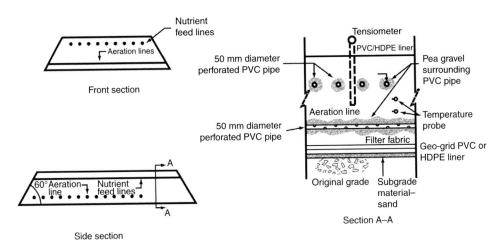

Fig. 11.8. Outline design of a typical biopile. HDPE, high-density polyethylene; PVC, polyvinyl chloride.

Fig. 11.9. View of a typical biopile (from Biogenie).

material, potentially decreasing their bio-availability to the microflora. This is likely to be the case with more slowly degradable organic material such as bark chippings. An example of this is shown in Table 11.3 (Bewley, 2000), in which degradation of TPH and polyaromatic hydrocarbons (PAH) was compared in nutrient-amended soil micro-cosms with and without the inclusion of bark chippings following a four-week period of incubation. Whilst reductions in total PAH of between 59 and 69% were achieved in the treatments with nutrient amendment only, less than 10% reduction was obtained in the

bark-amended treatment, presumably as a result of a reduction in bioavailability. The effects on the reduction of TPH were less pronounced, possibly as a result of the predominantly aliphatic compounds having a lower sorption capacity for the organic matrix.

A similar effect was observed in a field trial used to compare the effects of mechanical aeration and air injection in a prototype biopile as part of a pilot study to assess the feasibility of bioremediating soil contaminated with a mixture of several alkyl benzenes and naphthalene (Bewley *et al.*, 2005). For mechanically aerated treatments

Table 11.3. Extent of reduction in both total and leachable hydrocarbons in soil microcosms with (TPN and TPR) or without (TPS) addition of bark.

	Time (weeks)	TPN	TPR	TPS
Bark added		–	–	+
Nutrients added		+	+	+
TPH[a] (mg/kg)	0	4366	771	988
TPH[b] (mg/kg)	4	422 (90.3)	315 (59.1)	547 (44.7)
Leachable TPH (µg/l)	0	145	25	71
Leachable TPH (µg/l)	4	<10	<10	<10
PAH[c] (mg/kg)	0	2404	1559	1190
PAH[b] (mg/kg)	4	736 (69.3)	645 (58.6)	1081 (9.2)
Leachable PAH (µg/l)	0	127	46	75
Leachable PAH (µg/l)	4	1.0	2.0	0.7

TPH, total petroleum hydrocarbons; PAH, polyaromatic hydrocarbons.
[a]Aliphatic hydrocarbons < C10–44; [b]percentage reduction shown in parentheses; [c]sum of 16 United States Environmental Protection Agency (USEPA) priority pollutant PAH.

the respective differences in contaminant reduction between compost-amended and without, over approximately 3 months, were 94% versus 99% for 1,3,5-trimethylbenzene, 80% versus 98% for ethyl benzene, 90% versus 99% for xylenes and 93% versus 99% for naphthalene. For the biopiles, the respective reductions were 80% versus 79%, 25% versus 76%, 53% versus 64% and 64% versus 73%.

Such reductions in the degree of contaminant loss in treatments receiving mushroom compost could therefore possibly reflect a reduction in bioavailability in compost-amended treatments.

On the other hand, a more favourable effect of mushroom compost was observed in a field trial to assess the feasibility of bioremediation of relatively high concentrations of phenols. In this second trial, five treatments were established consisting of various permutations of mushroom compost (approx. 20% by volume), nutrients, an inoculum of commercially available bacteria ('Phenobac') together with a control. Two of the treatments were established as 'biopiles', with forced aeration, while two were subjected to periodic mechanical aeration.

From starting concentrations in the order of 600–700 mg/kg, the three treatments receiving mushroom compost achieved mean reductions in total phenols of 99% over an eight-week period. These compared with reductions of 30–48% in treatments receiving simply nutrients, with or without bioaugmentation, and less than 1% reduction in the control (although the latter showed significant variability).

Bioaugmentation appeared to increase the rate, but not the extent, of bioremediation: the mean concentration of total phenols in the compost-amended soil was reduced from 600 mg/kg to less than 20 mg/kg after two weeks, whereas in the non-inoculated, compost-amended soil the mean concentration was reduced from 710 to 320 mg/kg. After 8 weeks, however, mean concentrations ranged from 1 to 2 mg/kg in both treatments. This finding is consistent with other studies concerning bioaugmentation, where an increase in the rate of degradation has been reported (Cunningham and Philp, 2000), although the overall extent of contaminant reduction was similar to that of non-inoculated treatments (Bewley, 1996a).

Any reduction in bioavailability conferred by the compost in this second example therefore appears not to be significant. This may be as a result of the relatively polar nature of the phenolics reducing the degree of absorption as a result of the compost.

4.5 *In situ* systems

Whilst *ex situ* systems offer the greatest potential for optimization of the physico-chemical factors that control bioremediation, it is often impracticable to implement such schemes, particularly where the contamination exists at depth, or at an operational facility where excavation is not possible without significant disruption to on-site activities. In such cases *in situ* bioremediation may offer a potential solution. However, because *in situ* systems do not effect changes in the physical nature of ground conditions, a number of significant challenges need to be addressed.

In situ bioremediation of the unsaturated zone generally involves injection or extraction of air via a network of venting wells, so flushing oxygen through the interstitial pore spaces, a process referred to as bioventing, is utilized (Bewley, 1996a). The injection of air is more likely to result in channelling through preferential pathways, but it obviates the requirement for treatment of the vapour stream prior to atmospheric discharge. This may be required for systems that operate in extraction mode, depending upon the volatility of the contaminants involved. Bioventing systems should ideally be designed to supply sufficient oxygen for bioremediation of the contaminants concerned rather than attempting to strip the compounds concerned through volatilization. In practice, however, most air-based systems (for both the unsaturated and saturated zones) will involve a combination of physical and biological removal (Brown, 1993).

Systems may be designed to incorporate both extraction and re-injection of the extracted air although, even here, the geological nature of the subsurface has important implications for successful design (Fig. 11.10).

The implicit assumption underpinning all such bioventing systems is that oxygen is the key factor limiting degradation, at least for hydrocarbons and their derivatives. Whilst this may be true in many cases, (apart from more weathered contamination where bioavailability assumes greater significance), nutrient limitation may also be a significant factor, as has been confirmed in pilot-scale studies (Breedveld *et al.*, 1995). The efficient delivery of appropriate quantities of nitrogen, phosphorus and other key elements in liquid form will, however, be significantly limited by subsurface geology, especially if this is via passive delivery through pipes installed close to the surface.

This also applies where inocula are introduced by passive mechanisms. An example of this concerned a field trial for evaluating microbially mediated reduction of hexavalent chromium (Cr(VI)) to the less toxic trivalent form (Cr(III)), through the introduction of the bacterium *Shewanella alga* Br-Y that had demonstrated this capability at bench scale (Bewley *et al.*, 2000). The bacterium was introduced by passive infiltration into the test plot, facilitated by previously spiking the soil to enable permeation of the liquid suspension. Downward migration of the liquid continued through the saturated zone to the depth of the water table. In the saturated zone, where the presence of groundwater allowed for more rapid lateral dispersion of the inoculum, there was evidence of a significant shift in Cr(VI) from an average of 35 to 10% of the total Cr following treatment. No such changes were observed in the shallower (unsaturated zone) soil, potentially because the intervening areas between the point sources where injection had occurred had not been sufficiently colonized. (Whilst redox conditions may have been more amenable to the shift in Cr(VI) taking place at depth, the reduction process is facultative in nature, so this was probably a secondary issue.)

Introduction of solutions or suspensions of supplements into the unsaturated zone may therefore be of limited effectiveness. One possible method for achieving more efficient delivery, at least for nitrogen, is to supply this as gaseous ammonia. Application of gaseous ammonia to an operational bioventing site where a diesel spillage had taken place produced appreciable increases in microbial activity as observed through enhanced oxygen uptake (Marshall, 1995). For contaminants

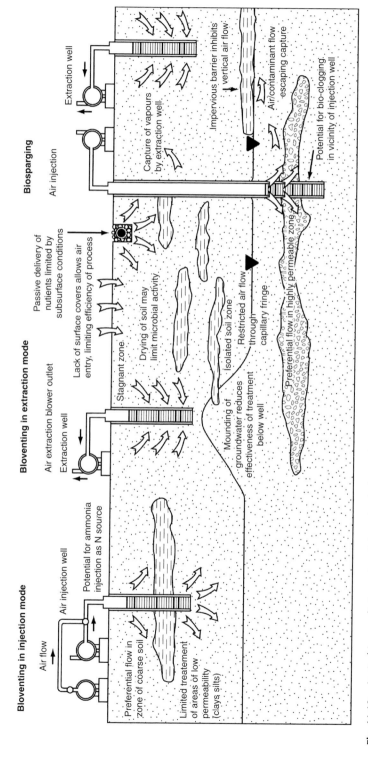

Fig. 11.10. *In situ* bioremediation using air-based techniques, showing potential limitations.

aerobically degraded by a co-metabolic process such as some of the lesser chlorinated compounds, including vinyl chloride and dichloroethene, such systems can also be used to deliver the primary substrate (e.g. methane or propane).

In the case of the saturated zone, remedial objectives for bioremedial systems are typically focused on the treatment of groundwater rather than soil, although a holistic approach is necessary to address the sorbed fraction of the contaminants of concern that provide an ongoing source to the groundwater. Indeed, this represents the key advantage of true *in situ* remediation over pump and treat, which requires a significant number of changes in pore space volume to allow for the slow partitioning of hydrophobic compounds into the aqueous phase. Because there is a continuum of the aqueous phase within the saturated zone, electron acceptors, nutrients – and, if necessary, inoculum – can be readily distributed throughout the zone of treatment.

Treatment of the saturated zone may take place through 'static' systems in which the groundwater is not actively displaced, or 'active' systems involving groundwater recirculation. Of the former, biosparging represents an air-injection approach to delivery of oxygen, and may also be used to strip volatile compounds that can later be captured through a soil vapour extraction system in the unsaturated zone. The permeability of the formation is often a key constraint (Brown, 1993), but such systems may also be vulnerable to geological heterogeneities as illustrated in Fig. 11.10, particularly where the presence of impermeable soil horizons may result in the spreading of contamination.

Static systems for groundwater also include more passive approaches that depend on the slow release of electron acceptors or donors throughout the contaminated zone. Commercially, these are referred to as oxygen release or hydrogen release compounds, respectively, in order to enhance natural attenuation processes typically relating to hydrocarbons or chlorinated compounds (Koenigsberg and Norris, 1999a, b). These comprise magnesium or calcium peroxide for oxygen release and a polylactate ester formulated for the release of lactic acid upon hydration, which can then undergo fermentation to produce hydrogen for reductive dechlorination. Such processes rely on dissolution and diffusion of the dissolved gases, and design considerations need to account for the overall 'efficiency' of utilization of these compounds rather than simply the stoichiometric requirements.

'Active' systems for *in situ* treatment of the saturated zone involve recirculatory approaches, in which groundwater is abstracted hydraulically down-gradient of the source, supplemented with nutrients, electron acceptors and inocula and then re-injected up-gradient by either injection wells or infiltration trench. Examples of these have been successfully carried out for hydrocarbons at a former lubricating oil terminal (Bewley et al., 2001b) and at a former gasworks currently occupied by a bitumen products facility (Bewley and Webb, 2001); whilst the remedial objectives were in each case based on groundwater quality criteria, *in situ* treatment ensured that any 'bounce back' in groundwater concentrations was limited.

Such systems may also be combined with classic pump and treat approaches, in order further to accelerate contaminant removal by surface treatment prior to re-injection, particularly where there are regulatory concerns regarding the re-infiltration of 'contaminated' groundwater.

4.6 Passive approaches to unsaturated zone treatment

Soil bioremediation is almost universally viewed as an 'active' process, in contrast to groundwater, where monitored natural attenuation has been gaining widespread acceptance by regulatory authorities as a potential remedial strategy either alone or as a 'polishing' technique following active remediation (Carey et al., 2000). Phytoremediation, the use of plants to remove, contain or render harmless environmental pollutants, does, however, represent a more passive approach for the unsaturated zone, especially in combination with bioremediation. The rhizosphere offers significant potential for such transformations, particularly as a result of the enhanced microbial population. This may both accelerate contaminant degradation and, through synergistic interactions, result in more complete degradation of highly complex molecules (Cunningham *et al.*, 1996). Additionally, the physico-chemical heterogeneity of the rhizosphere may also provide particularly conducive conditions for degradation of the more recalcitrant xenobiotics such as the highly chlorinated organics, where a two-phase, anaerobic–aerobic treatment may provide the most effective means of achieving mineralization.

Field application of such concepts has not as yet been implemented for full-scale remediation, although in fact the creation of aerobic microsites is a key feature of artificial reed beds that are used in the treatment of waste water effluents (McEldowney *et al.*, 1994), for which there are now a significant number of applications (Nuttall *et al.*, 1997). It is clear from preliminary studies that the interactions between the plant root, microbes and contaminant are highly complex and that indirect effects may be more important than direct acceleration of degradation. This was apparent from several studies by Johnson and co-authors (2004, 2005), including a field trial (Johnson and McGrath, 2004). Planting an industrially contaminated site with ryegrass/clover and a rhizobial inoculum enhanced the dissipation of chrysene and phenanthrene, two key PAH. However, rhizoremediation was not enhanced through the application of selected root exudates and defoliation. The increased degradation appeared to be brought about by enhanced root growth through the rhizobial inoculum, which in turn selectively enhanced populations of microorganisms with the capability of PAH degradation. The lack of effect with root exudates suggests that the benefits of increased root growth were attained through the improvement of soil physical conditions

promoting the ingress of oxygen into the soil.

5 Conclusions

One of the greatest challenges to the successful implementation of bioremediation, and indeed other remedial treatment technologies, is that of contaminant heterogeneity. *Ex situ* approaches enable a much greater element of control to be designed into such systems, through the achievement of at least a degree of homogenization. This results in a redistribution of contaminants, microorganisms, nutrients and electron acceptors, as well as facilitating the management of the bioremediation process through a better understanding of actual changes in contaminant concentration as the treatment proceeds.

In practical terms, however, *in situ* approaches may often be required due to operational and depth constraints. For saturated soils, the presence of water as a continuum enables the successful application of oxygen through biosparging or alternatively using chemically based approaches for the delivery of electron acceptors or donors through slow-release compounds. For the unsaturated zone, the problems of soil heterogeneity may be manifested at both the micro and macro scale in terms of electron acceptor and nutrient delivery.

The rhizosphere may offer potential for the treatment of contaminants through the creation of a mosaic of microsites of varying physico-chemical properties, which may facilitate degradation where this is dependent upon changes in redox potential, for example. Such approaches are likely to be seen as long-term, 'polishing' techniques designed to promote a general amelioration in soil quality rather than achieving the contaminant mass removal required to mitigate more immediate risks.

For all types of approach, however, significant improvements in the application of remedial technology will be heavily dependent upon our understanding of spatial heterogeneity in soil, at all levels.

References

Balba, M.T. and Bewley, R.J.F. (1991) Organic contaminants and micro-organisms. In: Jones, K.C. (ed.) *Organic Contaminants in the Environment*. Elsevier Applied Science, London, pp. 237–274.

Bewley, R.J.F. (1996a) Field implementation of *in situ* bioremediation: key physico-chemical and biological factors. In: Stotzky, G. and Bollag, J.M. (eds) *Soil Biochemistry*, vol. 9. Marcel Dekker, Inc., New York, pp. 473–541.

Bewley, R.J.F. (1996b) Best practice in soil bioremediation. In: *Wider Application and Diffusion of Bio-remediation Technology; The Amsterdam '95 Workshop*. Organization for Economic Cooperation and Development (OECD), Paris, pp. 369–379.

Bewley, R.J.F. (2000) Practical aspects of implementing bioremediation. In: Luo, Y.M. (ed.) *Proceedings of the International Conference of Soil Remediation (SoilRem 2000)*, Hangzhou, China, pp. 120–123.

Bewley, R.J.F. and Stotzky, G. (1983) Effects of cadmium and zinc on microbial activity in soil; influence of clay minerals. Part I: Metals added individually. *Science of the Total Environment* 31, 41–55.

Bewley, R.J.F and G. Webb, G.H. (2001) *In situ* bioremediation of groundwater contaminated with phenols, BTEX and PAHs using nitrate as electron acceptor. *Land Contamination and Reclamation* 9, 335–347.

Bewley, R.J.F., Ellis, B., Theile, P., Viney, I. and Rees, J.F. (1989) Microbial clean up of contaminated soil. *Chemistry and Industry* 23, 778–783.

Bewley, R.J.F., Ellis, B. and Rees, J.F. (1990) Development of a microbial treatment for restoration of oil contaminated soil. *Land Degradation and Rehabilitation* 2, 1–11.

Bewley, R.J.F., Jeffries, R. and Bradley, K. (2000) *Chromium Contamination: Field and Laboratory Remediation Trials*. Project Report 39, Construction Industry Research and Information (CIRIA), London.

Bewley, R.J.F., Jeffries, R., Watson, S. and Granger, D. (2001a) An overview of chromium contamination issues in the south-east of Glasgow and the potential for remediation. *Environmental Geochemistry and Health* 23, 267–271.

Bewley, R.J.F., Alexander, J.G. and Webb, G.H. (2001b) *Ex situ* and *in situ* bioremediation of former oil distribution terminals. *Land Contamination and Reclamation* 9, 175–189.

Bewley, R.J.F., Madawee, K., Clay, J. and Porter, J. (2005) Observations on the application of mushroom compost and bioaugmentation on the bioremediation of phenols and aromatic hydrocarbons. In: *Abstracts of the 156th Meeting, Society for General Microbiology*, Reading, UK, p. 57.

Bossert, I. and Bartha, R. (1984) The fate of petroleum in soil ecosystems. In: Atlas, R.M. (ed.) *Petroleum Microbiology*. Macmillan Publishing Company, New York, pp. 435–473.

Boswell, G.P., Jacobs, H., Davidson, F.A., Gadd, G.M. and Ritz, K. (2002) Functional consequences of nutrient translocation in mycelial fungi. *Journal of Theoretical Biology* 217, 459–477.

Breedveld, G.D., Olstad, G., Briseid, T. and Hauge, A. (1995) Nutrient demand in bioventing of fuel oil pollution. In: Hinchee, R.E., Miller, R.N. and Johnson, P.C. (eds) *In situ Aeration: Air Sparging, Bioventing, and Related Processes*. Battelle Press, Columbus, Ohio, pp. 391–399.

Brown, R. (1993) Treatment of petroleum hydrocarbons in ground water by air sparging. In: Norris, R.D. *et al.* (eds) *Handbook of Bioremediation*. CRC Press, Inc., Boca Raton, Florida, pp. 61–85.

Carey, M.A., Finnamore, J.R., Morrey, M.J. and Marsland, P.A. (2000) *Guidance on the Assessment and Monitoring of Natural Contaminants in Groundwater*. Environment Agency R&D Publication 95, Environment Agency, Bristol, UK.

Craul, P.J. (1999) *Urban Soils: Applications and Practices*. John Wiley and Sons, New York.

Craul, P.J. and Klein, C.J. (1982) Characterization of streetside soils of Syracuse. In: Craul, P.J. (ed.) *Urban Forest Soils: a Reference Workbook; Proceedings of a Workshop on Urban Forest Soils*, University of Maryland, College Park, Maryland.

Crumbling, D.M., Griffith, J. and Powell, D.M. (2003) Improving decision quality: making the case for adopting next-generation site characterization processes. *Remediation* 13, 91–111.

Cunningham, C.J. and Philp, J.C. (2000) Comparison of bioaugmentation and biostimulation in *ex situ* treatment of diesel contaminated soil. *Land Contamination and Reclamation* 8, 261–269.

Cunningham, S.D., Anderson, T.A., Schwab, A.P. and Hsu, F.C. (1996) Phytoremediation of soils contaminated with organic pollutants. *Advances in Agronomy* 56, 55–114.

Dec, J. and Bollag J.M. (1997) Determination of covalent and non-covalent binding interactions between xenobiotic chemicals and soil. *Soil Science* 162, 858–874.

Department of the Environment, Transport and the Regions (DETR) (2000) *Environmental Protection Act 1990: Part IIA Contaminated Land*. The Stationery Office, London.

Domenico, P.A. and Schwartz, F.W. (1998) *Physical and Chemical Hydrogeology*. John Wiley & Sons, Inc., New York.

Evans, D., Jefferis, S.A., Thomas, A.O. and Ciu, S. (2001) *Remedial Processes for Contaminated Land – Principles and Practice*. Construction Industry Research and Information (CIRIA), London.

Farmer, J.G., Graham, M.C., Thomas, R.P., Licona Mazur, C., Paterson, E., Campbell, C.D. *et al.* (1999) Assessment and modelling of the environmental chemistry and potential for remediative treatment of chromium contaminated land. *Environmental Geochemistry and Health* 21, 331–337.

Finnecy, E. (1987) Impacts of soils related to industrial activities: Part II. Incidental and accidental soil pollution. In: Barth, H. and L'Hermite, P. (eds) *Scientific Basis for Soil Protection in the European Community*. Elsevier Applied Science, London and New York, pp. 259–280.

Johnson, D.L. and McGrath, S.P. (2004) Rhizoremediation of land contaminated with persistent organic pollutants. In: *Biological Interactions in the Root Environment, Final Grantholders Workshop*. Biotechnology and Biological Sciences Research Council, Swindon, UK, p. 41.

Johnson, D.L., Maguire, K.L., Anderson, D.R. and McGrath, S.P. (2004) Enhanced dissipation of chrysene in planted soil: the impact of a rhizobial inoculum. *Soil Biology and Biochemistry* 36, 33–38.

Johnson, D.L., Anderson, D.R. and McGrath, S.P. (2005) Soil microbial response during the phytoremediation of a PAH contaminated soil. *Soil Biology and Biochemistry* 37, 2334–2336.

Koenigsberg, S.S. and Norris, R.D. (1999a) *Accelerated Bioremediation Using Slow Release Compounds*. Selected Battelle Conference Papers 1993–1999. Regenesis, San Clemente, California.

Koenigsberg, S.S. and Norris, R.D. (1999b) *Accelerated Bioremediation of Chlorinated Compounds in Groundwater*. Selected Battelle Conference Papers 1999–2000. Regenesis, San Clemente, California.

Lethbridge, G. and Wenham, M.F. (2004) *Addressing Soil Contamination in IPPC Site Baseline and Site Closure Reports*. SAGTA Reports Series, The Soil and Groundwater Technology Association/Shell Global Solutions.

Marshall, T.R. (1995) Nitrogen fate model for gas-phase ammonia enhanced *in situ* bioventing. In: Hinchee, R.E., Miller, R.N. and Johnson, P.C. (eds) *In situ Aeration: Air Sparging, Bioventing, and Related Processes*. Battelle Press, Columbus, Ohio, pp. 351–359.

McEldowney, S., Hardman, D. and Waite, S. (1994) *Pollution: Ecology and Biotreatment*. Longman Scientific & Technical, Harlow, UK.

Nuttall, P.M., Boon, A.G. and Rowell, M.R. (1997) *Review of the Design and Management of Constructed Wetlands*. Construction Industry Research and Information (CIRIA), London.

Reid, B.J., Jones, K.C. and Semple, K.T. (2000) Bioavailability of persistent organic pollutants in soils and sediments – a perspective on mechanisms, consequences and assessment. *Environmental Pollution* 108, 103–112.

Semple, K.T., Morris, A.W.J. and Paton, G.I. (2003) Bioavailability of hydrophobic organic contaminants in soils: fundamental concepts and techniques for analysis. *European Journal of Soil Science* 54, 809–818.

Smith, M.A. and Ellis, A.C. (1986) An investigation into methods used to assess gas works sites for reclamation. *Reclamation and Revegetation Research* 4, 183–209.

12 Biological Interactions within Soil Profiles Engineered for Sport and Amenity Use

Mark Bartlett* and Iain James

1 The Soils of Sports Surfaces

Land management practices and associated maintenance systems for sports surfaces, amenity horticulture and urban green spaces are different from those of primary production-orientated agricultural grassland systems. They too are revenue-generating systems, but principally dependent upon their playing quality and not upon yield of grass produced. In comparison with other grassland environments, the turfgrass found on sports fields represents varying degrees of 'naturalness'. Up until the 1980s, in professional football (soccer) and rugby, which are played during the winter in the northern hemisphere, it was common for players to change their shirts at half-time because they had become covered in mud. Modern football pitches are neatly striped, smoother and more durable, allowing greater precision and opportunity for skilled footballers to perform and entertain. The principal reason for this change is the level of engineering and design that is involved in the construction of the soil environment of modern sports pitches. This is driven by both economics and the need to provide facilities to allow communities to achieve health benefits from sports partici-

pation. For recreational football in a local park, the changes may be less apparent, but at the top end of the game a poorly performing pitch is considered an exception rather than the rule, and half-time muddy shirt changes are a thing of the past.

Despite the wide range of sports surfaces, there are relatively few differences in the internal architecture of their soils. However, the standard of the pitch is directly related to the cost of construction and maintenance. At one end of the spectrum, where budgetary constraints are tight, the sports surface may be little more than a consolidated meadow. Modern constructions of sports pitches have their internal architecture engineered to achieve optimal surface performance and grass aesthetics. Engineering designs have focused on two contrasting systems, one optimized to be free draining and the other to have a hard, mechanically resilient surface, depending on the concomitant requirements of each individual sport. Despite these optimizations for their mechanical behaviour, the playing surface must still be capable of supporting grass, which is an important component of player–surface interactions. The engineering constraints often produce a soil environment that is far from optimal for the free soil

*Corresponding author: mark.bartlett@scotts.com

biota, or indeed grass growth. Grass-covered sports systems therefore require intensive and continued manipulation and management, which include nutrient additions and the use of agrochemicals in order to maintain the playing characteristics that they were initially designed to deliver. Through concurrent manipulation of the abiotic and biotic components of these soils, there is an opportunity to reduce the management inputs to a sports surface, improving plant nutrient cycling and disease resistance, whilst still maintaining its overall quality.

Each different sport requires different characteristics from the playing surface. Within the sports industry there is a clear understanding that the pitch conditions will affect the performance of the athletes (Baker and Bell, 1986). In natural turf pitches the interaction between the player and the pitch can have a significant effect on the player's biomechanics (Stiles et al., 2009). To minimize unsafe and unfair variations between pitches they are commonly constructed using imported root zone soils. The internal architecture of these sports pitches is highly prescribed and tightly specified. Within sports pitches there are two common root zones used. One is sand-dominated, with typically 85–95% sand (USGA Green Section Staff, 2004). The resulting internal architecture of the soil is one of high porosity and low tortuosity.

In contrast, some root zones have higher clay contents (greater than ~30% clay), with correspondingly lower infiltration rates and a greater tortuosity within the pore network. For cricket, where clay-based root zones are used, the pitch must be hard and durable enough to last for a whole game, which could be as long as five days. This means that the wicket area, approximately 30 m^2, must be able to withstand cricket ball impacts off the surface, with the ball travelling up to 145 km/h in world class cricket. In this situation, the surface must be designed to minimize energy absorption on ball impact and to deliver the ball reliably from bowler to batsman (Carré et al., 2000). The soil characteristics and inner space of a cricket pitch are considerably different from a golf green. In golf, straight ball-

roll and rapid surface drainage (to allow a high throughput of golfers in poor weather) are primary business drivers in surface construction; root zones are therefore engineered from high-sand content soils. Both a cricket pitch and a golf green are designed to support turfgrass and promote its growth, but these two sports represent a dichotomy of built sports facilities. With increased engineering of the soil root zone, the flexibility of use of surfaces for different sports declines. If a golf green were to be constructed to the specifications of a cricket pitch the game would be virtually impossible to play and potentially dangerous to the participants. A cricket wicket constructed to the specifications of a golf green would be equally as dangerous and challenging to play on – and to the detriment of the sport.

The turf areas used for horse racing are also closely managed, but the root zones of these sports surfaces are not normally so significantly modified from surrounding soils. The soil matrix and the plant–soil relationships are important for both competition revenue and safety in horse racing (Dufour and Mumford, 2008). Horse injuries have been shown to be closely related to high soil strength and associated structural properties. These soil conditions can be desirable as harder surfaces result in faster, more competitive races (Henley et al., 2006).

A principal driving force for facilities of this nature is to allow access to the largest numbers of individuals, in order to increase participation in sports. In professional stadiums the drivers are different. The surfaces are engineered to produce the highest-quality playing surface with regard to both sports performance and safety for players, who are also valuable assets. The result is a pitch being founded on an anthropogenic soil profile and the sward composition being radically altered with a much lower species diversity compared with natural grassland. At Lord's Cricket Ground in London, these man-made soil profiles and grass species construction represent the whole facility. The soil in the area of the facility where the cricket wickets are sited is an imported heavy-clay soil, which can be consolidated to provide a suitable surface for ball bounce, while the outfield is constructed from a sandy soil so that it drains rapidly, thus minimizing the time lost to rain delays over a five-day test match.

Nutrient and water resource management at high-level, professional sports facilities is different from agricultural systems. Typically, a full range of both macro and micro nutrients is applied, not to promote a crop yield but in order to enhance plant colour, health, resistance to disease and tolerance to frequent mowing. The net result is that both plant and soil organisms are rarely nutrient limited, despite the frequent use of free-draining, low-cation exchange capacity, sandy soils. At the highest-quality sports facilities the turfgrass plants are rarely in water deficit, due to frequent irrigation. While this allows the turfgrass to continue to grow, it has a significant influence on the size, function, abundance and efficiency of the biological community within the soil.

An important difference between sports surfaces and agricultural soil environments is that, for most sports surfaces, some level of soil compaction is a desirable characteristic. Another differentiation is the intensity of surface water drainage in a modern pitch design. The grass species and their diversity, common in sports turf, are also different from natural grasslands. In the UK and other temperate climates, the turf ground cover is commonly maintained as one or two species of fine-leaved grasses such as *Festuca* spp. and *Agrostis* spp., in contrast to the diverse community assemblages found in natural grasslands (Balogh and Walker, 1992).

These differences mean that direct comparison and application of the knowledge of other grassland ecosystems are not always appropriate. Strict comparisons with other agricultural or horticultural cropping systems also have limited meaning. The principal driver in these latter systems is crop yield; for sports surface management, crop or grass yield is unimportant. In many cases increasing grass yield is undesirable as it increases maintenance requirements and costs.

It is difficult to argue that sand-construction sports surfaces are in practice environmentally sustainable – they are both resource intensive and financially costly. This lack of sustainability stems from the emphasis in design for hydraulic conductivity and drainage rather than environmental management. Our discussion here focuses on the extent to which the other functions of the soil inner space are affected, either by design or as a consequence of such design, and the extent to which surfaces can be redesigned with a knowledge of soil inner space to improve sustainability without significant detriment to required function.

2 Sports Surface Design and Root Zone Architecture

Modern, professional sports surfaces are constructed inside the stadium, from the bottom upwards. A natural turf pitch, constructed from a natural clay soil, might have an infiltration rate of 10 mm/h whereas a modern sand-construction pitch is designed to have an infiltration rate of at least 400 mm/h. This is achieved by constructing pitches from coarse sand root zones, overlying gravel and using a closely spaced (often less than 10 m) drainage network. The specification of the particle size of the sand and the gravel is important because they must be matched to create a hydraulic break and a resultant suspended water table. The aim is to move water from the surface as quickly as possible but also to retain a sufficient soil water content in the root zone to supply plant demand. Hydraulic performance of this system is dependent upon material selection, surface construction and surface maintenance. This radical alteration to the drainage characteristics of the pitch has major implications for the habitat space within the soil, allowing greater pore sizes and more connectivity, both within the soil matrix and between the soil and atmospheric system. A pertinent question is whether, with a better understanding of the inner space of soils, a more sustainable system can be designed that does not compromise drainage but reduces resource consumption.

The hydraulic conductivity of a soil is a function of pore size distribution and inter-pore connectivity. Any manipulation of the size, arrangement or biological occupancy of inner space must preserve sufficient hydraulic conductivity (and infiltration rates) to maintain surface playing performance. The original specifications for sand root zone pitches were designed for climate zones with high rainfall and high evapotranspiration. In the UK, lower-hydraulic conductivity material selections could be made, which would permit increased water and nutrient retention using soils with increased fine sand, silt and clay contents. Reducing pore size increases capillary rise above the hydraulic break and it is important that the suspended water table does not encroach on the playing surface, which would adversely affect player–surface interactions. Fine-grained soil material can also result in hydraulic continuity between drainage layers if particles migrate into gravel layers, removing the suspended water table. A 'blinding layer' of coarse sand between root zone and gravel is essential.

One strategy for increasing water retention, soil strength and cation exchange capacity is to increase organic matter by amendment of the sand root zone. Organic matter can have an adverse effect on the mechanical properties of a surface, increasing energy absorption and reducing ball rebound resilience. This is already practised to an extent by the amendment of sand root zones with peat, although the sustainability of this strategy is challenged both because of the use of (essentially non-renewable) peat-soils and because this organic matter is actually readily oxidized within the root zone. Studies have demonstrated that organic amendments decompose over time, with resultant obstruction of pores and reduced hydraulic conductivity (McCoy, 1992). To prevent this phenomenon, more resilient, inorganic amendments such as the zeolitic mineral clinoptilolite, which has a very high specific surface, have also been trialled but with varying success – in particular in the attempt to increase cation exchange capacity and water retention (Li et al., 2000; Waltz and McCarty, 2005).

Gas diffusion through soils has been studied extensively in both natural and agricultural systems (Stepniewski and Glinski, 1985); however, little attention has been paid to the role of the soil atmosphere in the health and functionality of sports turf. More densely compacted soils have reduced pore space, smaller average pore radii and reduced connectivity between pores, all of which directly affect the conductivity of both gas and water in soils (Stepniewski and Glinski, 1985; Horton et al., 1994; Lipiec and Hatano, 2003). Clay–loam cricket soils have a target dry bulk density in excess of 1.6 g/cm^3 (Baker and Adams, 2001), achieved by rolling with a heavy, smooth-wheeled roller. As soil becomes more compacted and total porosity decreases, the range of soil water potential that can be tolerated by the plant also becomes smaller and gas diffusivity reduces. Rooting has been shown to increase the proportion of larger pores, resulting in greater drainage potential in clay soils. The action of plant roots in the soil increases the number and connectivity of soil pores and, ultimately, total porosity (Whalley et al., 2005).

3 The Effects of Management on the Inner Space of the Root Zone

The management of sports fields also has a considerable impact on the associated below-ground ecology. A narrow range of grass species are used for sports facilities, in practice drawing upon only several species from seven genera. Different growth habits, leaf shape and turf colours are all factors considered in species selection, and different grass traits are required for different sports. The grass at any sports facility is typically grown in monoculture or, in some cases, a maximum of three-species mixtures and such monocultures can result in rapid disease infestations (Danneberger, 1993).

The success of sand-based pitches in football is unquestionable – Arsenal Football Club's Emirates Stadium (London) being a leading example. With valuable television rights comes the responsibility to provide aesthetic surfaces that guarantee that games will not be cancelled. However, in the root zone of these pitches the grass plant roots play a critical role in maintaining the overall functionality of a pitch. Plant roots influence soil mechanical properties that are important for player performance. In a sand-based root zone the grass roots play a role in reinforcing the sand matrix, increasing shear strength; without the grass, the sand simply 'divots' when kicked by the player. The presence of roots in sand-dominated root zones at least doubles resistance to shear stresses (Adams *et al.*, 1985). The effect on shear strength varies with grass species, a strong effect being seen in sports pitches planted with perennial rye grass (*Lolum perenne*) whereas those planted with brown top bent grass (*Agrostis tenuis*) showed no increase in shear strength (Adams and Jones, 1979). The rooting structure within the soil is critical in reinforcing the surface strength. The roots of mown cool season grasses predominate in the top 100 mm of the soil and rarely exceed 300 mm (Emmons, 2000), and this is in part due to the mechanical impedance within the soil. The grass surface is the interface between the pitch and the player and this is where the stresses on the soil are greatest. Research shows that vertical player loading of a pitch reduces rapidly with depth and is rarely observed below 100 mm (Dixon *et al.*, 2008). This is consistent with the suspended water table design that aims to hold water below this depth but within the rhizosphere. At the surface the pitch must provide sufficient traction to allow the player to manoeuvre without sliding uncontrollably, but not provide excessive grip so that the foot locks on the surface causing lower leg muscular–skeletal injuries to the player. A study of players running on natural turf showed that increasing the sand content of a pitch significantly increased its stiffness, compared with clay-dominated soils, allowing players to run more efficiently and therefore more quickly over the surface (Guisasola *et al.*, 2009). The same study has shown that the shear strength of sand soils, an important factor in player traction, is less sensitive to changes in water content than clay soils. Such clay soils are common in public park pitches, where surface performance can

change with small changes in water content; the change in shear strength of sands is much less. In a laboratory study to evaluate key mechanical properties from the soil, the presence of roots within the sample did not significantly affect the shear strength of a sandy root zone, but significantly reduced elastic moduli. The presence of roots within the soil matrix is therefore instrumental in increasing the stiffness of the surface (Guisasola *et al.*, 2010).

Although virtually all growing roots experience mechanical impedance to some degree (Bengough and Mullins, 1990), the engineering of sports surfaces to provide high soil strength means that the turf roots frequently experience these stresses. The internal soil environment of a cricket wicket, for example, is a high-impedance root zone but turfgrass cultivars used for cricket wickets are well adapted to such an environment. The root system reduces pitch disintegration from heavy wear or excess drying, thus reducing erratic and potentially dangerous ball bounce. The turfgrass also functions as a moisture removal mechanism. Transpiration by the plant is a necessary process for the even drying of the pitch soil profile to achieve optimum hardness of the surface, and also to increase density through shrinkage (Baker *et al.*, 1998; Adams *et al.*, 2001).

Mowing at sports facilities is a key process in maintaining both the aesthetics and surface performance of pitches. On a bowling green, for example, fine-leaved grasses are used, as these can be cut low to the ground, with the grass sometimes being as short as 3 mm (Brown, 2005). The short grass leaves result in a lower coefficient of friction between the surface and the ball, resulting in a smoother, more predictable roll (Hartwigera *et al.*, 2001). At the opposite extreme, grasses on a horse racecourse are typically maintained between 60 and 100 mm. The increased vigour in plant growth both above and below ground is essential in maintaining the quality of the track surface, helping to reduce damage caused by hoofs, cumulatively amounting to several tonnes of horse pounding the surface over a short time (Henley *et al.*, 2006).

Another component of the soil system that helps to minimize the impact of the horses on the running surface is 'thatch'. In classic soil taxonomy, thatch can be described as an O horizon at the surface, above the first true soil material from the A horizon. This soil feature is more commonly associated with woodland soils (Hodgson, 1997), but is a result of intensive management of turfgrass (Turgeon, 2002). Thatch in sports surfaces is a layer of living and dead grass stems, roots and crowns that has developed between the vegetation on the playing surface and the surface of the soil. It is the facility management that leads to the formation of thatch, predominantly turfgrass species selection and mowing regime, that affect the manner in which the plant grows. The plant response to 'damage' through mowing is to shed root mass not required to support the above-ground photosynthetic parts of the plant, reducing root depth and density and consequently soil strength. Frequent mowing also causes imbalances in synthesis of plant growth regulators, such as auxins and gibberellic acids which increase root division and cause cellular elongation. In horse racing the presence of this thatch layer helps to knit together the surface of the racecourse and also functions as a shock absorber, especially important in jump racing. In other sports, the presence of this thatch layer is highly undesirable, or even unsafe. If thatch is allowed to develop to a significant depth on a cricket square, energy absorption will increase, resulting in unpredictable and often dangerous ball bounce during bowling (McIntyre, 2001).

Thatch is a substrate-rich environment for many soil organisms, which in a less managed system would probably result in a feedback loop of self-regulation and subsequent decomposition. The use of agrochemicals and mowing in sports facilities distorts this regulation, as many of the soil flora and fauna that represent diseases or pests controlled by these chemicals also promote the degradation of this organic material (Potter et al., 1990). The disturbance in these processes by management can sometimes result in the thatch reaching a depth of 100–120

mm before the start of the mineral soil matrix, and implies that decomposers within the microbial community have been inhibited. To investigate the relationships between the activity of the soil microbiota and the management of sport surfaces, the functional diversity with respect to C cycling of soils taken from each of the key landscape features of a golf course (i.e. tee, fairway, green and rough) was investigated. Samples were taken from a gradient of management intensity, in the decreasing order: greens, tees, fairways and roughs. A multiple, substrate-induced respiration assay (after Degens and Harris, 1997) was used to assess the respiratory response of the soils to seven different carbonaceous substrates representing a range of molecular complexity, plus a grass–homogenate suspension in water. Soil respiration was monitored every 6 min for 16 h and the rates from each substrate were integrated to determine the net response on a substrate-by-substrate basis (Ritz et al., 2006). Differences in respiratory profiles were then analysed using principal components analysis. The rough zones of the golf course showed a significantly different response from the more managed areas ($p < 0.01$; Fig. 12.1). The loadings associated with the analysis indicated that the ability to use grass clippings as a substrate was causal in defining these relationships (Fig. 12.1b). The size of the response by the soil of the rough to the grass-clipping substrate was fivefold greater than for the other three areas of the course (data not shown). Bardgett and McAlister (1999) observed that fertilizer applications in managed agricultural grasslands influence soil microbiota and, while this effect is evident here, there is also clear indication that the management of the grass plants and associated agrochemical applications plays a fundamental role in influencing the nutrient cycling and community composition within the soil matrix.

The soil microflora and -fauna are critical in any productive nutrient cycling ecosystem, and sports surfaces are no exception. In order to maintain healthy turfgrass, an active soil microbial community must be present to cycle essential minerals and

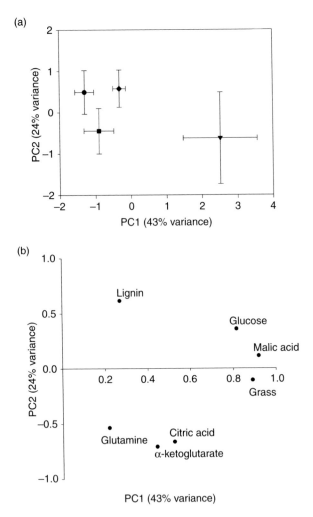

Fig. 12.1. Microbial functional diversity in different zones of a golf course. Principal component analysis of multiple substrate-induced respiration profiles of microbial communities associated with tees (■), fairways (♦), greens (•) and rough (▼). (a) mean coordinates in PC1 and PC2 for each area of a golf course; $n = 5$ per treatment; whiskers show standard error; $p < 0.01$ in PC1 only; (b) associated loading values from the analysis highlighting the substrates causing significant differentiation in the community structure.

nutrients. As well as imbalances in nutrient cycling, the size of the microbial community is also suppressed in the most intensively managed sports turf (Bartlett *et al.*, 2008a). In a study of sports surfaces within a single sports complex, on similar soil types, all the grass areas used for sport supported a significantly smaller microbial biomass compared with surrounding non-sport amenity grassland (Fig. 12.2). On the areas of the golf course (tee, fairway, green and rough) the management practices that had been adopted had a direct effect on the size of the soil microbial community. Where management was intensive the microbial community was suppressed: the microbial biomass C concentrations on these surfaces were significantly smaller than on the fairways or the rough ($p < 0.01$; Fig. 12.2). Different areas of the golf course received

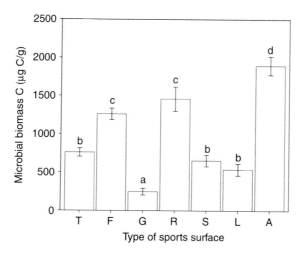

Fig. 12.2. Soil microbial biomass associated with different types of sports surfaces from within a single sports complex are significantly different ($p < 0.01$). T, golf tee; F, golf fairway; G, golf green; R, golf rough; S, football (soccer); L, lawn tennis; A, non-sport amenity turfgrass. Whiskers show standard error; letters indicate homogeneous groups by LSD (least significant difference).

different quantities and frequencies of chemical inputs, particularly fertilizers. Each tee and green, which must be visited by every individual playing the game, receive much higher nutrient inputs in comparison with the rest of the golf course (Bartlett *et al.*, 2009). The soils of the lawn tennis courts and football pitches, managed at a similar intensity to golf tees, also had a similar microbial biomass.

The physical characteristics of the soil also significantly affect the phenotypic expression of the microbial community structure. Changing the physical character- istics of the soil has a demonstrable effect on soil biology, an effect that has been shown to be geographically consistent. In a study of the phenotypic microbial commu- nity structure of golf course soils, differ- ently managed surfaces at a range of golf courses across a gradient of soil types indi- cated consistent community structures (Bartlett *et al.*, 2007). On some golf courses, where for whatever reason the greens can- not be used, temporary greens are sometimes constructed. These greens are made by mowing an area of the fairway very closely (c. 5 mm), over-seeding with fine-leaved grasses and agrochemicals applied, both to

promote growth and prevent grass diseases. Once suitable for play, these greens can be maintained for as long as they are required, often between 6 and 12 months. The result of these modifications to soil internal envi- ronment causes a significant shift in the structure of the microbial community (Bartlett *et al.*, 2007). These findings are consistent with the community structure of pasture lands at different geographical loca- tions, where the effects of management practice – and not their physical location – dominate the microbial community struc- ture (Steenwerth *et al.*, 2002). Changes in vegetation management in agricultural sys- tems, such as radically altering cultivation practices, have been shown to have similar impacts on the associated microbial com- munity (Hedlund, 2002; Clegg, 2006).

Of the hundreds of thousands of differ- ent species of micro-, meso- and macroflora and fauna supported by the soils of amenity turf, less than 1% are detrimental to plant health (Grewal, 1999). These soil biota have adverse effects on either the health of the grass plants or performance of the surface; the action of these soil-borne biota results in non-smooth, non-level, poorly performing pitches. Some species of fungi found within the soil environment are pathogenic to turf-grass. Diseases such as *Microdochium nivale*, a pathogen of winter cereals and turfgrass, can result in patches of dead grass. Spores of this fungus are capable of surviving warm, dry weather within the soil matrix and germinating in cool, wet conditions. The fungi subsequently feed on the turfgrass, which ultimately results in plant death (Haigh *et al.*, 2009). This damage to the turfgrass is as a result of the pathogen taking advantage of the soil habitat of the sports surfaces and proliferating when the turfgrass health is weak and environmental parameters favourable. Other meso- and macrofauna that are detrimental to grass health also take advantage of the habitat space provided by the soil. The soil supports a large number of insects and larvae, for example the larvae of *Phyllopertha horticola*, *Melolontha melolomtha* and *Rhizotrogus* spp. (chafer beetles), which can cause damage to the turfgrass, feeding on the plants roots. Turfgrass pests of this nature, when present in high numbers, can cause considerable damage to the turfgrass, having a significant impact on the playing quality of the surface.

Mann (2003) suggests that the most important pest species in UK amenity turf are earthworms and leatherjackets (Tipulid larvae). Where thatch build-up is frequently a management problem, earthworms are capable of breaking this organic root debris down and redistributing it through the soil (Potter *et al.*, 1990). The net result is an increase in available nutrients and a faster rate of oxidation of organic material. Where earthworm populations are suppressed, greater problems of thatch build-up and surface compaction can lead to ponding, resulting in both turf and surface degradation. All of these beneficial actions can be related to endogenic and anecic earthworms. Despite these many benefits to the health of the amenity turf, it is the action of anecic earthworms that determines why they are considered as pests in sports root zones.

The ecology of anecic earthworm species is to maintain permanent or semi-permanent burrows, significantly altering the internal space of the soil. The net result

of these modifications is that the material moved from within the soil matrix must be deposited somewhere and, in the case of anecic earthworms, this is as casts on the soil surface. On playing surfaces where ball roll (e.g. golf and bowls) or bounce (e.g. cricket and tennis) is important, earthworm casts can be problematic because they can result in an uneven playing surface. This can cause the ball to deviate unpredictably, either interfering with play or resulting in potentially dangerous ball–surface interactions (Binns *et al.*, 1999; Baker *et al.*, 2000). Surface earthworm casts also present problems with mowing (Fig. 12.3a). Most high-performance sports turf surfaces are cut very close to the ground, with mowers typically being set to a height of 3–20 mm. Earthworm casts, which can be up to 25 mm in height (Lee, 1985), will stand proud of the grass and may damage and blunt mower blades as they pass. The presence of earthworms within the soil can also have a major effect on the quality of the playing surface and its gross architecture because of the behaviour of their vertebrate predators. Mammals such as badgers (*Meles meles*) and moles (*Talpa europaea*) that inhabit the soil feed extensively on earthworms, which can account for up to 62% of their diet (Goszczynski *et al.*, 2000). Larger earthworm species such as *Lumbricus terrestris* are an important component of a badger's diet, and they are known to excavate large areas searching for earthworms (Fig. 12.3b).

Leatherjacket is the generic name given to the turfgrass tissue-feeding larvae of *Tipula paludosa* and *Tipula oleracea* (European crane fly). They are considered a pest in both amenity turf and some agricultural systems, as the larvae feed with rasping mouthparts on the plant roots and have significant effects on soil structure and plant growth habit (Dawson *et al.*, 2002). *Tipula paludosa* and *T. oleracea* are most frequently found in areas of open grassland such as sports facilities, as these systems have relatively low disturbance to their life cycle and their larvae have a habitat preference for sandy soils (Brown, 2005). The two species have different life cycles and mate

(a)

(b)

Fig. 12.3.   The impact of soil biology on the playing quality of sports surfaces. (a) The scale of a single earthworm cast to a golf ball and the typical grass length of a fine turf surface (b) and soil disruption caused by badger (*Meles meles*) foraging on a sports field (images courtesy of S. O'Hara).

at different times of year, but both pupate in the summer. Females from both species lay eggs at the base of grass leaves and, once hatched, they feed on the roots of the plant (Dawson, 1932). *Tipula oleracea* will feed on root and other organic material during the spring, growing to approximately 30 mm after which it burrows deeper into the soil to pupate. This activity modifies the soil below the surface, changing its structure and critically reducing surface hardness and resulting in an uneven surface for play. As well as the damage to the surface

these larvae can cause considerable damage to the grass plants. Initial stages of an infestation are signified by patches of turf that wilt in full sun because of the root system damage; in major infestations the larvae can destroy the whole sward. Turf can also be damaged by higher-trophic level feeders such as birds foraging through the turf surface for the larvae, thus causing even more surface damage and disruption. Either condition results in the prevention of the sports facility from being able to function as it was designed (Brown, 2005). The cranefly larva has a natural nematode parasite (*Steiner-nema feltiae*) that has been deployed as a successful biological control vector in some situations (Peters and Ehlers, 1994). This nematode feeds both on the tissues of the larvae and also on bacteria which thrive on the decomposing cranefly corpse. Other workers have shown that a significant reduction in cranefly population can be achieved by introducing the bacterium *Bacillus thuringiensis*, although control is never as effective as through chemical means (Waalwijk *et al.*, 1992).

4 The Soil Biota as a Tool for Optimization in Sports Root Zones

Modern sports pitches that have been designed from an engineering standpoint to have uniform mechanical behaviour across the whole surface still have an intrinsic variability. Where the internal architecture of these soils is so similar within a single pitch, it is the effects of the soil biota that are responsible for many of the inconsistencies in playing quality. Engineering solutions have been proposed in attempting to eliminate the biotic component of sports pitches, resulting in the development of synthetic turf. The most significant advances in optimization and engineering of sports root zones have been driven by soil–water physics problems (Guisasola *et al.*, 2009) or the need to optimize player–surface biomechanics (Stiles *et al.*, 2009). The resulting systems require intensive management and are resource inefficient. Applied research has identified ways to optimize water, fertilizer and agrochemical applications within these systems (Blume *et al.*, 2009; Hindahl *et al.*, 2009; Rossi and Grant, 2009).

Manipulations of the soil environment attempt to control earthworm populations by physical intervention; one method proposed sought physically to exclude earthworms from colonizing the soil using geotextile membrane barriers (Bartlett *et al.*, 2008b). This method showed some limited success in controlling the surface casting of anecic earthworms. Approaches like this,

that are predominantly physical, are common. The deliberate manipulation of the biological component of the soil has rarely been considered as a mechanism for optimization of the playing quality An alternative to engineering the soil structure could be to engineer the earthworm community within the soil. Earthworms from the anecic ecological grouping that form permanent and semi-permanent burrows can be discouraged from colonizing soil when there is high structural disturbance within the matrix (Lee, 1985). Earthworms from the endogeic ecological grouping do not maintain permanent burrows and rarely cast on the surface (Sims and Gerard, 1999), and consequently they are responsible for considerable bioturbation within the soil matrix (Lavelle, 2001). The actions of this ecological group could potentially be harnessed to control anecic earthworm activity. Increasing the population of endogeic earthworms within the soil could increase burrow disturbance to a level that forces the anecic earthworms from this feeding niche. Field-scale research would be required to confirm these ecological relationships and concepts, founded on research from other soil environments.

Manipulating the soil earthworm community to be dominated by endogeic earthworms could also stimulate the breakdown of thatch material. Earthworm feeding has been closely associated with the organic matter in the soil, although it is governed by the associated microbial community (Bonkowski *et al.*, 2000; Pilar Ruiz *et al.*, 2006). The response of substrate addition to golf course soils suggests that stimulating the size of the microbial community could help to mitigate thatch accumulation (Fig. 12.2). The suppression of the size of the soil's microbial community means that the magnitude of the nutrient transfer within the soil system is smaller than for natural or semi-natural grasslands. Applying labile carbon sources with nitrogen fertilizers may help to redress the imbalances in nutrient supply for the microbial community caused by turfgrass management. Stimulation of the size of the microbial community in this way would theoretically lead to a more rapid nutrient turnover, including the degradation of excessive organic material in the soil. Manipulation of the soil biota in this way, rather than changing the structure of the soils, could also lead to a reduction in turfgrass disease incidence (Mazzola, 2004).

Heat sterilization of soils prior to construction is a current approach to control weeds and diseases in newly built sports pitches. The effect on both the soil structure and the whole of the microbial community is considerable (Trevors, 1996) and, while this has the effect of initially controlling some soil organisms, it also creates a 'playing-in' period while the pitches perform suboptimally. Further investigations are required to determine the interactions between the biotic–abiotic system that result in the optimal performance of playing surfaces for sport.

Numerous beneficial synergies have been reported between grasses and arbuscular mycorrhizal (AM) fungi, improving nutrient uptake efficiency, rooting depth, speed of establishment of the turf and disease and drought resistance (Sylvia and Burks, 1988; Gemma and Koske, 1989; Augé et al., 1995). Despite this, the size of the mycorrhizal community associated with sports pitches is typically small (Koske et al., 1997a; Barry et al., 2005). However, sports pitches have also been reported to support a relatively high diversity of AM fungi (Koske et al., 1997b). The intensive management of soil rather than soil physical and structural characteristics is likely to be responsible. Application of fungicides to sports surfaces is significantly greater than for other grassland systems, but a limited toxicity to AM fungi from these chemicals has been reported (Barry et al., 2005). However, the high frequency of mowing of the grass means that such frequent foliage removal is likely to reduce the capacity of the plant to direct assimilated carbon to the fungi (Jakobsen et al., 2002). Another contributing factor is likely to be through the frequent application of fertilizer, most notably phosphorus: it is well known that AM fungi become less abundant in high-P soil environments (Demiranda and

Harris, 1994). The growth habit of common meadow grass (Poa annua) makes it a problematic weed species in many sports pitches; however, research has shown that stimulating the size of the AM fungal community can act as a biological control against P. annua by increasing the competitive advantages of the desired grass species (Gange, 1998; Gange et al., 1999). Therefore the functional requirements of the pitch dictate the composition of this component of the soil biota rather than directly by the architecture of the soil.

5 The Biology of Synthetic Turf Pitches

An increasingly common alternative to a natural soil pitch is a synthetic turf pitch. Synthetic turf has developed beyond the infamous pitches of the 1980s, where ball bounce was uncontrollably high and the risk of player injury was unacceptably large (Baker and Wollacott, 2005). Player perceptions of this kind of pitch are still closely linked with the overall playing environment (Flemming et al., 2005). Using a programme structured to drive product innovation, the Fédération Internationale de Football Association (FIFA), the international governing body for football, aims to emulate elite natural turf pitches using synthetic surfaces with longer pile and granulated rubber infill materials. These modern surfaces are termed third- or fourth-generation synthetic surfaces. This type of sports surface is generally selected for its high usage capacity and 'most-weather' suitability. A recent survey of independent schools in the UK highlighted that synthetic turf pitches commonly sustain 44 hours of play per week, whereas natural turf pitches supported only 4.1 hours before there was significant surface degradation (McLeod and James, 2007). Some sports, such as field hockey at National League level, are exclusively played on these types of constructed surfaces (FIH Equipment Committee, 2008). The physical structure of these surfaces, the materials used for infill and the maintenance of the surface create unique environments that are essentially soil analogues in the context of sports surfaces. Biological interactions with these materials have barely been considered.

Synthetic turf pitches are constructed using a tufted or woven synthetic turf fibre carpet and a backing material. In second-generation synthetic turf pitches, the most commonly used design, the fibres are 22–25 mm long and the infill material is rounded sand (Fig. 12.4a). For third-generation pitches the pile depth is longer, between 60 and 90 mm, depending on the sport that is to be played. The infill material is also different, being a combination of sand and crumb rubber (Fig. 12.4b).

The infill matrix has a number of functions. The primary role is to keep the synthetic turf fibres upright; without the infill, the fibres would simply flatten, giving a slippery surface with high ball roll that would not drain. The infill matrix also holds down the carpet. In a second-generation synthetic turf pitch each square metre of turf has 20–25 kg of sand infill (McLeod and James, 2007). The infill is also designed to act as a filter, stopping particulate material from migrating into the carpet backing where it would inhibit surface drainage.

The internal environment and physical construction within these surfaces represents a unique habitat for many microorganisms; a complex biological system appears

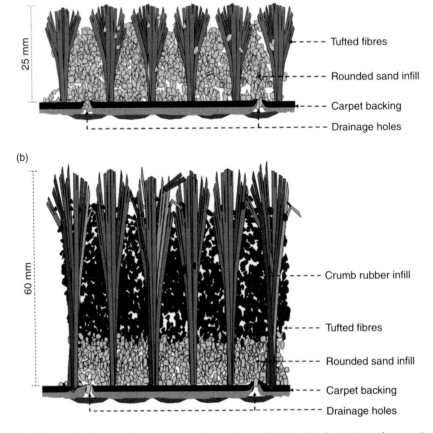

Fig. 12.4. Schematic diagram of a cross-section through a synthetic turf pitch. (a) Second-generation pitch with sand infill; (b) third-generation pitch with sand and crumb rubber infill.

to develop, dominated by bacterial communities (A. McNitt, Pennsylvania, 2008, personal communications). Where the synthetic pitch is sand-filled, the pore spaces and other soil physical characteristics are similar to those of a sand-based root zone. They provide a habitat for microbial colonization and biological succession, and are probably linked with pedogenic processes. Some synthetic turf pitches are frequently irrigated, while others retain high moisture contents, commonly resulting in algal and cyanobacterial biofilms forming on the playing surface. They may also provide conditions for pathogenic bacterial survival and in vivo growth. These biological components are typically controlled with biocides. Moss and higher plants can also become established in artificial turf, this normally being observed when the required mechanical maintenance associated with these surfaces is not carried out. In third-generation constructions, where crumb rubber is also used as an infill material, the dimensions and complexity of the internal habitat of the playing surface are considerably expanded because of the high porosity of each rubber crumb. The rubber crumb potentially provides additional carbon substrates for appropriately adapted microorganisms. The material from which the turf carpet is constructed may also represent a substrate for some microorganisms within the matrix. Their activity may lead to more rapid wearing of the turf fibres, or increased susceptibility to damage when surface maintenance is carried out.

The use of these playing surfaces is still relatively new and, as such, associated challenges are only just beginning to be addressed. Any questions regarding the scale of microbial activity within these surfaces remain unanswered. The closest comparable systems are sand dunes and newly constructed sand root zone golf greens, where research has indicated microbial biomass is between 70 and 120 µg/g dry weight of soil (Bartlett et al., 2008a), although with the limited input from photosynthetic organisms the microbial biomass is likely to be even smaller. It is impossible to infer data on the level of activity and even

functionality of the microorganisms present. However, research carried out in the USA has indicated that synthetic turf surfaces may present health risks to players. One study showed that turf abrasions suffered by American Football players could be directly linked to the transmission of methicillin-resistant Staphylococcus aureus (Kazakova et al., 2005).

6 Summary

The popular approaches to design of modern sports surfaces and facilities in general have been driven by maximizing both their capacity and enjoyment for use and their ability to generate revenue. The manipulations and modifications made to sports root zones confer major biomechanical advantages to athletes, and this in turn enhances the entertainment value for either participant or spectator. The impact and implications of these design solutions for the biotic system within the soil have often been considered only superficially, if at all. The consequence of the design optimization of the soil structure for playing performance has led to highly managed surfaces that are resource intensive. Recent research has highlighted the environmental impact of this kind of management, showing that the contribution to global climate change from the management of a UK parkland golf course caused a release of approximately 10.75 Mg C/year, with a net carbon balance of −33 Mg C/year, due to the level of the management of the soil required (Bartlett and James, 2011). Such significant implications to the environmental impact from the management of sports surfaces force us to question the current approaches to optimization, focusing on only one single facet of design.

The underlying question remains: what are the desirable characteristics of the matrix that supports the pitch? If the overall design aim is a surface on which players can perform consistently, which has predictable mechanical properties and so results in uniform biomechanical interactions from the athletes playing on it, then removing the biological interactions as far as possible from within the system is desirable. A third-generation synthetic turf pitch represents this mechanically optimized system. While there are mechanical similarities between synthetic and an optimized natural pitch system, the unique interactions of the turfgrass, the soil biological community and the structure of the soil are what make playing sport on natural turf unique. The variability of the (bio)mechanical interactions of the player with the surface of a natural turf pitch is often what makes playing on these pitches enjoyable to participants. The variability means that a talented athlete could outperform a less skilled player who had a physical strength advantage alone. Natural turf pitches, like any soil-based ecosystem, need to have interactions within and between the whole system of soil matrix, root network and free biota. Neglecting or biasing towards the soil matrix creates a system that needs more support from the other two components. While the propensity for designing sports pitches to optimize their mechanical and drainage properties prevails, the biological components of the system are likely to remain neglected. Contemporary resource-intensive methods of management and overall impact on the soil biota are likely to remain in maintaining playing performance. A priority for future research should be to approach the design of these soil systems to include, rather than exclude, soil biology. As our understanding of how roots, soil organisms and microbes interact within a soil system increases and which abiotic parameters influence each, it should be possible to engineer soils that support playing surfaces that require fewer resources but remain mechanically optimized for sport.

References

Adams, W.A. and Jones, R.L. (1979) The effects of particle size composition and root binding on resistance to shear of sports turf surfaces. *Rasen Turf Gazon* 2, 48–53.

Adams, W.A., Tanavud, C. and Springsguth, C.T. (1985) Factors influencing the stability of sports turf root zones. *International Turfgrass Society Research Journal* 5, 391–399.

Adams, W.A., Young, R.J. and Baker, S.W. (2001) Some soil and turf factors affecting the playing characteristics of premier cricket pitches in Britain. *International Turfgrass Society Research Journal* 9, 451–457.

Augé, R.M., Stodola, A.J.W., Ebel, R.C. and Duan, X. (1995) Leaf elongation and water relations of mycorrhizal sorghum in response to partial soil drying: two *Glomus* species at varying phosphorus fertilization. *Journal of Experimental Botany* 46, 297–307.

Baker, S.W. and Adams, W.A. (2001) *Classification and Provisional Guidelines for Moisture Content and Soil Bulk Density Values for First-class Squares*. The Sports Turf Research Institute, Bingley, UK.

Baker, S.W. and Bell, M.J. (1986) The playing characteristics of natural turf and synthetic turf surfaces for association football. *Journal of the Sports Turf Research Institute* 62, 9–35.

Baker, S.W. and Wollacott, A.R. (2005) Comparison of the playing performance of 3rd generation artificial grass with natural turf used for professional soccer. *International Turfgrass Society Research Journal* 10, 15–26.

Baker, S.W., Cook, A. and Binns, D.J. (1998) The effect of soil type and profile construction on the performance of cricket pitches. 1. Soil properties and grass cover during the first season of use. *Journal of the Sports Turf Research Institute* 79, 80–92.

Baker, S.W., Firth, S.J. and Binns, D.J. (2000) The effect of mowing regime and the use of acidifying fertiliser on rates of earthworm casting on golf fairways. *Journal of Turfgrass Science* 76, 2–11.

Balogh, J.C. and Walker, W.J. (1992) *Golf Course Management and Construction: Environmental Issues*. Lewis, Boca Raton, USA.

Bardgett, R.D. and McAlister, E. (1999) The measurement of soil fungal:bacterial biomass ratios as an indicator of ecosystem self-regulation in temperate meadow grasslands. *Biology and Fertility of Soils* 29, 282–290.

Barry, F., Gange, A.C., Crane, M. and Hagley, K.J. (2005) Fungicide levels and arbuscular mycorrhizal fungi in golf putting greens. *Journal of Applied Ecology* 42, 171–180.

Bartlett, M.D. and James, I.T. (2011) Are golf courses a source or sink of atmospheric CO_2: A modelling approach. *Proceedings of the Institution of Mechanical Engnieers, Part P: Journal of Sport Engineering and Technology* 225, 75–83.

Bartlett, M.D., James, I.T., Harris, J.A. and Ritz, K. (2007) Interactions between microbial community structure and the soil environment found on golf courses. *Soil Biology and Biochemistry* 39, 1533–1541.

Bartlett, M.D., James, I.T., Harris, J.A. and Ritz, K. (2008a) Size and phenotypic structure of microbial communities within soil profiles in relation to play surfaces on a UK golf course. *European Journal of Soil Science* 59, 835–841.

Bartlett, M.D., James, I.T., Harris, J.A. and Ritz, K. (2008b) The relationship between profile construction and casting behaviour of earthworms in golf course soils. In: Crews, D. and Lutz, R. (eds) *Science and Golf V.* Energy in Motion, Inc., Mesa, Arizona, pp. 489–495.

Bartlett, M.D., James, I., Harris, J. and Ritz, K. (2009) Microbiological profiles of golf course soils. *International Turfgrass Society Research Journal* 11, 949–957.

Bengough, A.G. and Mullins, C.E. (1990) Mechanical impedance to root growth: A review of experimental techniques and root growth responses. *European Journal of Soil Science* 41, 341–358.

Binns, D.J., Baker, S.W. and Piearce, T.G. (1999) A survey of earthworm populations on golf course fairways in Great Britain. *Journal of Turfgrass Science* 75, 36–44.

Blume, C., Christians, N.E. and Joo, Y.K. (2009) Nitrogen release timing of organic fertilisers applied to turf. *International Turfgrass Society Research Journal* 11, 957–966.

Bonkowski, M., Griffiths, B. and Ritz, K. (2000) Food preferences of earthworms for soil fungi. *Pedobiologia* 44, 666–676.

Brown, S. (2005) *Sports Turf and Amenity Grassland Management.* The Crowood Press, Marlborough, England.

Carré, M.J., Haake, S.J., Baker, S.W. and Newell, A.J. (2000) Predicting the dynamic behaviour of cricket balls after impact with a deformable pitch. In: Subic, A.J. and Haake, S.J. (eds) *The Engineering of Sport – Design, Development and Innovation.* Blackwell Science, Oxford, UK, pp. 177–184.

Clegg, C.D. (2006) Impact of cattle grazing and inorganic fertiliser additions to managed grassland on the microbial community composition of soils. *Applied Soil Ecology* 31, 73–86.

Danneberger, T.K. (1993) *Turfgrass Ecology and Management.* Franzak and Foster, Cleveland, Ohio.

Dawson, L.A., Grayston, S.J., Murray, P.J. and Pratt, S.M. (2002) Root feeding behaviour of *Tipula paludosa* (Meig.) (Dipetra: Tipulidae) on *Lolium perenne* (L.) and *Trifolium repens* (L.). *Soil Biology and Biochemistry* 34, 609–615.

Dawson, R.B. (1932) Leather jackets. *Journal of the Board of Green Keeping Research* 2, 183–195.

Degens, B. and Harris, J. (1997) Development of a physiological approach to measuring the catabolic diversity of soil microbial communities. *Soil Biology and Biochemistry* 29, 1309–1320.

Demiranda, J.C.C. and Harris, P.J. (1994) Effects of soil phosphorus on spore germination and hyphal growth of arbuscular mycorrhizal fungi. *New Phytologist* 128, 103–108.

Dixon, S.J., James, I.T., Blackburn, D.W.K., Pettican, N. and Low, D. (2008) Influence of footwear and soil density on loading within the shoe and soil surface during running. *Journal of Sports Engineering and Technology* 222, 1–10.

Dufour, M.J.D. and Mumford, C. (2008) GoingStick technology and electromagnetic induction scanning for naturally-turfed sports surfaces. *Sports Technology* 1, 125–131.

Emmons, R.D. (2000) *Turfgrass Science and Management.* Delmar, Kentucky, USA.

FIH Equipment Committee (2008) *Handbook of Performance Requirements for Synthetic Turf Hockey Pitches Incorporating Test Procedures.* International Hockey Federation, Lausanne, Switzerland.

Flemming, P.R., Young, C., Roberts, J.R., Jones, R. and Dixon, N. (2005) Human perceptions of artificial surfaces for field hockey. *Sports Engineering* 8, 121–136.

Gange, A.C. (1998) A potential microbiological method for the reduction of *Poa annua* L. in golf greens. *Journal of Turfgrass Science* 74, 40–45.

Gange, A.C., Lindsay, D.E. and Ellis, L.S. (1999) Can arbuscular mycorrhizal fungi be used to control the undesirable grass *Poa annua* on golf courses? *Journal of Applied Ecology* 36, 909–919.

Gemma, J.N. and Koske, R.E. (1989) Field inoculation of America beachgrass (*Ammophila breviligulata*) with VA mycorrhizal fungi. *Journal of Environmental Management* 29, 173–182.

Goszczynski, J., Jedrzejewska, B. and Jedrzejewski, W. (2000) Diet composition of badgers (*Meles meles*) in a pristine forest and rural habitats of Poland compared to other European populations. *Journal of Zoology* 250, 495–505.

Grewal, P.S. (1999) Factors in the success and failure of microbial control in turfgrass. *Integrated Pest Management Reviews* 4, 287–294.

Guisasola, I., James, I.T., Llewellyn, C., Bartlett, M.D., Stiles, V. and Dixon, S. (2009) Human–surface interactions: an integrated study. *International Turfgrass Society Research Journal* 11, 1097–1106.

Guisasola, I., James, I.T., Llewellyn, C., Stiles, V. and Dixon, S. (2010) Quasi-static mechanical behaviour of soils used for natural turf sports surfaces and stud force prediction. *Sports Engineering* 12, 111–122.

Haigh, I.M., Jenkinson, P. and Hare, M.C. (2009) The effect of a mixture of seed-borne *Microdochium nivale* var. *majus* and *Microdochium mivale* var. *nivale* infection on *Fusarium* seedling blight severity and subsequent stem colonisation and growth of winter wheat in pot experiments. *European Journal of Plant Pathology* 124, 65–73.

Hartwigera, C.E., Peacock, C.H., DiPaolac, J.M. and Casseld, D.K. (2001) Impact of light-weight rolling on putting green performance. *Crop Science* 41, 1179–1184.

Hedlund, K. (2002) Soil microbial community structure in relation to vegetation management on former agricultural land. *Soil Biology and Biochemistry* 34, 1299–1307.

Henley, W.E., Rogers, K., Harkins, L. and Wood, J.L.N. (2006) A comparison of survival models for assessing risk of racehorse fatality. *Preventive Veterinary Medicine* 74, 3–20.

Hindahl, M.S., Miltner, E.D., Cook, T.W. and Stahnke, G.K. (2009) Surface water quality impacts from golf course fertiliser and pesticide applications. *International Turfgrass Society Research Journal* 11, 19–30.

Hodgson, J.M. (1997) *Soil Survey Field Handbook. Soil Survey Technical Monograph No. 5.* Soil Survey of England and Wales, Silsoe, UK.

Horton, R., Ankeny, M.D. and Allmaras, R.R. (1994) Effects of compaction on soil hydraulic properties. In: Soane, B.D. and van Ouwerkerk, C. (eds) *Soil Compaction in Crop Production.* Elsevier, Amsterdam.

Jakobsen, I., Smith, S.E. and Smith, F.A. (2002) Function and diversity of arbuscular mycorrhizae in carbon and mineral nutrition. In: van der Heijden, M.G.A. and Sanders, I.R. (eds) *Mycorrhizal Ecology.* Springer-Verlag, Berlin, pp. 75–92.

Kazakova, S.V., Hageman, J.C., Matava, M., Srinivasan, A., Phelan, L., Garfinkel, B. *et al.* (2005) A clone of methicillin-resistant *Staphylococcus aureus* among professional football players. *New England Journal of Medicine* 352, 468–475.

Koske, R.E., Gemma, J.N. and Jackson, N. (1997a) A preliminary survey of mycorrhizal fungi in putting greens. *Journal of Turfgrass Science* 73, 2–8.

Koske, R.E., Gemma, J.N. and Jackson, N. (1997b) Mycorrhizal fungi associated with three species of turfgrass. *Canadian Journal of Botany* 75, 320–332.

Lavelle, P. (2001) *Soil Ecology,* Kluwer Academic Publishers, Dordrecht, The Netherlands.

Lee, K.E. (1985) *Earthworms: Their Ecology and Relationships with Soils and Land Use.* Academic Press, Sydney, Australia.

Li, D., Joo, Y.K., Christians, N.E. and Minner, D.D. (2000) Inorganic soil amendment effects on sand-based sports turf media. *Crop Science* 40, 1121–1125.

Lipiec, J. and Hatano, R. (2003) Quantification of compaction effect on soil physical properties and crop growth. *Geoderma* 116, 107–136.

Mann, R.L. (2003) A survey to determine the spread and severity of pest and disease on golf greens in the UK and Ireland. *Journal of Turfgrass and Sports Surface Science* 80, 2–19.

Mazzola, M. (2004) Assessment and management of soil microbial community structure for disease suppression. *Annual Review of Phytopathology* 42, 35–59.

McCoy, E.L. (1992) Quantitative physical assessment of organic materials used in sports turf root zone mixes. *Agronomy Journal* 84, 375–381.

McIntyre, K. (2001) *Cricket Wickets: Science v Fiction.* Horticultural Engineering Consultancy, London.

McLeod, A.J. and James, I.T. (2007) The effect of particulate contamination on the infiltration rates of synthetic turf surfaces. In: Fleming, P., Carré, M., James, I., Walker, C. and Dixon, S. (eds) *Science, Technology and Research into Sports Surfaces.* Loughborough University, Loughborough, UK, pp. 35–42.

Peters, A. and Ehlers, R.U. (1994) Susceptibility of leatherjackets to the entomopathogenic nematode *Steinernema feltiae. Journal of Invertebrate Pathology* 63, 163–171.

Pilar Ruiz, M., Ramajo, M., Jesus, J.B., Trigo, D. and Diaz Cosin, D.J. (2006) Selective feeding of the earthworm *Hormogaster elisae* (Oligochaeta, Hormogastridae) in laboratory culture. *European Journal of Soil Biology* 42, S289–S295.

Potter, D., Powell, A. and Smith, M.S. (1990) Degradation of turfgrass thatch by earthworms (Oligochaeta: Lumbricidae) and other soil invertebrates. *Entomological Society of America* 32, 205–211.

Ritz, K., Harris, J.A., Pawlett, M. and Stone, D. (2006) *Catabolic Profiles as an Indicator of Soil Microbial Functional Diversity.* Science Report SC040063/SR, Environment Agency, Bristol, UK.

Rossi, F.S. and Grant, J.A. (2009) Long term evaluation of reduced chemical pesticide management of golf course putting turf. *International Turfgrass Society Research Journal* 11, 77–90.

Sims, R.W. and Gerard, B.M. (1999) *Synopses of the British Fauna (New Series): Earthworms.* Dorset Press, Dorchester, UK.

Steenwerth, K.L., Jackson, L.E., Calderon, F.J., Stromberg, M.R. and Scow, K.M. (2002) Soil microbial community composition and land use history in cultivated and grassland ecosystems of coastal California. *Soil Biology and Biochemistry* 34, 1599–1611.

Stepniewski, W. and Glinski, J. (1985) *Soil Aeration and its Role for Plants.* CRC Press Inc., Boca Raton, Florida.

Stiles, V., James, I.T., Dixon, S. and Guisasola, I. (2009) Natural turf surfaces: The case for continued research. *Sports Medicine* 39, 65–84.

Sylvia, D.M. and Burks, J.N. (1988) Selection of vesicular-arbuscular mycorrhizal fungus for practical inoculation of *Uniola paniculata. Mycologia* 80, 565–568.

Trevors, J.T. (1996) Sterilization and inhibition of microbial activity in soil. *Journal of Microbiological Methods* 26, 53–59.

Turgeon, A.J. (2002) *Turfgrass Management.* Pearson Education, Inc., Upper Saddle River, New Jersey.

USGA Green Section Staff (2004) *USGA Recommendations for a Method of Putting Green Construction.* United States Golf Association, Far Hills, New Jersey.

Waalwijk, C., Dullemans, A., Weigers, G. and Smit, P. (1992) Toxicity of *Bacillus thuringiensis* variety *Israelensis* against Tipulid larvae. *Journal of Applied Entomology* 114, 415–420.

Waltz, F.C. and McCarty, L.B. (2005) Field evaluation of soil amendments used in root zone mixes for golf course putting greens. *International Turfgrass Society Research Journal* 10, 1150–1158.

Whalley, W.R., Riseley, B., Leeds-Harrison, P.B., Bird, N.R.A., Leech, P.K. and Baker, S.W. (2005) Structural differences between bulk and rhizosphere soil. *European Journal of Soil Science* 56, 353–360.

Index